Ergebnisse der Physiologie

Biologischen Chemie und experimentellen Pharmakologie

Reviews of Physiology

Biochemistry and Experimental Pharmacology

69

Herausgeber / Editors

R. H. Adrian, Cambridge · E. Helmreich, Würzburg
H. Holzer, Freiburg · R. Jung, Freiburg · K. Kramer, München
O. Krayer, Boston · F. Lynen, München · P. A. Miescher, Genève
H. Rasmussen, Philadelphia · A. E. Renold, Genève
U. Trendelenburg, Würzburg · K. Ullrich, Frankfurt/M.
W. Vogt, Göttingen · A. Weber, Philadelphia

With 15 Figures

Springer-Verlag Berlin Heidelberg GmbH 1974

ISBN 978-3-662-30958-2 ISBN 978-3-540-37810-5 (eBook)
DOI 10.1007/978-3-540-37810-5

Ursprünglich erschienen bei Springer-Verlag Berlin Heidelberg New York 1974
Softcover reprint of the hardcover 1st edition 1974

Library of Congress Catalog Card Number 73-14479.

Inhalt/Contents

The Adrenergic Innervation of the Gastrointestinal Tract. By J. B. FURNESS and M. COSTA, Victoria/Australia. With 5 Figures 1

Regulation of Fatty-Acid Synthesis in Higher Animals. By S. NUMA, Kyoto/Japan. With 6 Figures 53

Amino Acid Transmitters in the Mammalian Central Nervous System. By D. R. CURTIS and G. A. R. JOHNSTON, Canberra City/Australia. With 4 Figures 97

Author Index 189

Subject Index 216

Mitarbeiter/List of Contributors

COSTA, M., Dr., Department of Zoology, University of Melbourne, Parkville 3052, Victoria/Australia

CURTIS, D. R., Prof. Dr., Department of Pharmacology, John Curtin School of Medical Research, P.O. Box 334, Canberra City, A.C.T. 2601/Australia

FURNESS, J. B., Dr., Department of Zoology, University of Melbourne, Parkville 3052, Victoria/Australia

JOHNSTON, G. A. R., Prof. Dr., Department of Pharmacology, John Curtin School of Medical Research, P.O. Box 334, Canberra City, A.C.T. 2601/Australia

NUMA, S., Prof. Dr., Department of Medical Chemistry, Kyoto University, Faculty of Medicine, Yoshida, Sakyo-ku, Kyoto 606/Japan

The Adrenergic Innervation of the Gastrointestinal Tract

J. B. FURNESS and M. COSTA*

Contents

I. Introduction . . . 1
II. Adrenergic Nerves and the Inhibition of Gastrointestinal Movement . . . 2
A. Release of Noradrenaline by Stimulation of Extrinsic Nerves . . . 2
B. Positions of Adrenergic Cell Bodies . . . 3
C. Ramifications of the Terminals . . . 5
D. Ultrastructural Identification of the Adrenergic Axons . . . 12
E. Sites of Action of the Neurally Released Transmitter . . . 13
F. Receptor Mechanisms . . . 18
G. Non-Adrenergic Inhibitory Neurones . . . 20
III. Adrenergic Innervation of the Sphincters . . . 21
IV. Adrenergic Nerves and Gastrointestinal Vasculature . . . 23
A. Effects of Nerve Stimulation . . . 23
B. Influence of Blood Flow on Secretion and Absorption . . . 26
C. Blood Flow and Motility . . . 27
V. Adrenergic Activity in Vivo . . . 27
A. Resting and Peak Discharge Rates in Adrenergic Nerves . . . 27
B. Reflex Changes in Adrenergic Discharge . . . 30
1. Gastrointestino-Gastroinstestinal Inihibitory Reflexes . . . 30
2. Excitation of Afferent Nerves outside the Intestine . . . 33
3. Adynamic Ileus . . . 36
VI. The Functions of the Adrenergic Nerves which Supply the Gastrointestinal Tract . . . 37
VII. Summary . . . 39
References . . . 40

I. Introduction

This article is intended to provide the reader with a clear understanding of the relationship between adrenergic nerves and gastrointestinal function, drawn from an analysis of the available physiological, pharmacological and anatomical data. It is not intended to be an historical review. We direct attention to earlier work where we feel it is of particular interest or has been important in shaping current concepts, but we concentrate on presenting the most recent and, as far as we can judge, the most relevant and reliable information. For an appreciation of the historical development of this subject, the reader is referred to the following: BUNCH, 1898; STARLING, 1902; MÜLLER, 1924; MCSWINEY, 1931; GARRY, 1934;

* Department of Zoology, University of Melbourne, Parkville, 3052, Victoria, Australia.

MCDOWALL, 1935; ALVAREZ, 1948; YOUMANS, 1949, 1952; KUNTZ, 1953; GRIM, 1963; GRAYSON and MENDEL, 1965; KOSTERLITZ, 1968.

The survey of literature pertaining to this review was concluded in July, 1972.

II. Adrenergic Nerves and the Inhibition of Gastrointestinal Movement

A. The Release of Noradrenaline by Stimulation of Extrinsic Nerves

In the second half of last century, it was observed that stimulation of the thoracolumbar outflow, of splanchnic, or of mesenteric nerves inhibits the movements of the gastrointestinal tract (LUDWIG, 1852; PFLÜGER, 1857; LISTER, 1858; BASCH, 1873; BRAAM-HOUKGEEST, 1874; LANGLEY and DICKINSON, 1889; LANGLEY and ANDERSON, 1895a; COURTADE and GUYON, 1897; BUNCH, 1898; BAYLISS and STARLING, 1899). In the following 40 years the similarity of this inhibition and of other effects of the stimulation of sympathetic nerves to those caused by adrenaline was noted and it was suggested that adrenaline or a similar substance was released at the nerve endings (ELLIOTT, 1904b, 1905; STARLING, 1906[1]; CANNON and ROSENBLUETH, 1937; DALE, 1938). FINKLEMAN (1930) found that the inhibition caused by stimulation of the mesenteric nerves to the isolated duodenum of the rabbit was mimicked by adrenaline ($0.5–2 \times 10^{-7}$ g/ml), and that both effects were similarly antagonized by ephedrine or by desensitization of the muscle to adrenaline. FINKLEMAN showed that during stimulation of the mesenteric nerves a substance was released which could cause relaxation of a second, unstimulated, segment of intestine. From these experiments he concluded that the sympathetic nerves act by liberating an inhibitory substance similar to adrenaline. The relaxations of the intestine caused by stimulation of mesenteric nerves and by sympathomimetic amines were compared again by ASTRÖM (1949), who concluded from his own experiments and from assays which showed that the mesenteric nerves contain 12–15 times more noradrenaline than adrenaline (EULER, 1951), that noradrenaline was released at the endings of mesenteric nerves within the gut wall. MANN and WEST (1951), who used a bioassay technique, deduced that about 75% of the catecholamine released when the inferior mesenteric nerves of the cat were stimulated was noradrenaline. The venous outflow from colons of cats and rabbits has been collected during mesenteric nerve stimulation and analyzed by chromatographic separation followed by bioassay (MIRKIN and BONNYCASTLE, 1954). In rabbits, 95% of the catecholamine in the effluent was noradrenaline, and in cats this amine accounted for 90%. Assays of tissue homogenates of rabbit small intestine (SHORE, 1959) of rat stomach and small intestine (BRODIE et al., 1964) and of guinea-pig small intestine (GOVIER et al., 1969) and large intestine (BAUMGARTEN et al., 1970) reveal high levels of norad-

1 In a footnote, STARLING writes: "Adrenaline, the hormone manufactured by the suprarenal bodies, seems to be necessary for the normal display of the functions of the sympathetic system, and its motor or inhibitory effects on the gut are produced through this system".

renaline. GOVIER et al. (1969) were unable to detect any adrenaline in the longitudinal muscle and myenteric plexus of the guinea-pig ileum. Fluorescence histochemical studies also suggest that noradrenaline, rather than adrenaline, is associated with nerves in the stomach and intestines. GILLESPIE and MACKENNA (1961) showed that the inhibitory responses of the colon and ileum to stimulation of mesenteric nerves were abolished if the rabbits were pretreated with reserpine, and that the inhibition was restored if noradrenaline or its precursors (dopa, dopamine) were included in the organ bath for a short while before stimulation. These and many subsequent experiments provide unequivocal evidence that mesenteric inhibitory nerves to the gastrointestinal tract are adrenergic. Because noradrenaline is the actual transmitter substance, it might be more accurate to refer to the nerves as noradrenergic. However, the term "adrenergic" was introduced to describe these nerves before the exact chemistry of the transmitter was determined and has continued in general use. It is therefore used in this article.

B. Positions of Adrenergic Cell Bodies

Most of the cells which give rise to the adrenergic fibres innervating the gastrointestinal tract are located in the prevertebral ganglia of the coeliac, superior mesenteric, inferior mesenteric and pelvic plexuses (Fig. 1). This was first demonstrated by experiments in which nicotine was used to block transmission through the ganglia and has since been confirmed by studies using histochemical means to locate adrenergic nerves.

LANGLEY and DICKINSON (1889) observed that in cats, dogs and rabbits the application of nicotine to the coeliac ganglia prevented the inhibition of the movements of the stomach which is normally caused by stimulation of the splanchnic nerves, but did not prevent the inhibition caused by stimulation of the mesenteric nerves beyond the ganglia. Nicotine applied to the splanchnic nerves alone was ineffective. Similar results were obtained when transmission of impulses through the superior mesenteric ganglia to the intestine were examined. These results indicate that the cell bodies of mesenteric inhibitory (adrenergic) nerves to the stomach and small intestine are in the coeliac and superior mesenteric ganglia respectively. With similar methods, ELLIOTT (1904a) showed that the sympathetic pathway to the ileo-colic sphincter has a ganglionic interruption in the coeliaco-mesenteric plexus. LANGLEY and ANDERSON (1895b) used nicotine to show that most cell bodies of sympathetic fibres ending within the colon and rectum in cats and rabbits are in the inferior mesenteric ganglia, but they also pointed to the undoubted existence of a small number of more peripherally located cells. In the cat, stimulation of the hypogastric nerves emerging from the inferior mesenteric ganglia caused a contraction of the internal anal sphincter, which was weakened, although not blocked, by nicotine. This suggests the presence of sympathetic post-ganglionic (adrenergic) neurones in the pelvic plexus, and the presence of a large number of efferent myelinated axons passing through the inferior mesenteric ganglia to run in the hypogastric nerves is consistent with this postulate (LANGLEY and ANDERSON, 1894/95; M'FADDEN et al., 1935).

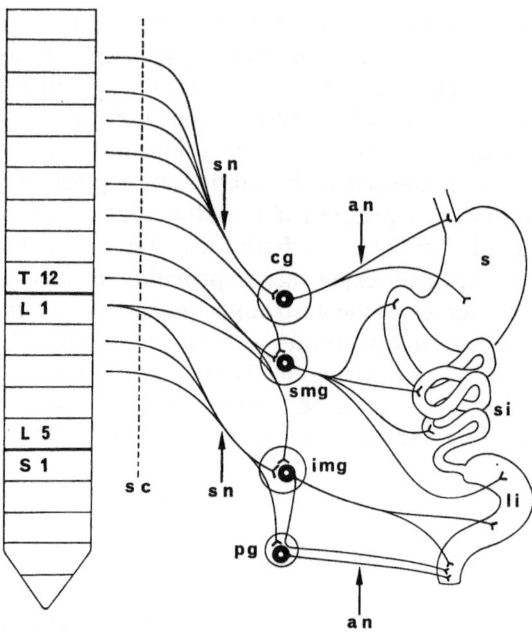

Fig. 1. The positions of neurones which supply the adrenergic innervation of the gastrointestinal tract and the levels of the spinal cord from which cholinergic nerves activating the adrenergic neurones arise. The adrenergic cell bodies are located predominantly in the ganglia of the prevertebral plexuses, i. e., the coeliac (*cg*), superior mesenteric (*smg*), inferior mesenteric (*img*) and pelvic (*pg*) ganglia. The position of the sympathetic chains is indicated by the dotted line (*sc*). *an* adrenergic nerves, *li* large intestine, *s* stomach, *si* small intestine, *sn* splanchnic nerves. This diagram has been drawn to give an indication of the general arrangement found in most mammals. It is based on information in KUNTZ (1953), MITCHELL (1953), MONNIER (1968), and other references given in the text. Some of the differences between species are also dealt with in the text

These are almost certainly pre-ganglionic sympathetic nerves which synapse with adrenergic cells in the pelvic ganglia. The existence of a few groups of cells scattered along the mesenteric nerves beyond the prevertebral ganglia has been confirmed by KUNTZ and JACOBS (1955) who found small ganglia in paravascular nerves supplying the gut in cat, rat and man.

Experiments employing the fluorescence histochemical technique recently developed for the detection of adrenergic nerves have confirmed the results of the early experiments with nicotine. Most of the cell bodies in the prevertebral ganglia contain histochemically demonstrable levels of noradrenaline (HAMBERGER and NORBERG, 1963; NORBERG and HAMBERGER, 1964). The fluorescence histochemical method was first used to examine the adrenergic innervation of the gut by NORBERG (1964) and JACOBOWITZ (1965), who found adrenergic fibres, but no adrenergic cell bodies, within the gut wall. Studies in which mesenteric nerves were interrupted or prevertebral ganglia were removed several days prior to examination demonstrate that the adrenergic nerves to the stomach, small intestine and colon of the cat, the colon of the dog, the small intestine and distal colon of the guinea-pig, the small intestine of the rabbit and the human sigmoid colon arise in extrinsic ganglia (see FURNESS and COSTA, 1971).

The only part of the gut of any mammal known to contain a substantial number of adrenergic cell bodies is the proximal colon of the guinea-pig (COSTA et al., 1971; FURNESS and COSTA, 1971). In this area, there are approximatelly 10000 adrenergic cell bodies scattered throughout the myenteric plexus. It is likely that adrenergic cells occur is small numbers in other parts of the gut which receive the major part of their adrenergic innervation from cells in prevertebral ganglia. For example in the last 10 cm of the distal colon of the guinea-pig, extrinsic denervation usually results in complete disappearance of adrenergic terminals (FURNESS, 1969a; 1970a), although we have occasionally observed a few adrenergic cells in the myenteric plexus (COSTA and FURNESS, 1973a). Thorough examination of the adrenergic innervation of the gastrointestinal tract in rats, guinea-pigs and rabbits has not revealed any significant number of adrenergic cells in parts other than the guinea-pig proximal colon (COSTA and GABELLA, 1971). No adrenergic cell bodies have been observed in the submucous plexuses.

The fluorescence histochemical method has recently been used in experiments to determine the source of adrenergic nerves to the rectum and anal sphincter in the guinea-pig (COSTA and FURNESS, 1973a). It was found that some fibres run to the internal sphincter from the sacral sympathetic chains via nerves following the rectal arteries and that other fibres arise in or pass through the posterior pelvic plexuses, but that no adrenergic axons reach the sphincter from the inferior mesenteric ganglia.

C. Ramifications of the Terminals

The ramifications of the adrenergic terminals within nonsphincter areas of the gut wall as revealed by the fluorescence histochemical method will be described in this section, but it is worthwhile first to consider some of the earlier work in relation to recent observations. One of the features which is immediately apparent from the fluorescence histochemistry is that many adrenergic fibres supply the ganglia of the myenteric plexus but that few fibres supply the muscle (NORBERG, 1964; JACOBOWITZ, 1965). Recent articles have mistakenly suggested that this is a novel observation, although many earlier investigators had observed that the sympathetic fibres actually run to the ganglionated plexuses of the intestine (HENLE, 1879; OPENCHOWSKI, 1889; CAJAL, 1893, 1911; DOGIEL, 1895; LA VILLA, 1898; MÜLLER, 1920, 1924; CARPENTER, 1924; WADDELL, 1929; BAUMANN, 1949). The impression that the fibres run only, or principally, to the muscle seems to have arisen from the work of LANGLEY and his collaborators who found that nicotine did not block the inhibition caused by stimulation of mesenteric nerves. However, it is quite clear that LANGLEY himself was not firm in this view. He states (LANGLEY and ANDERSON, 1895b):

"It is hardly conceivable that some of the sympathetic nerve fibres should not become connected with the numerous nerve cells which are present in these plexuses, and the observations of RAMON Y CAJAL, V. GEHUCHTEN, V. KÖLLIKER and DOGIEL show fairly conclusively that such a connexion exists.

We may then, at any rate, conclude that these connexions are of a different nature from those which exist between pre-ganglionic fibres and sympathetic nerve cells; and that the nerve cells of the plexuses of Auerbach and Meissner do not belong to the sympathetic system but are of a different class".

The length of intestine supplied by the adrenergic nerves following any one artery as it enters the gut wall is approximately coextensive with the area of supply of the artery and its branches. If the nerves following a particular artery are interrupted the adrenergic terminals in the area supplied by its branches degenerate, but there is little change in the adrenergic innervation of adjacent areas. Similarly, if the nerves following one mesenteric artery are left intact and those on either side are interrupted the terminals in the middle section remain, while those on either side degenerate. There is only a small area of overlap between areas supplied by paravascular nerves following adjacent arteries and there is no histochemical evidence that non-terminal adrenergic fibres contribute to nerves ascending or descending within the gut wall. DOWNMAN (1952) studied the areas of intestine under the control of adrenergic fibres arising from nerves accompanying individual arteries in the cat mesentery. He found that vasoconstriction and inhibition of intestinal movements were confined to the area supplied by each artery and that adjacent fields overlapped by a maximum of 20%. He found no evidence for the presence of ascending or descending adrenergic fibres within the gut.

After penetrating the serosa, paravascular nerves give rise to small filaments which supply adrenergic terminals to various intramural structures (Fig. 2). With the fluorescence technique adrenergic fibres appear smooth and moderately fluorescent in their non-terminal parts and are varicose and brightly fluorescent in their terminal parts. It is believed that noradrenaline is released from the varicosties along the entire length of the terminal portion when an action potential passes down the axon (MALMFORS, 1965; FURNESS, 1970b).

The myenteric (AUERBACH'S) plexus lies between the longitudinal and circular muscle coats of the gut wall and is composed of small ganglia, sometimes referred to as nodes, which are joined by bundles of nerve fibres, the internodal strands. Adrenergic terminals ramify extensively amongst the nodes and also contribute to the internodal strands (Fig. 3a–c). There seems to be no basic difference in the adrenergic supply to the myenteric ganglia of stomach and intestines in different species of mammals although the size and shape of the ganglia do differ. A few adrenergic terminal fibres branch from those supplying the ganglia without coming into any close relationship with other neurones. The adrenergic terminals do not form true pericellular endings around the ganglion cells of the myenteric plexus; i. e. each does not branch into a series of short processes which envelop a ganglion cell. We have traced the paths of individual fibres in laminar preparations in which the fluorescence of only a few terminals had survived chemical depletion of catecholamines (COSTA and FURNESS, 1971) or partial denervation (Fig. 4 and unpublished). Many of the varicose fibres run for long distances, several mm and occasionally up to 1 cm, without branching. They course close to several ganglion cells in the nodes and then may follow an internodal strand to an adjacent ganglion where they again come close to several cells (Fig. 4d). Some axons branch within the plexus and come close to ganglion cells. In preparations counterstained for dehydrogenases, the fluorescent fibres and the ganglion cells can be observed at the same time (COSTA

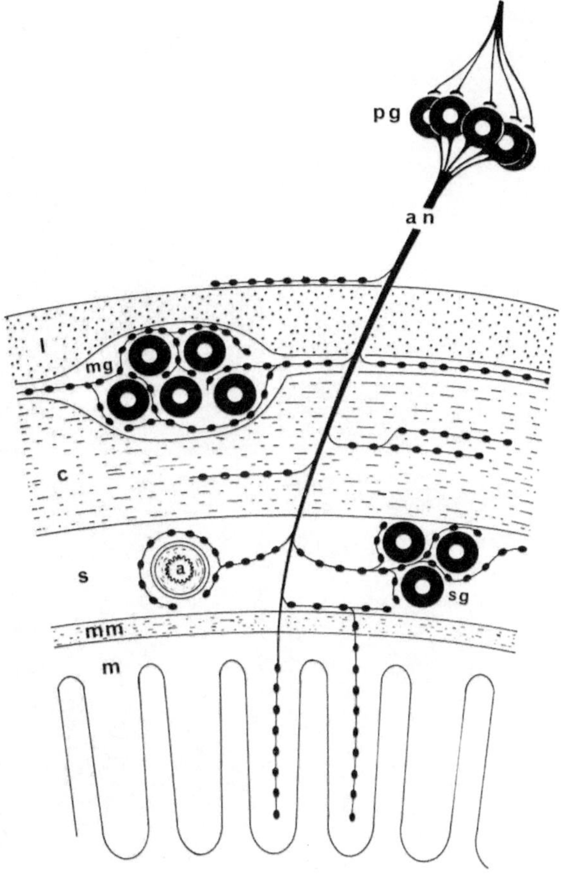

Fig. 2. The distribution of the terminals of adrenergic nerves (drawn as varicose fibres) within the wall of the stomach and intestine. Most of the terminals ramify amongst neurones of the myenteric (*mg*) and submucous (*sg*) ganglia or form plexuses around arteries (*a*). A few fibres innervate the circular muscle and the muscularis mucosae and some axons extend into the villi. A few adrenergic axons contribute to the subserous plexus where present. *an* adrenergic nerve trunk, *c* circular muscle, *l* longitudinal muscle, *m* mucosa, *mm* muscularis mucosae, *pg* prevertebral ganglion, *s* submucosa

and FURNESS, 1973b). Such preparations show that all the ganglion cells in the nodes are invested by the network of adrenergic fibres (Fig. 3c), but that there are a few extraganglionic neurones which are apparently not innervated. We have also observed bright, ring-like formations which appear to surround a small number of ganglion cells in the myenteric plexus. They are connected with the adrenergic fibres (they disappear after section of the mesenteric nerves and their fluorescence is similarly affected by drugs) but we do not know if they represent specialized relationships between adrenergic axons and particular ganglion cells. At an ultrastructural level, adrenergic axons have been identified by their content of granulated vesicles or by the autoradiographic localization of noradrenaline and found to form synapses with enteric neurones (HONJIN et al., 1965; TAXI

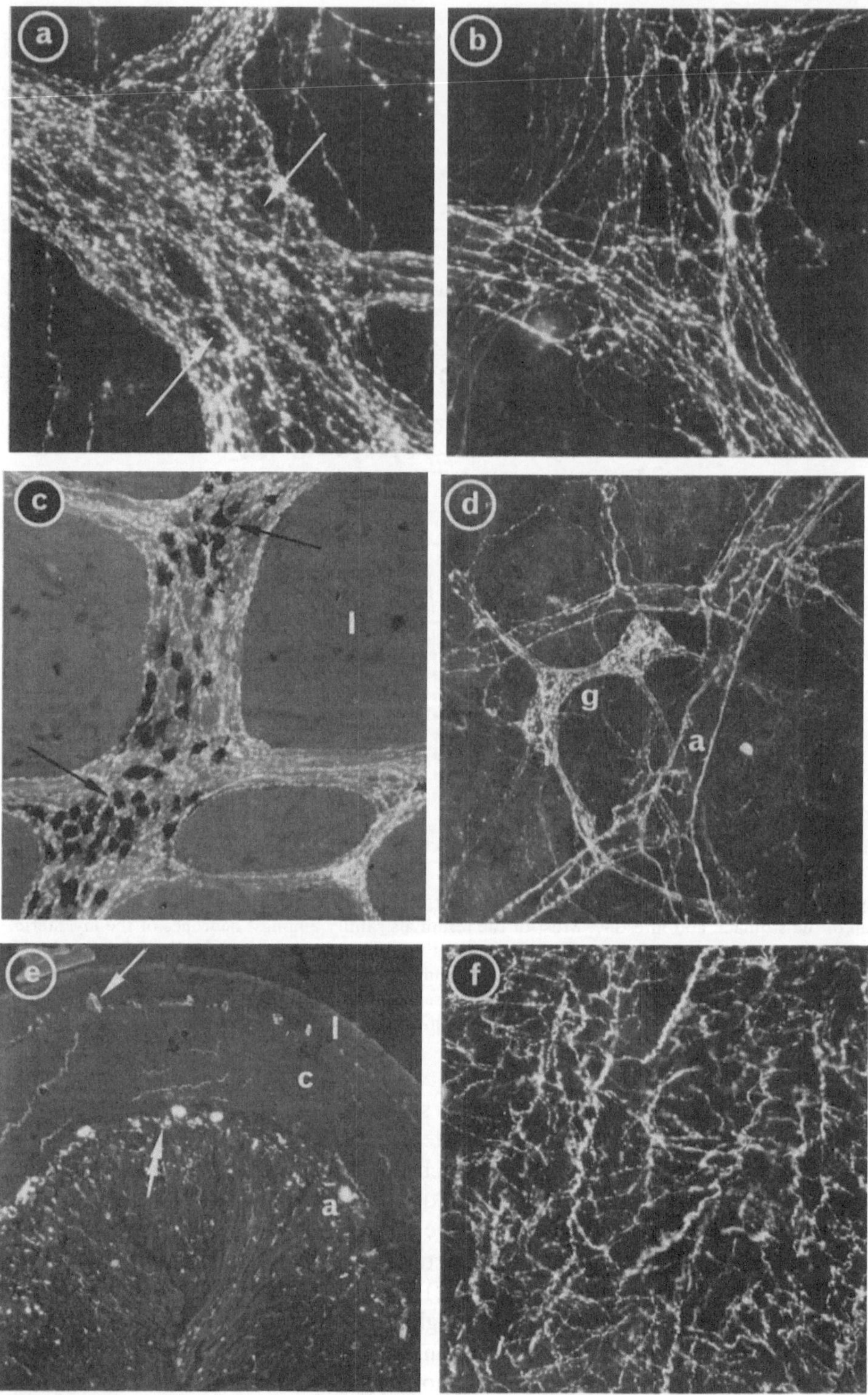

Fig. 3a–f

and DROZ, 1969; GABELLA, 1971). ROSS and GERSHON (1970), who used 6-hydroxydopamine to identify adrenergic nerves, reported that most adrenergic terminals in the myenteric plexus in guinea-pigs were remote from ganglion cells but were close to the smooth muscle, although GABELLA'S (1972) description of this plexus indicates that none of the axons apposing the smooth muscle have the ultrastructural characteristics of adrenergic nerves.

In comparison with the myenteric plexus, the innervation of the muscularis externa (other than at the sphincters) is sparse (Fig. 3e). In small mammals (monkeys, rats, guinea-pigs, cats and rabbits) there are practically no adrenergic fibres within the longitudinal muscle of the small intestine (NORBERG, 1964; JACOBOWITZ, 1965; GABELLA and COSTA, 1967; COSTA and GABELLA, 1971; SILVA et al., 1971) and the situation in the large intestine is similar (GILLESPIE, 1968; READ and BURNSTOCK, 1969; FURNESS, 1970a; COSTA and GABELLA, 1971), except that adrenergic fibres are found in the longitudinal muscle where it is gathered into taeniae (ÅBERG and ERÄNKÖ, 1967; BENNETT and ROGERS, 1967) and in the longitudinal muscle of the rectum, at least in guinea-pigs, cats and dogs (HOWARD and GARRETT, 1973; FURNESS and COSTA, 1973b). Although it is sparse, there is some adrenergic innervation of the circular muscle of both the small and large intestine (GABELLA and COSTA, 1967; FURNESS, 1970a; COSTA and

Fig. 3a–f. The arrangement of adrenergic axons within the gut wall, revealed by the fluorescence histochemical method. In these micrographs the adrenergic nerves appear white against a dark background. Figs. a to c show preparations of the myenteric plexus and longitudinal muscle which were separated from the other layers of the wall and laid flat on a glass slide. The myenteric plexus appears as a lamina in these figures. Fig. d is a similar preparation of the isolated submucosa. Figs. e and f are of sections 15 μm thick, taken from paraffin-embedded tissue.

a. The adrenergic axons in a ganglion of the myenteric plexus from the rabbit ileum. Note that the adrenergic axons form a dense meshwork. The positions of non-fluorescent neurones of the ganglia can often be recognized as "holes" in the meshwork, such as those indicated by arrows. ×230

b. The adrenergic axons which ramify within a ganglion of the myenteric plexus of the sheep ileum. ×195

c. An area of the myenteric plexus of the guinea-pig duodenum which has been treated with the fluorescence histochemical method to reveal adrenergic axons and with a NADH dehydrogenase method which stains many of the neurones of the myenteric ganglia. It can be seen that the varicose adrenergic axons ramify amongst the neurones (arrowed) and that they also run in the strands between ganglia. Note that there are no adrenergic axons in the background longitudinal muscle (*l*). ×130

d. Adrenergic axons in the submucosa of the guinea-pig duodenum. There are many adrenergic terminals within submucous ganglia (*g*) and in the plexuses surrounding submucous arteries (*a*). ×125

e. Transverse section through the ileum of the guinea-pig. There are no adrenergic axons in the longitudinal muscle (*l*), but there are a few in the circular muscle (*c*). Adrenergic terminals are found in the ganglia of the myenteric plexus (single arrow) and submucous plexus (double arrow). The large fluorescent areas in the submucosa are the perivascular plexuses of submucosal arteries (*a*). At the magnification used in this figure, the individual fibres of the plexus cannot be distinguished. Enterochromaffin cells appear as small white dots scattered throughout the submucosa and mucosa. Some adrenergic axons also extend into the villi. This micrograph can be compared with the schematic representation of the innervation shown in Fig. 2. ×50

f. The dense adrenergic innervation of sphincter muscle. This micrograph is of a section taken through the circular muscle of the internal anal sphincter of the guinea-pig. ×140

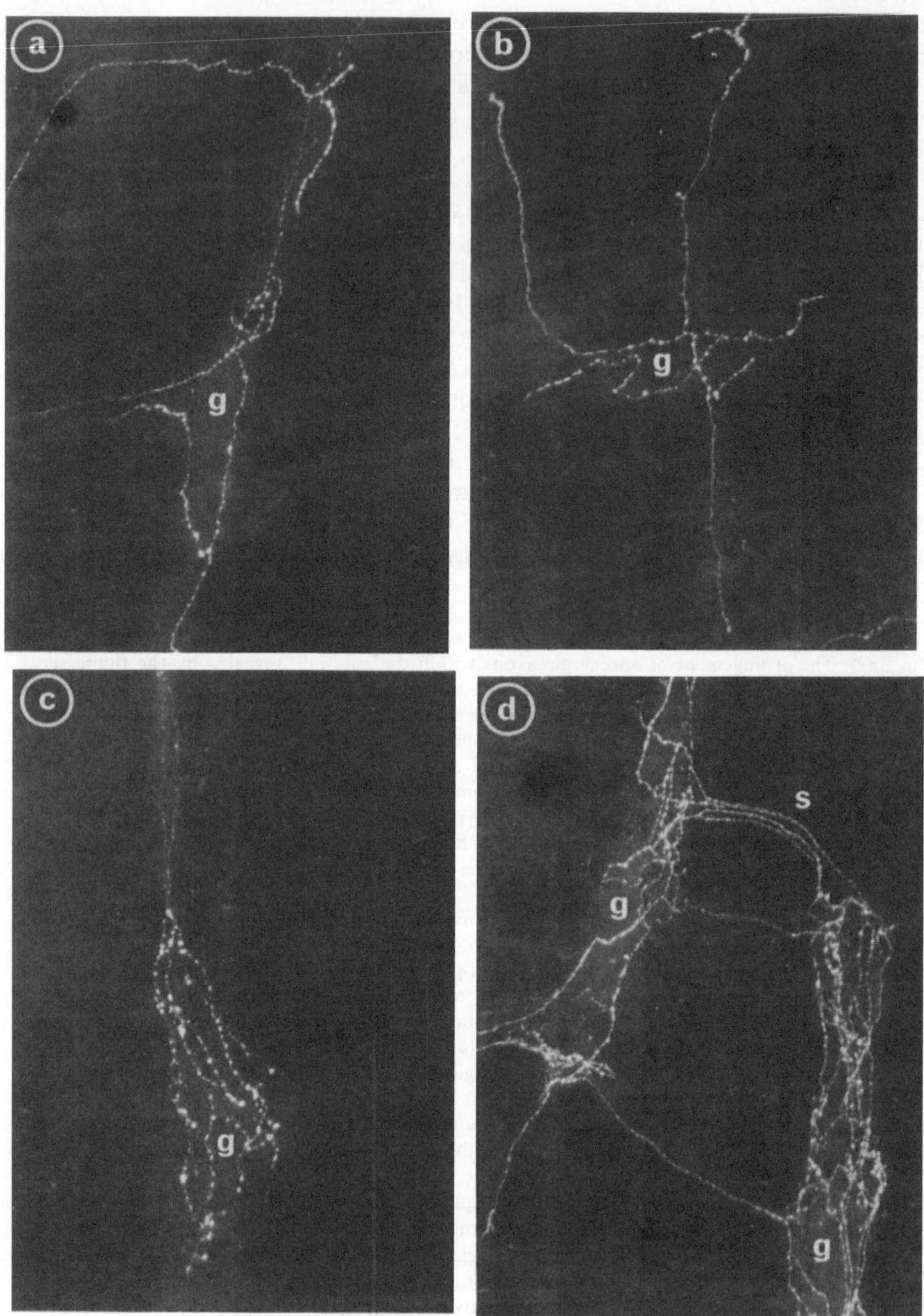

Fig. 4a–d

GABELLA, 1971). The dense network of nerve fibres in the inner layer of the circular muscle (CAJAL, 1911; LI, 1940) is probably mostly non-adrenergic, but SILVA et al. (1971) have shown it to include a substantial number of adrenergic fibres in the cat small intestine. We have found some regional differences in the innervation of the stomach. In guinea-pigs, mice and rats there was a particularly dense innervation of the muscle of the lesser curvature. The innervation of the greater curvature was similar to that in other non-sphincter regions in the intestine.

In the human, there is a sparse adrenergic innervation of the longitudinal and circular coats of the muscularis externa of the stomach and small intestine, and a slightly denser innervation towards the anal end of the large intestine (BAUMGARTEN, 1967; CAPURSO et al., 1968).

The ganglia of the submucosa belong to two plexuses, that of Meissner (the submucous plexus proper) and that of Henle (plexus entericus internus), which is adjacent to the circular layer of the muscularis externa [2]. In the small intestine, Henle's plexus is poorly developed and almost all ganglia belong to Meissner's plexus (LI, 1940; GUNN, 1968), but in the colon and rectum both plexuses can be clearly identified (RINTOUL, 1960; GUNN, 1968). Fluorescence histochemical studies show only one ganglionated plexus in the submucosa of the small intestine and this is well supplied with adrenergic terminals (Fig. 3d). In the rectum, however, two plexuses can be distinguished and adrenergic terminals ramify amongst the ganglion cells of both (FURNESS and COSTA, 1973b). The submucous ganglia of the stomach in rats (MURYOBAYASHI et al., 1968) and in cats, guinea-pigs and rabbits (personal observations) are not supplied by adrenergic fibres.

There is a sparse adrenergic innervation of most of the muscularis mucosae, although where this muscle is thicker, as in the cardia and anal canal, a greater number of fibres is present (BAUMGARTEN and LANGE, 1969a; COSTA and GABELLA, 1971). A few fibres extend into the villi and a small number of adrenergic axons form an open network about the bases of the mucosal glands (GABELLA and COSTA, 1967, 1968; HÅKANSON et al., 1969; RATZENHOFER et al., 1969; FURNESS, 1970a).

2 The division of the submucous ganglia into the plexuses of HENLE and MEISSNER dates from the work of SCHABADASCH (1930) and STÖHR (1930). SCHOFIELD (1968) has emphasized that this division does not correspond to the original descriptions given by MEISSNER and HENLE. This distinction between the two plexuses has been retained in the present article although in view of the little we know of their function, it would probably be more realistic to group together all the ganglia and their interconnections under the term "submucous plexus".

Fig. 4a–d. Adrenergic axon terminals in partially denervated preparations of the myenteric plexus of the guinea-pig small intestine

a and b. Most of the adrenergic fibres have degenerated. The remaining varicose adrenergic axons run through the ganglia (*g*). They seldom branch and do not form true pericellular endings. Both ×125

c. Surviving adrenergic axons which branch and ramify amongst the cells of a myenteric ganglion (*g*). ×240

d. An area of the myenteric plexus with more axons remaining after denervation. Varicose axons run in the nerve strands (*s*) between the ganglia (*g*). The axons run for considerable distances in the ganglia and can be traced from one ganglion to another. ×150

The innervation of the vasculature of the gastrointestinal tract is described in Section IV of this review.

D. The Ultrastructural Identification of Adrenergic Axons

Criteria for the identification of adrenergic axons in tissue examined under the electron microscope have been developed over the last decade (WOLFE et al., 1962; RICHARDSON, 1962; BURNSTOCK and ROBINSON, 1967; MACHADO, 1967; TAXI and DROZ, 1967; TRANZER and THOENEN, 1967; HÖKFELT, 1968; VAN ORDEN et al., 1970). The principal criterion, the presence of granular vesicles of characteristic size within the axon, has been derived from studies of nerves outside the gastrointestinal tract; as discussed below, the identification of adrenergic nerves within the gut wall is more complicated. For organs other than the gastrointestinal tract, the vesicles are very numerous in adrenergic terminals and have diameters of 300–700 Å with an average of about 500 Å. Adrenergic axons also contain a proportion (usually 2–4%) of large granular vesicles of average diameter about 1 000 Å. The small granular vesicles appear slightly larger after glutaraldehyde than after OsO_4 fixation (MACHADO, 1967). Comparison of the work of VAN ORDEN et al. (1966, 1970) also indicates a dependence of vesicle diameter on fixation technique; the small granular vesicles in axons of the rat vas deferens ranged from 600–800 Å in diameter if glutaraldehyde was used as a fixative but were 400–700 Å following fixation with $KMnO_4$.

Within the gastrointestinal tract the situation is complicated by two factors. First, the presence of non-adrenergic axons containing large opaque vesicles, 800–1600 Å in diameter, and, second, the probable presence of two classes of adrenergic axon. Axons in which most vesicles are 800–1600 Å in diameter and contain electron-opaque cores which almost fill the vesicle occur throughout the intestine (HAGER and TAFURI, 1959; TAXI, 1965; NAGASAWA and MITO, 1967; ONO, 1967; PICK, 1967; TAXI and DROZ, 1967; BAUMGARTEN et al., 1970). These axons are not adrenergic, as borne out by the fact that they cannot be labelled autoradiographically with catecholamines (TAXI and DROZ, 1967), and are not affected by the false transmitter substances 5- and 6-hydroxydopamine (BAUMGARTEN et al., 1970). The vesicles have been referred to as p-type (BAUMGARTEN et al., 1970), large opaque vesicles (LOV) (BURNSTOCK, 1972) and heterogeneous granulated vesicles (HGV) (GABELLA, 1972).

Electronmicroscopists have found it difficult to locate axons containing small granular vesicles within the wall of the alimentary canal, and TAXI (1969) suggested that the vesicles of adrenergic nerves supplying the intestine may have different properties from those of other peripheral adrenergic nerves. It now seems that there are two types of adrenergic axon within the gut wall, one containing a preponderance of medium-sized granular vesicles, 500–1 100 Å (average about 800 Å) in diameter, and the other a preponderance of small granular vesicles, 300–700 Å in diameter and thus similar to the vesicles of adrenergic axons supplying other organs. The difference in the sizes of the vesicles of these two axon types is greater than the differences which have been observed as a consequence of variations in preparative techniques. Furthermore, both types of axon have been found in tissue prepared by the same method (DERMIETZEL, 1971; GABELLA, 1972).

The medium-sized granular vesicles have an electron-lucent halo between the granule and the vesicle membrane. They have been found in the stomach, upper small intestine, and colon, but not in the ileum, of humans, monkeys, mice and guinea-pigs and throughout the gastrointestinal tract in rats (GRILLO and PALAY, 1962; TAFURI, 1964; HONJIN et al., 1965; PICK, 1967; BAUMGARTEN et al., 1970; DERMIETZEL, 1971). BAUMGARTEN et al. (1970) used 5- and 6-hydroxydopamine to identify adrenergic nerves in the large intestine of guinea-pig, monkey and man. They found that most of the vesicles of adrenergic axons were medium-sized, between 500 and 900 Å in diameter and contained electron-dense cores. GABELLA (1972) has confirmed this observation for the guinea-pig. DERMIETZEL (1971) examined the nerves innervating the inner circular muscle of the pylorus of the mouse, which can be shown histochemically to be densely innervated by adrenergic nerves, and found axons containing large numbers of granular vesicles 600–1 000 Å in diameter (average 830 Å). This is in excellent agreement with the observations of HONJIN et al. (1965) who found axons within Auerbach's plexus of the stomach, jejunum, colon, and rectum of mice containing granular vesicles with diameters of 600–1 200 Å (average 830 Å). Similar granular vesicles, with average diameters of 840 Å, have been reported to be the predominant type in certain axons of the enteric plexuses of rats; some of these axons form synapses with neurone somas in the myenteric plexus (GRILLO and PALAY, 1962). Medium-sized granular vesicles observed in the human jejunum have average diameters of 700 Å (PICK, 1967). TAFURI (1964) reported average diameters of 800 Å (guinea-pig duodenum) and 900 Å (guinea-pig colon).

The adrenergic nerves supplying intestinal blood vessels contain small granular vesicles of 300–700 Å and are therefore similar to other adrenergic nerves (DEVINE and SIMPSON, 1967; DERMIETZEL, 1971). Likewise, small granular vesicles with diameters in the range normally expected for adrenergic nerves have been found in some regions of the small intestines of rats and guinea-pigs. NAGASAWA and MITO (1967) found nerve fibres in Auerbach's plexus of the small intestine of the rat which contained small granular vesicles averaging 400 Å in diameter, and GABELLA (1972) reported that the small granular vesicles of putative adrenergic axons in the small intestine of the guinea-pig were 400–600 Å in diameter. GABELLA (1970) found small granular vesicles, 400–700 Å in diameter, in axons of the circular muscle of rat small intestine.

The division of the adrenergic axons supplying the gastrointestinal tract into two ultrastructurally distinct types is justified, but is based only on a few studies. Further investigation is required to determine whether the two types respond differently to drug treatment and whether they have different functions, and to compare their distributions.

E. Sites of Action of the Neurally Released Transmitter

The anatomical observation that adrenergic fibres ramify extensively in the ganglia of Auerbach's plexus and that few adrenergic fibres run in the musculature suggests that adrenergic nerves cause inhibition of gastrointestinal movement principally by their action on enteric ganglia; but this is not as clear-cut as the anatomical results might suggest; noradrenaline released within the myenteric plexus can act both on the enteric ganglia and on the muscle. Adrenergic nerves could inhibit motility by activating intrinsic inhibitory neurones or by antagonizing transmission from intrinsic excitatory neurones. With the possible exception of the recent work of OHKAWA and PROSSER (1972a, b), the evidence available strongly points to the second alternative. Stimulation of mesenteric nerves does not activate intrinsic inhibitory neurones because it causes only a slight hyperpolarization of the muscle (GILLESPIE, 1962; BENNETT et al., 1966a; FURNESS, 1969b), whereas stimulation of intramural (non-adrenergic) inhibitory nerves results in large transient hyperpolarizations (BENNETT et al., 1966b; BÜLBRING and TOMITA, 1967; KURIYAMA et al., 1967; FURNESS, 1969b; see also FURNESS and COSTA, 1973a). There is now a large body of evidence which shows that exogenous adrenaline or noradrenaline can reduce the amount of acetylcholine released from intrinsic cholinergic neurones (see LEE, 1970; FURNESS and BURNSTOCK, 1974). However, attempts to demonstrate that the activation of the adrenergic nerves supplying Auerbach's plexus can cause a similar reduction have not always been successful. GERSHON (1967) examined the effect of stimulation of adrenergic nerves on the resting output of acetylcholine from the guinea-pig ileum and rabbit jejunum. In the guinea-pig ileum, 60% of the output is blocked by tetrodotoxin or hexamethonium and therefore depends on the activity of intrinsic cholinergic neurones (PATON and ZAR, 1968; PATON et al., 1971). The resting output in the ileum measured by GERSHON was 53.6 ng/g/min, that by PATON et al. (1971) was 51 ng/g/min; therefore the proportion of the output arising from active neurones was likely to be similar in the two experiments. If the adrenergic nerves act on the ganglia to inhibit acetylcholine release, then reductions up to 60% in output can be expected when paravascular nerves are stimulated. However, GERSHON (1967) found no difference in output with stimulus frequencies of 5 and 20 Hz. In the rabbit jejunum, stimulation at 5 and 10 Hz did not reduce the output although the movements of the jejunum were inhibited. There was some reduction in acetylcholine output at 30 and 40 Hz. A further attempt to show an inhibition of acetylcholine release was made by BEANI et al. (1969). These authors found that stimulation of mesenteric nerves at a frequency of 50 Hz for 10 minutes caused an insignificant reduction in the output of acetylcholine from the guinea-pig colon during stimulation of pelvic or intramural nerves at 10 Hz but that the output during stimulation of pelvic or intramural nerves at 1 Hz was slightly but significantly reduced. BEANI et al. (1969) pointed out that stimulation of mesenteric nerves in reserpine-pretreated animals causes the release of acetylcholine and that this acetylcholine may have masked an inhibition of the output from neurones in the gut wall. Further studies have been made using the guinea-pig ileum and rabbit small intestine (KNOLL and VIZI, 1971; VIZI and KNOLL, 1971). In the guinea-pig ileum it was found that amphetamine reduced the spontaneous release of acetylcholine by 62% and that pretreatment with reserpine in a dose which would deplete the adrenergic nerves of noradrenaline (COSTA and FURNESS, 1971) abolished this effect of amphetamine. VIZI and KNOLL (1971) found that stimulation of mesenteric nerves at 10 Hz reduced the resting output of acetylcholine from rabbit jejunum by 45%. They also found that the release of acetylcholine at rest and during stimulation was reduced by a brief exposure to guanethidine (4×10^5 M). The authors argue that this action of guanethidine is due to its displacing noradrenaline from adrenergic nerves and the noradrenaline so released acting on cholinergic neurones.

Taken together, the results of *in-vitro* experiments described above show that noradrenaline released from nerve endings within the gut wall reduces the output of acetylcholine from the neurones of the myenteric plexus but, as will be discussed latter, additional modes of action can be demonstrated *in-vitro* or at high stimulus frequencies *in-vivo*. On the other hand, when the actions of adrenergic nerves are examined *in-vivo*, and their discharge is elicited reflexly or with low (< 10 Hz) frequencies of stimulation, it is found that their action is almost entirely confined to the inhibition of the activity of enteric cholinergic ganglia (KEWENTER, 1965; JANSSON and MARTINSON, 1966; HULTÉN and JODAL, 1969; JANSSON, 1969b; JANSSON and LISANDER, 1969; JANSSON et al., 1969). In KEWENTER'S (1965) experiments, adrenergic nerves supplying the ileum were tonically active and only weak excitation occurred when the vagus nerves were stimulated. If adrenergic transmission was blocked by ergotamine or guanethidine, the effect of vagal stimulation was markedly enhanced. The adrenergic nerves were not acting on the muscle; contractions in response to acetylcholine were not dependent on the level of activity of adrenergic nerves. JANSSON and his colleagues have demonstrated that the situation in the cat stomach is similar. JANSSON and MARTINSON (1966) showed that the activation of adrenergic nerves during an intestino-gastric inhibitory reflex did not affect the basic myogenic tone of the stomach but that it did inhibit contractions in response to stimulation of vagal cholinergic nerves. In contrast, the gastric contraction caused by exogenous acetylcholine was unaffected by the reflex firing of adrenergic nerves to the stomach. Direct stimulation of the adrenergic nerves at up to 10 Hz produced prompt inhibition of movement when vagal fibres were active but gave only weak inhibition after the animals were treated with atropine. In this context, it should be pointed out that the stomach of the cat is still capable of substantial relaxation to appropriate stimuli after the administration of atropine (JANSSON, 1969a). Supporting evidence is given by the experiments of JANSSON and LISANDER (1969), who showed that stimulation of adrenergic nerves supplying the stomach, in which there was an abnormally elevated activity of cholinergic nerves following chronic vagotomy, markedly inhibited the cholinergic contractions even at low frequencies of stimulation. After the cats had been given atropine, similar stimulation of adrenergic nerves failed to relax the gastric muscle. It has also been shown that the activation of adrenergic nerves by stimulation of the hypothalamus (JANSSON et al., 1969) or of somatic pressor afferents (JANSSON, 1969b) inhibits vagal excitation of the cat stomach, but does not affect contractions caused by exogenous acetylcholine. HULTÉN and JODAL (1969) have brought forward evidence for a ganglionic site of action of ihnibitory adrenergic fibres supplying the colon of the cat. They found that low frequencies of stimulation of adrenergic nerves suppressed contractions caused by pelvic or vagus nerve stimulation or by spontaneous activity of enteric neurones but that neither the myogenic contraction in response to distension of the colon nor the contractions caused by exogenous acetylcholine were antagonized. WATT (1971) has shown that stimulation of mesenteric nerves is more effective in reducing the contraction caused by a single stimulus pulse applied to intrinsic cholinergic nerves than in reducing the contraction to exogenous acetylcholine in the guinea-pig ileum *in-vitro*. In contrast to the results of JANSSON and his colleagues, REED and SANDERS (1971) found that stimulation of splanchnic nerves equally reduced the excitatory effects of vagal stimulation and close intra-arterial injection of acetylcholine in the cat stomach, suggesting an action at the muscle in their experiments. REED and SANDERS stimulated the distal ends of the severed splanchnic nerves in cats under ether anaesthesia at a frequency of 10 Hz, although the experiments of JANSSON and LISANDER (1969) suggest that stimuli at frequencies of 2 or 4 Hz act through the ganglia, but those at 8 or 16 Hz also act directly on the muscle. The direct action on the muscle could be a result of release from the few adrenergic fibres which run in the musculature, or overflow of noradrenaline from fibres supplying the myenteric plexus, or a combination of these effects.

When cholinergic effects are abolished by muscarinic blocking agents (hyoscine or atropine), stimulation of paravascular nerves with frequencies as low as 1–2 Hz still inhibits intestinal movements *in-vitro* (GILLESPIE, 1960; BURNSTOCK et al., 1966; CAMPBELL, 1966). This implies that noradrenaline released from nerves within the musculature is more effective *in-vitro* and/or noradrenaline diffuses from the terminals in the plexus to the muscle. Such diffusion is feasible; BOULLIN et al. (1967) have shown that radioactive label appears in the venous outflow from the intestine preincubated with H^3 -noradrenaline even when adrenergic nerves are not stimulated. CELANDER (1959) suggested that the inhibition of intestinal muscle which is observed when mesenteric or splanchnic nerves are stimulated is due to the overflow of noradrenaline from nerves supplying blood vessels. It now seems that release from nerves in the myenteric plexus is more plausible. In the proximal colon of the guinea-pig, the perivascular adrenergic nerves arise in extrinsic ganglia, while most of the adrenergic terminals of the myenteric plexus come from intrinsic adrenergic cell bodies (FURNESS and COSTA, 1971). Adrenergic terminals are not found in the longitudinal muscle (COSTA and GABELLA, 1971) but when mesenteric nerves are stimulated in the presence of hyoscine, the longitudinal muscle relaxes. The relaxation is blocked by guanethidine, bretylium or by a combination of phentolamine and propranolol, indicating that it is due to the release of noradrenaline; it is also antagonized (and in some experiments blocked) by pentolinium, indicating that there is a ganglionic interruption between the point of stimulation and the points of release of noradrenaline (COSTA and FURNESS, unpublished). Therefore the adrenergic relaxation of the longitudinal muscle depends on the release of noradrenaline from terminals in Auerbach's plexus and not from vasomotor nerves. GERSHON (1967) found that an adrenergic inhibition of the guinea-pig stomach could be obtained with stimulus frequencies of 5 Hz even though excitation of intramural ganglia was blocked by hexamethonium and the action of acetylcholine on the muscle was blocked by hyoscine. This inhibition is probably due to noradrenaline released both from the myenteric plexus and from the nerves within the muscular layers of the guinea-pig stomach (see Section II C). Although the experiments cited above indicate that the overflow of noradrenaline from the nerves of the myenteric plexus to the muscularis externa occurs during low frequencies of stimulation *in-vitro*, there is no evidence that it happens *in-vivo* with the possible exception of high-frequency discharge of adrenergic fibres during the high-threshold intestino-intestinal inhibitory reflex (see Section V). In the normal animal, noradrenaline which is not reabsorbed by the nerves is possibly swept up into the bloodstream before it can act on the muscle.

CREMA et al. (1970) have shown that the stimulation of adrenergic nerves *in-vitro* antagonizes the peristaltic reflex in segments of colon from cats and guinea-pigs. In such segments the introduction of a bolus into the lumen induces contraction of both muscle coats and causes a peristaltic wave which moves the bolus aborally. Stimulation of the mesenteric inhibitory nerves at 2 Hz significantly reduced the contractions and slowed the movement of the bolus. The reflex was completely blocked by stimulation at 8 Hz or less, but this did not antagonize the contractile effect of the intra-arterial injection of acetylcholine.

The foregoing arguments indicate that noradrenaline released from nerve fibres within the myenteric plexus acts on intrinsic cholinergic neurones to reduce their excitatory effect on the muscle and it is only *in-vitro*, or at high frequencies of stimulation *in-vivo*, that noradrenaline diffuses from nerve terminals in the plexus to act directly on the muscle. The observed distribution of adrenergic nerves is clearly consistent with this mechanism. Although we consider that diffusion of noradrenaline to the non-sphincter muscle of the gut usually makes only a small contribution to the inhibition of motility which occurs when these nerves act *in-vivo*, diffusion of transmitter from a localized plexus to a muscular effector which is relatively free of nerves seems important in other situations. For instance, most of the adrenergic fibres which supply mammalian arteries form a network between the connective tissue of the adventitia and the medial smooth muscle, and yet these nerves act very effectively to constrict the vessels. A further example is transmission from cholinergic nerves to the guinea-pig ileum; although these nerves are almost completely confined to the myenteric plexus, a single shock releases sufficient acetylcholine to contract the longitudinal muscle (PATON and ZAR, 1968; AMBACHE and FREEMAN, 1968).

Just as the recent anatomical demonstration of adrenergic nerves innervating intramural ganglia is a revival of earlier observations (Section C, above), the conclusion of modern physiologists that mesenteric inhibitory nerves act on intramural neurones was arrived at many years ago. In 1858, writing of the effects of these nerves on the rabbit intestine, Lord Lister said, "the inhibitory influence does not operate directly on the muscular tissue, but on the nervous apparatus by which its contractions are, under ordinary circumstances, elicited". Lister had found that the propagation of contraction away from a point of local irritation of the intestine, which, he had concluded from other experiments, depended on the intramural ganlionated plexuses, was prevented by stimulation of the thoracic source of the splanchnic nerves.

A new approach to the study of the innervation of the intestine is to record directly from the enteric ganglion cells. OHKAWA and PROSSER (1972a), who used extracellular electrodes to make such recordings, reported that stimulation of mesenteric nerves caused no significant change in the spontaneous firing of enteric neurones of the cat small intestine. On the other hand, adrenaline and noradrenaline both increased the discharge in some enteric neurones (OHKAWA and PROSSER, 1972b). More work needs to be done to determine the type of cells sampled by these authors (there are cholinergic, non-adrenergic inhibitory, sensory and possibly other neurone types in enteric ganglia). There is no reason at the moment to regard their findings as contradicting the concept of an inhibitory action of adrenergic nerves on the activity of enteric cholinergic neurones.

F. Receptor Mechanisms

Since the work of AHLQUIST (1948), receptor sites for the action of noradrenaline have come to be divided into two groups, α and β, first on the basis of the relative potencies of directly-acting sympathomimetic amines and more recently on the basis of the selective antagonism of the action of catecholamines by certain drugs. These two categories are adhered to, although more refined analysis shows that each group could be subdivided (FURCHGOTT, 1972).

Our knowledge of the receptor types involved in the inhibition of gastrointestinal movement by exogenous catecholamines has been reviewed recently (FURNESS and BURNSTOCK, 1974). The following generalization was made: inhibition of the release of acetylcholine from intrinsic cholinergic neurones is mediated through α-adrenoceptors; direct inhibition of the muscle is mediated through both α-and β-receptors. The relative importance of the two receptor populations on the muscle is different in different segments of the alimentary canal. Beta-receptors on the muscle are nearly always inhibitory, but α-receptors often mediate excitation, in some cases in areas where the dominant response is inhibition.

The mechanism by which noradrenaline released from adrenergic nerves and acting through α-receptors inhibits the release of acetylcholine from intrinsic neurones is not known. Present evidence indicates that the action could be on the cell body, on the cholinergic axon, or on both.

Noradrenaline probably has an hyperpolarizing action on the ganglion cells, thus raising the threshold for their activation and inhibiting ganglion cell firing; this would be consistent with the mechanism of inhibition via α-receptors which is encountered in other autonomic ganglia (for review see TRENDELENBURG, 1967). Depolarization block is unlikely to occur, as this would involve an initial excitation of cholinergic neurones by noradrenaline which is not in fact observed. PATON and VIZI (1969) suggest that noradrenaline might inhibit the output of acetylcholine by hyperpolarizing the cholinergic nerve endings. However, it seems more likely that depolarization of the endings would be caused by noradrenaline because it is generally observed that hyperpolarization of nerve terminals enhances, whereas depolarization reduces, the release of transmitter consequent upon the arrival of an action potential or on local suprathreshold stimulation (ECCLES, 1961, 1964; KATZ, 1966; LLINAS, 1966).

JENKINSON and MORTON (1967) examined the effects of noradrenaline and isoprenaline on the uptake and loss of radioactive K from the guinea-pig taenia coli and the antagonism of the catecholamine effects by α-and β-adrenoceptor blocking agents. Their work showed that α-receptors were involved in increasing the membrane permeability to K but that activation of β-receptors inhibited the muscle by a separate mechanism, possibly a limitation of the availability of Ca for interaction with contractile proteins. JENKINSON and MORTON's conclusions were confirmed by BÜLBRING and TOMITA (1969a, b, c) who further analyzed the mechanism of catecholamine inhibition of the taenia coli in experiments using a double sucrose gap technique. BÜLBRING and TOMITA (1969a) showed that the reversal potential for adrenaline was 10 to 20 mV more negative than the resting membrane potential. This is well short of the equilibrium potential for K, which is 35–37 mV below the resting membrane potential (CASTEELS, 1966; CASTEELS and KURIYAMA, 1966), suggesting that adrenaline increases the conductance to other ions as well as K. The effects of altered Cl and Na concentrations in the bathing solution indicate an increased conductance of Cl but not of Na. The actions of adrenaline and noradrenaline in decreasing membrane resistance and causing hyperpolarization of the muscle are blocked by phentolamine, indicating an action through α-receptors (BÜLBRING and TOMITA, 1969b). However, as also shown by BÜLBRING and TOMITA (1969b), the action of catecholamines on β-receptors causes suppression of action potentials without any change in membrane conductance. Both α- and β-effects are anatagonized by removal of Ca ions from the external solution and BÜLBRING and TOMITA (1969c) have speculated that interaction of catecholamines with α-receptors might facilitate the binding of Ca to membrane sites in such a way as to increase K conductance and that interaction with β-receptors might antagonize the removal of Ca from the membrane which is normally involved in the generation of pacemaker potentials. Thus the action through α-receptors would be to increase Ca binding, that through β-receptors to preserve Ca already bound. In a great variety of tissues, effects mediated through β-receptors have been shown to involve the activation of adenyl cyclase and it has been proposed that adenyl cyclase is intimately connected with, or may be part of, the β-receptor (see ROBISON et al., 1971). Stimulation of β-receptors results in the formation of cyclic AMP (adenosine-3′-5′-monophosphate) and phosphorylase a and in the utilization of ATP. In the intestine significant increases in phosphorylase a were found when β-receptors were stimulated (DIAMOND and BRODY, 1966; BRODY and DIAMOND, 1967), and BUEDING et al. (1966) showed that levels of cyclic AMP increased by 40% in response to 2.5×10^{-8} M adrenaline. ANDERSSON and MOHME-LUNDHOLM (1970) found that intestinal relaxation mediated through β-receptors was preceded by an increase in cyclic AMP and phosphorylase a and a decrease in ATP. In addition, the relaxant and metabolic effects of β-receptor activation are mimicked by cyclic AMP and its dibutyryl analogue (KIM et al., 1968; ANDERSSON and MOHME-LUNDHOLM, 1970; BOWMAN and HALL, 1970). The relaxation of intestinal smooth muscle mediated through β-receptors is enhanced by theophylline, diazoxide and puromycin (antagonists of the hydrolysis of cyclic AMP by phosphodiesterase) and is reduced by imidazole, which enhances phosphodiesterase activity (ANDERSSON and MOHME-LUNDHOLM, 1969; WILKENFELD and LEVY, 1969; BOWMAN and HALL, 1970). The relaxations mediated through α-receptors were not modified by phosphodiesterase inhibition in the experiments of ANDERSON and MOHME-LUNDHOLM (1969) and WILKENFELD and LEVY (1969), but BOW-

MAN and HALL (1970) found that concentrations of theophylline greater than those used by the former groups antagonized the relaxation mediated through α-receptors. Imidazole is not specific in its action; it also reduces α-effects of catecholamines (BÜLBRING, 1967; ANDERSSON and MOHME-LUNDHOLM, 1969; BOWMAN and HALL, 1970), although WILKENFELD and LEVY (1969) found it antagonized only the β-receptor-mediated relaxation in the rabbit ileum. In contrast to other investigators, BÜLBRING and TOMITA (1969b) found that imidazole antagonized the effect of adrenaline (α-action) but did not affect that of isoprenaline (β-action) in the guinea-pig taenia coli. The increase in ATP levels noted by BUEDING et al. (1967) when the taenia coli is relaxed by adrenaline could arise through other mechanisms and cannot be taken as inconsistent with the mediation of β-effects through the activation of adenyl cyclase. For example, BUEDING et al. (1967) did not block the interaction of adrenaline with α-receptors, although it has been found that activation of α-receptors is often associated with decreased cyclic AMP levels (ROBISON et al., 1971) which suggests that tissue ATP might increase. Relaxation itself might be anticipated to increase tissue concentrations of ATP by decreasing demand on its usage. However, BUEDING et al. (1967) have shown that the increase in ATP induced by adrenaline is independent of relaxation. The lack of correlation between phosphorylase activation and relaxation in response to adrenaline in the guinea-pig taenia coli (BUEDING et al., 1962, 1966) might have arisen because the relaxation was principally mediated through α-receptors, as is suggested by the results of BÜLBRING and TOMITA (1969b) for this tissue. It is also worth considering that increased adenyl cyclase activity does not involve increased phosphorylase activity in the taenia coli of guinea-pig, although it does in other intestinal muscle.

In summary, the stimulation of α-receptors on intrinsic cholinergic neurones reduces the release of acetylcholine, probably by hyperpolarizing the cell bodies and possibly also by depolarizing the terminals. In the muscle, stimulation of α-receptors causes a Ca-dependent increase in the permeability to K and Cl in areas in which α-receptors mediate inhibition. It is tempting to speculate that α-receptors in areas in which catecholamines contract intestinal muscle are involved in similar cellular changes; increase in the proportion of the active current carried by chloride, at the expense of that carried by potassium, could change inhibition to excitation[3]. Activation of β-receptors on the muscle causes a stabilization of the membrane, again dependent on Ca. This inhibition is an energy-requiring process and involves the formation of cyclic AMP.

G. Non-Adrenergic Inhibitory Neurones

The characteristics and distribution of inhibitory nerves which innervate gastrointestinal muscle but release some substance other than noradrenaline have been

3 BÜLBRING et al. (1968) have suggested that a change in the relative proportions of current carried by potassium and chloride might cause the change from relaxation (in the virgin) to contraction (in the pregnant animal) when myometrial α-receptors of cat are stimulated. They speculate that an increase in potassium permeability predominates in the virgin, but that an increase in chloride permeability predominates during pregnancy.

reviewed recently (BORTOFF, 1972; BURNSTOCK, 1972; FURNESS and COSTA, 1973a). Our present knowledge indicates that all inhibitory pathways which are anatomically sympathetic are adrenergic, and that vagal and intrinsic inhibitory pathways are non-adrenergic. The non-adrenergic inhibitory fibres form the efferent links in a cascade of descending inhibitory reflexes that aid the passage of material along the alimentary canal (FURNESS and COSTA, 1973a).

Most cell bodies of the non-adrenergic inhibitory neurones are in Auerbach's plexus, but the terminals of adrenergic nerves which ramify amongst the ganglion cells do not seem to have any effect on the non-adrenergic neurones (JANSSON and MARTINSON, 1966; GERSHON, 1967).

III. Adrenergic Innervation of the Sphincters

The sphincters of the gastrointestinal tract are the regions whose specialized function is to provide a mechanical hindrance to the mixing of the contents lying proximal and distal to them. The clearly identifiable sphincters are the lower oesophageal (cardiac) sphincter, the sphincter of Oddi (choledocho-duodenal), the ileo-colic sphincter and the internal anal sphincter. The gastro-duodenal junction (pylorus) cannot be regarded as a true sphincter because its functional activity cannot be separated from the rest of the stomach and it is not associated with a region of raised intraluminal pressure (EDWARDS and ROWLANDS, 1968).

Although modern methods have been applied to investigations of the histochemistry and pharmacology of their adrenergic innervation in very few instances, we can deduce the way in which activation of adrenergic nerves affects the sphincters from less direct evidence.

The specific fluorescence histochemical method has been used in only a small number of studies, so that a generalized pattern of the adrenergic innervation of sphincter muscle cannot be presented, but it can be stated that the circular muscle of the sphincter regions is adrenergically innervated and that the density of innervation is often markedly greater than in adjacent non-sphincter muscle (Fig. 3f). In the lower oesophageal sphincter of the guinea-pig there is a very dense network of adrenergic fibres (COSTA and GABELLA, 1971), while in the monkey, cat and rat there are few adrenergic fibres in the musculature, no more than in adjacent non-sphincter areas (BAUMGARTEN and LANGE, 1969a; GILLESPIE and MAXWELL, 1971). The circular musculature of the pyloric region of the rat and guinea-pig is densely supplied by adrenergic fibres (GILLESPIE and MAXWELL, 1971; COSTA and GABELLA, 1971). Intramuscular adrenergic fibres are present in the common bile duct and the sphincter of Oddi of all species that have been examined, that is, in cat, guinea-pig, monkey and rabbit (BAUMGARTEN and LANGE, 1969b; MORI et al., 1971; PERSSON, 1971). There is no significant increase in the density of the adrenergic innervation in the region of the ileo-colic junction of guinea-pig whereas the ileo-colic sphincter in rat and rabbit is more densely innervated than is the adjacent nonsphincter muscle (unpublished observations). The internal anal sphincter is richly supplied by adrenergic nerves in guinea-pig (Fig. 3f), cat and dog (COSTA and GABELLA, 1971; HOWARD and GARRETT, 1973; FURNESS and COSTA, 1973b) but is less heavily innervated in rat and man (BAUMGARTEN, 1967; GILLESPIE and MAXWELL, 1971).

The origins of the adrenergic fibres supplying the sphincters has not been extensively investigated with modern methods, but they do seem to correspond with the sources of adrenergic fibres to adjacent non-sphincter muscle (Fig. 1). The lower oesophageal sphincter receives sympathetic (presumably adrenergic) fibres which follow the gastric arteries from the coeliac ganglion (MITCHELL, 1953). The cell bodies of the adrenergic nerves to the sphincter of Oddi are located in the coeliac plexus judging by denervation experiments in cats (BAUMGARTEN and LANGE, 1969b), while those innervating the ileo-colic sphincter are in the superior mesenteric ganglia (ELLIOTT, 1904a). The internal anal sphincters of cat and dog receive sympathetic nerves running in the colonic and hypogastric nerves. LANGLEY and ANDERSON (1895b) used nicotine to show that a substantial proportion of the nerves forming efferent pathways to the internal anal sphincter of the cat pass through the inferior mesenteric ganglia and form synapses with more peripherally located neurones. In contrast, in the dog almost all of the efferent pathways to the anal sphincter appear to have a ganglionic interruption in the inferior mesenteric ganglion (LEARMONTH and MARKOWITZ, 1929). In guinea-pig, we found that no or very few adrenergic fibres to the internal anal sphincter originate in the inferior mesenteric ganglion, but adrenergic axons run from the sacral sympathetic chains and from the posterior pelvic plexuses to the sphincter (COSTA and FURNESS, 1973a). It is not clear whether adrenergic fibres arise from cells in the posterior pelvic plexuses or whether part of the innervation from the sympathetic chains passes through these plexuses to the sphincter.

With few exceptions, those investigators who were able to record mechanical responses from sphincter muscle following stimulation of the sympathetic supply reported a constrictor effect which was mimicked by adrenaline and noradrenaline. The lower oesophageal sphincter is contracted by adrenaline or noradrenaline and by sympathetic nerve stimulation in cat, rabbit, dog, monkey, and man (BOTHA, 1962). Although CARLSON et al. (1922) reported relaxation of this sphincter on sympathetic stimulation in rabbit, BRÜCKE and STERN (1938) observed contraction. The sphincter of Oddi is contracted by stimulation of the splanchnic nerves (DOYON, 1894; WESTPHAL, 1923) but a reflex relaxation to distension of the gall-bladder mediated by a pathway through the coeliac ganglion has been reported in the cat (WYATT, 1967). There are both α-excitatory and β-inhibitory receptors for catecholamines in the muscle and exogenous noradrenaline can cause relaxation or contraction, depending on the conditions (BENZI et al., 1964; MORI et al., 1971; PERSSON, 1971). The response of the ileo-colic sphincter is clear-cut; there is general agreement that in a wide variety of mammals it contracts to sympathetic nerve stimulation and to adrenaline and noradrenaline. In dog and guinea-pig, stimulation of sympathetic nerves causes contraction of the distal ileum (SMETS, 1936a; MUNRO, 1953), so that this final part of the small intestine may function with the ileo-colic sphincter to inhibit transport between the ileum and the colon. The internal anal sphincter is contracted by stimulation of the sympathetic nerves in all mammals with the probable exception of the rabbit, in which it relaxes in response to sympathetic nerve stimulation and to adrenaline (LANGLEY and ANDERSON, 1895a; LANGLEY, 1901). The human internal anal sphincter contracts when exposed to adrenaline or noradrenaline (FRIEDMANN, 1968) but stimulation of the hypogastric nerves produces inhibition (SHEPHERD and WRIGHT, 1968). The type of nerve involved in this inhibition has not been identified.

In all the sphincters, contractions elicited by adrenaline and noradrenaline are blocked by α adrenoceptor-blocking drugs and are usually reversed to relaxation by β-adrenoceptor-blocking drugs. The β-adrenoceptor agonist isoprenaline usually relaxes the sphincters. The contraction of these sphincters to the stimulation of the adrenergic nerves can also be reversed to relaxation by α-adrenoceptor blocking drugs. The sphincter muscle therefore has α-excitatory and β-inhibitory receptors but it seems that that the α-receptors are the ones which are activated physiologically by noradrenaline released by the adrenergic nerves. The excitatory effect of adrenergic nerves at the sphincters appears to be due to a direct effect of noradrenaline on receptors in the muscle.

IV. Adrenergic Nerves and Gastrointestinal Vasculature

A. Effects of Nerve Stimulation

The primary action of adrenergic nerves on the vessels of the gastrointestinal tract is to decrease arterial calibre, thus elevating the resistance to flow through the splanchnic vasculature, and to constrict the larger veins, thus decreasing the volume of blood contained in this circuit (see reviews by GREEN and KEPCHAR, 1959; LACROIX, 1960; BRADLEY, 1963; GRIM, 1963; GRAYSON and MENDEL, 1965). The adrenergic nerves constrict the arteries of the mesentery and of the intestinal wall, causing blanching of the mucosa and a fall in intraluminal temperature. The mesenteric and portal veins are also constricted; with vigorous stimulation, the constriction of the large mesenteric and of the portal veins is so strong that the blood displaced from the splanchnic bed gives an initial increase in outflow, coincident with a decrease in inflow due to arterial constriction. The splanchnic vessels hold about 20% of the total blood volume of the body, most of it in the veins (LACROIX, 1960; BRADLEY, 1963) and some 35–40% of this blood can be displaced from the splanchnic bed by moderate stimuli (FOLKOW et al., 1964a).

The mesenteric arteries are surrounded by a dense plexus of adrenergic nerves at their adventitio-medial border, and the arteries within the gut wall are similarly innervated (FURNESS, 1971). The veins within the walls of the alimentary tract are very sparsely innervated, but the nerve supply increases as larger vessels are formed in the mesentery and becomes dense in the walls of the superior mesenteric and hepatic portal veins.

The decrease in intestinal blood flow in response to stimulation of adrenergic nerves declines after 1–3 minutes of maintained stimulation even at low frequencies (FOLKOW et al., 1964a; DRESEL and WALLENTIN, 1966; HULTÉN et al., 1969).

This phenomenon is known as autoregulatory escape[4]. When stimulation is stopped there is a further increase in flow, in this case to a rate greater than the control value (reactive hyperaemia). The escape is not specific to adrenergic nerve stimulation and occurs with a variety of constrictor drugs in different animals, but the degree of escape does depend on the species, drug, method of administration and concentration (DRESEL and WALLENTIN, 1966; SHEHADEH et al. 1969; ROSS, 1970, 1971c; HENRICH and LUTZ, 1971; FARA and ROSS, 1972). Autoregulatory escape occurs under conditions of constant flow *in vivo* and in isolated vascular segments *in vitro,* and thus it is not a consequence of the accumulation of vasoactive metabolites. There is some controversy as to whether the escape from the influence of adrenergic nerves and infused catecholamines involves a relaxation of the vessels originally constricted, which, in the absence of other changes, would mean that the relative magnitudes of flow through different layers of the gut would be maintained, or whether there is a preferential flow through some vessels, resulting in 'selective' underperfusion of parts of the mucosa. The evidence at present available indicates that there is a redistribution of blood flow but that this is not the cause of the autoregulatory escape.

FOLKOW et al. (1964b) used the distribution of Indian ink to assess relative changes in blood flow through the mucosa, submucosa, external muscle and serosa of the cat intestine. Segments of intestine were taken before nerve stimulation, during the peak resistance increase to stimulation at 8 Hz, and during the period of autoregulatory escape. The segments were immediately frozen and 100 μm sections were used for examination. Their published figure suggests that flow decreased markedly in all four areas during peak constriction and then recovered partially in the mucosa and substantially in the submucosa during autoregulatory escape, while flow in the muscle and serosa remained depressed or was reduced even further. FOLKOW et al. (1964b) deduced that there was a well-sustained, often profound diversion of blood flow from the mucosa even when total flow was little reduced during the autoregulatory escape from vasoconstrictor fibre stimulation. DRESEL and WALLENTIN (1966) compared the effects of infused noradrenaline and of the stimulation of adrenergic nerves and concluded that both act in the same manner, causing sustained underperfusion of the mucosa during the period of autoregulatory escape. ROSS (1971a) measured the uptake of ^{86}Rb into mucosa, submucosa and external muscle in the small intestine of cats, before and during infusion of noradrenaline, and found no significant difference in flow distribution between the control and escape periods. The apparent conflict probably arises because the diversion of blood is not from the entire mucosa but only from its deeper layers while flow increases in the lamina propria. During the period of autoregulatory escape, the ability of raised venous pressure to transfer fluid from the circulation into the extravascular space of the intestine is lessened as is the diffusion of ^{86}Rb into the tissue (DRESEL et al., 1966). However, the effect on fluid transfer is greater than that on ^{86}Rb transfer. These results are best explained by a redistribution of flow between capillaries, so that some are effectively closed off while others remain patent

4 ROSS (1971b) criticizes the use of the term 'autoregulatory escape' maintaining that autoregulation should only be used to describe the regulation of blood flow according to metabolic needs and that, for the escape of intestinal vessels from vasoconstrictor influences, this is not the case.

and carry a greater flow than normal (DRESEL et al., 1966). There is little likelihood that blood flow is diverted through non-nutritional shunts, because, although DRESEL et al. were able to detect artificial bypassing of as little as 5% of the blood, there was no indication of such shunting during the escape period. Moreover, the oxygen consumption of the intestine is greater during the escape period than it is during the initial constriction (BAKER and MENDEL, 1967). Other experiments make it clear that some capillaries are bypassed during the autoregulatory escape, because although blood flows may return to near control during this period, capillary filtration remains depressed (FOLKOW et al., 1964a, b; HULTÉN et al., 1969). The direct observation of mesenteric capillaries during noradrenaline infusion shows there is a redistribution of flow amongst capillaries arising from the one arteriole (RICHARDSON and JOHNSON, 1970).

The reactions of the large arteries running through the mesentery to supply the gut contribute to the autoregulatory escape. FARA (1971) examined the effects of nerve stimulation on isolated arterial rings from the cat mesentery and observed that the constriction decreased during maintained nerve stimulation.

It is concluded that during continued activation of adrenergic nerves there is a sustained diversion of blood flow from some vessels in the mucosa, and that there is an escape from the vasoconstriction which does not involve nutritional shunts but which principally arises from a relaxation of the innervated arteries and arterioles of the mesentery and submucosa.

RICHARDSON and ZWEIFACH (1970) used micropipettes to measure the pressures in different vessels of the cat mesentery. They found that most of the fall in pressure across this vascular bed occurs in vessels of 20–40 μm in internal diameter, and hence these vessels provide most of the resistance to flow at rest. When the nerves to the mesentery are stimulated at moderate frequencies (3–6 Hz), there is substantial constriction in the large mesenteric arteries, as well as in arterioles greater than about 20 μm in diameter (FURNESS and MARSHALL, 1973). This suggests that, during discharge of adrenergic nerves, a greater length of the arterial tree becomes important in providing resistance to flow and that it is not changes in the immediately precapillary vessels that provide the major increase in resistance.

The constrictor action of adrenergic nerves and of catecholamines is blocked or changed to slight dilatation by ergotoxin, dibenamine, azapetine and similar drugs (DALE, 1913; WOODS et al., 1932; Clark, 1934; BÜLBRING and BURN, 1936; FOLKOW et al., 1948; DEAL and GREEN, 1956, among others) and is mediated through α-receptors. The dilatation of intestinal vessels in response to nerve stimulation or to adrenaline after blockade of α-receptors suggests the presence of inhibitory β-receptors, and their presence is confirmed by more recent experiments (GREEN et al., 1955; ROSS, 1967). The autoregulatory escape from the constrictor effect of adrenergic nerve stimulation is not reduced by β-receptor blockade (ROSS, 1971c).

B. The Influence of Blood Flow on Secretion and Absorption

Before the experiments of JACOBSON et al. (1966) the evidence for a direct relationship between gastric secretion and gastric mucosal blood flow was largely intuitive. JACOBSON et al. used the clearance of aminopyrine to estimate mucosal blood flow. For reasons discussed in their publication, it is very probable that the rate at which aminopyrine is cleared from the blood into the gastric contents is primarily dependent on the rate of flow through mucosal vessels in diffusive communication with the contents, so long as gastric pH remains low. Histamine and gastrin both increased blood flow and secretion, and while these drugs were used to provide a background elevation of flow and secretion the effects of isoprenaline, the secretory stimulant urecholine, and the inhibitors vasopressin, adrenaline, and secretin were examined. In all cases there was a positive correlation between increase in gastric secretory rate and aminopyrine clearance except that isoprenaline at the lower of two doses tested significantly increased clearance without altering an already high secretory rate. JACOBSON (1967) has reviewed the available evidence suggesting that secretion and blood flow are directly linked. From this and later work (JACOBSON, 1970) it is concluded that any change in mucosal blood flow tends to change secretion in the same direction and that any change in secretion tends to involve a similar flow change. Situations in which secretion and blood flow do not appear to change in parallel manner have been found but these do not invalidate this generalization.

These experiments do not allow us to decide whether adrenergic nerves act on blood flow and secretory effects follow, whether the opposite occurs or whether the nerves have a primary effect on both, but the substantial adrenergic innervation of mucosal and submucosal arteries and the paucity of innervation of other mucosal structures commend the first alternative. THOMPSON and VANE (1953) measured total gastric blood flow and gastric secretion in the vivi-perfused stomach and found that both secretion and flow were decreased by stimulation of the splanchnic nerves, but their experiments give little indication of any causal relationship. A further assessment of the relationship between mucosal blood flow and secretion during adrenergic nerve stimulation has been made by REED et al. (1971), who showed that stimulation of splanchnic nerves caused an increase in systemic blood pressure but had little effect on mucosal blood flow and secretion. When a blood reservoir was used to eliminate the rise in pressure, mucosal blood flow and secretion were both depressed. If the adrenergic nerves did have a direct effect on secretion, it would seem unnecessary to change blood pressure to reveal it. On the other hand, the fact that secretion followed blood flow changes indicates a direct dependence in these experiments. These studies also showed that mucosal blood flow remained reduced during stimulation; in fact, the average mucosal blood flow and secretion were lower during the second than during the first half of a 20-minute period of stimulation at 10 Hz.

Thus there is almost no doubt that adrenergic nerves inhibit acid secretion in the stomach by reducing mucosal blood flow and that they have little or no direct effect on acid secretion. Early suggestions that stimulation of splanchnic nerves may increase mucus secretion (BABKIN, 1950) deserve to be reinvestigated, particularly now that periglandular adrenergic nerves have been demonstrated (Section IIC).

There is no evidence that adrenergic nerves have a direct influence on absorption in the gastrointestinal tract and it is only in extreme cases that mucosal blood flow is a limiting factor for absorption by diffusion, although active absorption is more readily affected (SCHANKER et al., 1957; VARRO et al., 1965). Passive absorption is not substantially diminished if blood flow to the intestine is halved, but active absorption is reduced by 30–40% (VARRO et al., 1965).

C. Blood Flow and Motility

Interdependence of blood flow and motility has been demonstrated in the gastrointestinal tract, but its nature and importance are imperfectly understood. Under different conditions, the correlation can be positive, negative or absent (SHEHADEH et al., 1969; SEMBA and FUJII, 1970). Thus, the possibility is raised that the constrictor action of adrenergic nerves may contribute to motility changes and their actions on gastrointestinal muscle might influence blood flow. Inhibition of motility by adrenergic nerves is not a consequence of the restriction of blood flow; BAYLISS and STARLING (1899) were able to inhibit the movements of the intestine by stimulation of the splanchnic nerves in a freshly-killed animal, and numerous investigations have demonstrated inhibition in nonperfused segments of stomach and intestine *in vitro*. In the anaesthetized cat, with the adrenals excluded and the nerves to the intestine cut, the blood supply to the intestine has to be reduced to 25% or less before any significant reduction of motility occurs (KOCK, 1959). Relaxation of intestinal muscle tends to increase flow (SEMBA and SASAKI, 1953; SIDKY and BEAN, 1958; SEMBA and FUJII, 1970) and might be supposed to contribute to the autoregulatory escape from vasoconstriction caused by nerve stimulation or catecholamines. However, SHEHADEH et al. (1969) demonstrated escape from the constrictor actions of both noradrenaline and angiotensin although noradrenaline caused relaxation and angiotensin caused contraction of the intestinal muscle. Moreover, in the case of noradrenaline, maximum vasoconstriction coincided with maximum inhibition of motility. In conclusion, there is probably no significant interaction between the actions of adrenergic nerves on blood flow and on motility.

V. Adrenergic Activity in Vivo

A. Resting and Peak Discharge Rates in Adrenergic Nerves

There is probably no tonic activity of adrenergic nerves causing inhibition of non-sphincter regions of the gastrointestinal tract, but there is some evidence which, although not fully corroborated, suggests that adrenergic nerves normally exert a tonic constrictor action on some sphincter muscle. There is very good evidence that the adrenergic nerves make some contribution to the maintenance of tone in splanchnic vessels.

In most cases where a tonic inhibitory action of adrenergic nerves on motility has been proposed, the registration of gastrointestinal movement has required exposure of the viscera or surgical interference at the time of observation. Laparotomy, exposure, or handling of the intestine causes a profound and long-lasting inhibition of gastrointestinal tone and even cutting the surface skin of an anaesthetized animal causes a significant suppression of motility (CANNON and MURPHY, 1906; OLIVECRONA, 1927; GARRY, 1934; JANSSON and LISANDER, 1969). CANNON (1906) avoided this difficulty by using an X-ray method to examine movements of the stomach and intestines of cats before and after sympathectomy. His experiments show that there is no tonic adrenergic inhibition of gastric movement or tonic adrenergic closure of the pyloric sphincter in rested cats. In the experiments of JANSSON and LISANDER (1969), once the effects of surgical interference had subsided, there was no evidence of continued adrenergic inhibition of gastric motility. MCCREA (1926) has described experiments in which X-ray examination before and after section of sympathetic nerves did indicate some tonic activity of these fibres in the dog. The transection resulted in acceleration of gastric emptying, which appeared to be due to increased gastric activity rather than to relaxation of the pylorus. CANNON'S (1906) experiments indicate that there may be tonic inhibition of small intestine motility induced by certain substances when present in the contents, because the passage of protein (lean beef) was markedly accelerated by severing the splanchnic nerves, although the passage of carbohydrate (mashed potato) was almost unaffected. The experiments of M'FADDEN et al. (1935), who also used an X-ray technique, show that removal of the inferior mesenteric ganglia of the cat causes little appreciable change in the activity of the colon and rectum.

Even with X-ray techniques, it is difficult to derermine whether the sympathetic nervous system contributes to the maintenance of sphincter tone. However, there is no evidence that tonic activity of adrenergic nerves is a normal requirement. For example, in laboratory animals with their entire sympathetic chains excised, there are no apparent changes in the digestive process and no long-lasting changes in sphincter activity (CANNON et al., 1929; MOORE, 1930). Division of sympathetic nerves does not usually cause any significant change in the tone of the lower oesophageal sphincter (OCHSNER and DEBAKEY, 1940; INGELFINGER, 1958). In the case of the ileo-caecal sphincter, activity in sympathetic nerves appears to contribute to tone in dog and rat, but not in cat (ELLIOTT, 1904a; HINRICHSEN and IVY, 1931). LANGLEY and ANDERSON (1895a) found that section of the lumbar splanchnic nerves reduced the tone of the internal anal sphincter in the cat and that there was an even greater reduction if the inferior mesenteric ganglia were removed.

There were several early suggestions of a continuous descharge of adrenergic fibres supplying splanchnic blood vessels (see BUNCH, 1899; IZQUIERDO and KOCH, 1930) but most of the experiments are open to the same criticism as above, namely that sympathetic activity may have been induced by the surgical interference of the experiment. The problem was approached in a new way by IZQUIERDO and KOCH (1930) and KREMER and WRIGHT (1932) who placed snares around the splanchnic nerves at operation and led out the ligatures through small incisions. After the animals had been allowed to recover for 1 h (by the first authors, who used rabbits) or $\frac{1}{2}$ to 7 h (by the second authors, who used cats), the ligatures were pulled to break the nerves. This caused an immediate fall in blood pressure of up to 25% (usually 0–15%) in the cat and 25–35% in the rabbit as the splanchnic vessels dilated. If the compensatory function of the reflexes initiated by carotid and aortic baroreceptors was inhibited, the fall in blood pressure was more pronounced (averaging about 50% in the cat and 70% in the rabbit). Besides this, there is a vast literature to show that sustained discharge of vasomotor nerves, including the adrenergic nerves to the gastrointestinal vasculature, contributes to the maintainence of the normal systemic blood pressure (FOLKOW, 1955; HEYMANS and NEIL, 1958). Stimulation of the sinus depressor nerve increases intestinal blood flow by removing the tonic action of adrenergic nerve fibres on the splanchnic vessels, not by increasing the discharge of dilator nerves supplying the vessels (CELANDER and FOLKOW, 1951); in fact it is unlikely that any dilator nerves supply these vessels (GRAYSON and MENDEL, 1965; KEWENTER, 1965). COHEN and GOOTMAN (1970) have examined the spontaneous firing of axons in the splanchnic nerves and have found several periodicities in the combined discharge: a cyclic variation in activity at 10/sec; sometimes locked 3:1 with the cardiac cycle; a discharge locked 1:1 with the cardiac cycle; and a rhythmic firing synchronized with respiration. These oscillations occur in anaesthetized or decerebrate cats and activity is almost completely abolished by high cervical section of the spinal cord. Spontaneous changes in splanchnic nerve activity are directly related to oscillations in systemic blood pressure (GOOTMAN and COHEN, 1970). NISIMARU (1971) recorded the grouped discharge in the central cut ends of nerve bundles accompanying the gastric arteries in anaesthetized open-chest cats. He found that the total activity consisted of a baroreceptor-sensitive and a baroreceptor-insensitive component. The baroreceptor-sensitive component was partly synchronized with the cardiac cycle. The combined discharge was blocked by hexamenthonium but was not influenced by bilateral vagus section. The results suggest that under the conditions used by NISIMARU there was continuous activity in adrenergic nerves controlling gastric motility as well as in vasomotor nerves.

An estimate of resting and peak rates of action-potential discharge in sympathetic fibres can be made by matching conditions at rest and during reflex activity with the effects of artifical stimulation of decentralized nerves or by recording from single-fibre preparations. The literature surveyed FOLKOW (1955) suggests that resting discharge rates in sympathetic vasomotor nerves are equivalent to nerve stimulation at 1–3 Hz and that maximum rates elicited reflexly are probaly equivalent to stimulation at 8–10 Hz. The effects on motility of activation of adrenergic nerves through central reflex pathways can be mimicked by direct stimulation at 2-8 Hz (JANSSON and MARTINSON, 1966; HULTÉN, 1969; JANSSON and LISANDER, 1969), although there is some indication that during peripheral reflexes higher average discharge rates might be achieved (see below). Recordings of unit activity in single pre- and postganglionic fibres indicate that individual axons carry impulse traffic at frequencies of less than one up to ten impulses/sec with an average of about 2 impulses/sec (IGGO and VOGT, 1960; KOIZUMI and SUDA, 1963; SATO, 1972). Groups of 2–3 action potentials may occur at 30/sec. Reflex activation of the sympathetic outflow elicits a fast initial burst of activity followed by a slower steady discharge if the reflexogenic stimulus is maintained.

Initial bursts lasting 60–300 msec at rates of 10–100 impulses/sec and steady firing rates of 2–25 impulses/sec have been recorded from individual fibres (IGGO and VOGT, 1960; SATO, 1972). It is unrealistic to expect artificial stimuli at frequencies greater than 10 Hz to mimic the physiological actions of adrenergic nerves supplying the gastrointestinal tract, especially in non-perfused tissue in which transmitter is not removed into the circulation.

B. Reflex Changes in Adrenergic Discharge

1. Gastrointestino-Gastrointestinal Inhibitory Reflexes

It has often been demonstrated that distension or mucosal irritation of one part of the gastrointestinal tract can inhibit the movements of other areas. Two types of reflex occur, the intestino-intestinal reflex, in which adrenergic nerves are involved, and the descending inhibitory reflex which is not mediated through adrenergic nerves but involves intrinsic non-adrenergic pathways (FURNESS and COSTA, 1973a). The intestino-intestinal inhibitory reflex, originally described in detail by PEARCY and LIERE (1926), in which distension of any section of the intestine inhibits movements throughout the gastrointestinal tract (YOUMANS, 1949, 1968), has as its afferent limb mesenteric nerves and as its efferent limb mesenteric, adrenergic nerves. HULTÉN (1969) has found that distension of the small intestine elicits a similar reflex inhibition of the proximal colon in cat, but that inhibition of the distal colon is weak or absent. Distension of the stomach prompts reflex relaxation which accommodates the increased volume, but this reflex is mediated by vagal afferents and vagal, non-adrenergic, efferents (ABRAHAMSSON, 1971). Reflexes which arise from intestinal distension and, through adrenergic nerves, contract sphincter muscle can propery be regarded as intestino-intestinal inhibitory reflexes because they oppose the normal movement of the digesta. Stimulation of the central end of the cut hypogastric nerve, the other hypogastric nerve being intact, causes contraction of the internal anal sphincter; this appears to be an axon-reflex and does not reveal a true reflex pathway (LANGLEY and ANDERSON, 1894).
The intestino-intestinal reflex has a low and a high threshold component, the former involving a spinal pathway and the latter an entirely peripheral pathway through prevertebral ganglia. KUNTZ and his collaborators examined the high threshold reflex in cats in which the prevertebral ganglia were decentralized and the intestine was transected between the points of stimulation and recording (KUNTZ, 1940; KUNTZ and BUSKIRK, 1941; KUNTZ and SACCAMANNO, 1944). KUNTZ and BUSKIRK (1941) found that distension of the distal ileum or of the colon caused inhibition of the proximal ileum and that the application of nicotine to the coeliac ganglia blocked the reflex. Following criticism that the decentralization might have been incomplete, KUNTZ and SACCAMANNO (1944) reported experiments which leave no doubt that the reflex was entirely peripheral. The authors performed one series of experiments on animals in which the entire spinal cord distal to the lower cervical region was removed (in some cases the vagi were also divided) and a second series in which the inferior mesenteric ganglia were

decentralized by bilateral removal of the lumbar sympathetic chains and section of the coeliac roots and hypogastric nerves a week or more before the experiment. At the time of the experiment, the colon was transected between the point of distension in the distal half and the recording point in the proximal half. In both series, the intestino-intestinal inhibitory reflex could be successfully elicited. A similar reflex was also demonstrated in the small intestine. KUNTZ and SACCAMANNO (1944) point out that the pressure required to elicit this reflex (200 cm H_2O or more) are much greater than those needed to elicit the intestino-intestinal inhibitory reflex through spinal pathways. SEMBA (1954a) initiated a reflex inhibition of the small intestine by the application of hypertonic solution (5% NaCl) to an adjacent, but unconnected, segment. In his experiments, the reflex was unaffected by adrenalectomy, vagotomy, abdominal sympathectomy or destruction of the spinal cord below t_1, but was eliminated when nicotine was applied to the coeliaco-mesenteric plexus. Synaptic connexions between nerves arising in the periphery, including presumed sensory fibres from the intestine, and cells of the prevertebral ganglia have been demonstrated electrophysiologically (JOB and LUNDBERG, 1952; MCLENNAN and PASCOE, 1954; CROWCROFT et al., 1971) and anatomically (KUNTZ, 1940; UNGVÁRY and LÉRÁNTH, 1970). By dividing mesenteric nerves and examining degenerative changes in their central and peripheral ends, ROSS (1958) found that 30% of the axons in nerves of mesentery of the small intestine arose from cells in the gut wall. It is likely that many of these axons are afferent processes involved in the peripheral intestino-intestinal inhibitory reflex, because the cell bodies of the afferent axons involved in the spinal reflex lie outside the intestine (see below). In experiments similar to those of ROSS, SCHOFIELD (1960) also found surviving axons in the peripheral stumps and, in addition, found signs of retrograde degeneration in neurones of the myenteric plexus, but not in neurones of the submucosa.

Reflex gastric inhibition, involving a peripheral pathway through the coeliac plexus, occurs when the jejunal or duodenal mucosa is exposed to acid or the jejunal mucosa is exposed to hypertonic solutions. This reflex has been given the name enterogastric. In dog, it is not significantly modified by vagotomy, section of the splanchnic nerves or removal of the thoraco-lumbar sympathetic chains, but it is abolished by application of nicotine to the coeliac plexus, resection of the plexus or division of the mesenteric nerves (SEMBA, 1954b; SHAPIRO and WOODWARD, 1959). In humans, no one appears to have tested for the reflex after coeliac ganglionectomy, but, as in dogs, the reflex persists after the other denervation procedures as well as after division of the intrinsic plexuses between the point of irritation and the stomach (SHAPIRO and WOODWARD, 1955). Thus, the enterogastric reflex pathway consists of afferent nerve fibres which arise in the upper small intestine and synapse with the cell bodies of efferent ihibitory (adrenergic) neurones in the coeliac plexus. Some authors have reported a partial reduction of the reflex after vagotomy, which proably arises because of the reduction in cholinergic activity for the adrenergic nerves to act against. The enterogastric reflex seems to delay gastric emptying principally by its inhibition of normal motility and is not significantly aided by constriction of the pylorus. The enterogastric reflex is probably constantly acting to regulate gastric emptying and restrict upper intestinal acidity and tonicity; in dog, decentralization of the coeliac plexus leaving the reflex intact has no significant effect, but destruction of the reflex pathway by removal of the coeliac plexus results in severe ulceration of the upper small intestine, bloody diarrhea and, in somes cases, death (IVY et al., 1950). An operation such as removal of the coeliac plexus will almost inevitably interfere with the blood supply to the gut and cause damage to the lymphatics, but it does not seem likely that these additional effects of surgery completely account for the observations of the several authors cited by IVY et al. (1950). A notable contrast to this finding is presented by the work of CANNON et al. (1929), who found that the coeliac ganglia could be removed from cats without disturbing their digestive tracts. It is not known wheter there is a species difference or whether differing experimental conditions (e.g. diet) could have influenced the result.

The low-threshold reflex inhibition, elicited in cat or dog by distension pressures of 20–100 cm H_2O, is abolished by section of the splanchnic nerves, by section of dorsal root fibres, by destruction or anaesthesia of the spinal cord or by guanethidine (MORIN and VIAL, 1934; CHANG and HSU, 1942; YOUMANS et al. 1942; HUKUHARA et al. 1959; JOHANSSON and LANGSTON, 1964; JANSSON and MARTINSON, 1966). Stimulation of medullary depressor areas or of hypothalamic sympatho-inhibitory areas activates descending fibre tracts which suppress the low threshold inhibitory reflex (JOHANSSON et al., 1965; JANSSON et al., 1969). This means that brain stem or spinal cord stimulation can increase intestinal motility by reducing the inhibitory action of adrenergic nerves, and it is important not to confuse such effects with the direct stimulation of excitatory pathways. The blockade of the spinal intestino-intestinal inhibitory reflex by stimulation of the medulla was often accompanied by a fall in blood pressure, but this was not a necessary condition to observe the effect. Transection of the cervical cord caused a reduction of intestinal motility and a decrease in the threshold pressure needed to elicit the reflex, suggesting that descending pathways exert a tonic restraint on the spinal intestino-intestinal inhibitory reflex under experimental conditions (JOHANSSON et al., 1968). These authors also demonstrated that the reflex could be enhanced by stimulation of suprabulbar areas. JANSSON et al. (1969) found that the reflex was facilitated by stimulation of the hypothalamic defence area. The experiments of CHANG and HSU (1942) show that, for reflexes initiated and recorded from the small intestine, the afferent limb is connected with the ipsilateral efferent outflow and that the reflex pathway is not eliminated when ascending and descending pathways are interrupted. The sensory endings of the afferent nerves involved in this reflex are embedded in the external muscle, probably in the longitudinal layer, and the cell bodies lie outside the intestine (HUKUHARA et al., 1959, 1960). The arrangement of the neurones involved in the low-threshold reflex, indicated by present data, is shown in Fig. 5. Anatomical and physiological studies both indicate that at least one spinal interneurone should be included in this pathway (FOERSTER et al.,1933; FRANZ et al., 1966; POMERANZ et al., 1968; PETRAS and CUMMING, 1971).

Comparison of the results of KOCK (1959) and those of JANSSON and MARTINSON (1966) indicates that the rate of discharge in the adrenergic nerves is greater in the high threshold peripheral reflex than in the low threshold spinal reflex. In KOCK's experiments, intestinal inhibition in response to carotid occlusion,

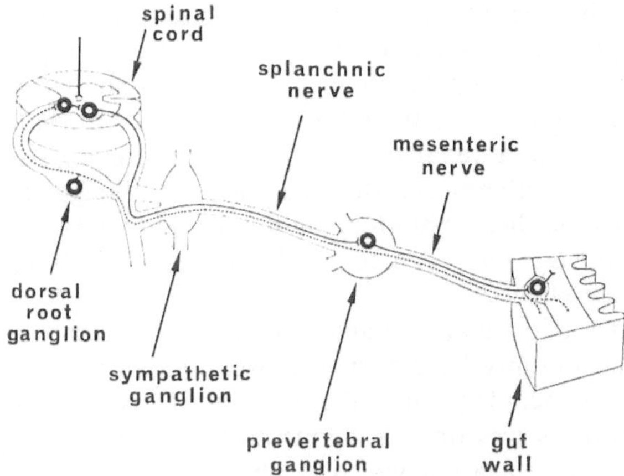

Fig. 5. Pathway of the spinal intestino-intestinal inhibitory reflex. The fibres of the afferent part of the circuit (dotted) have their endings in the external muscle of the intestine and have cell bodies in dorsal root ganglia. The efferent part of the circuit consists of cholinergic nerves with their cell bodies in the spinal cord, which pass via the sympathetic chains and splanchnic nerves to prevertebral ganglia and form excitatory synapses with adrenergic neurones. The adrenergic axons then follow mesenteric nerves to the gut wall where they form inhibitory contacts with enteric cholinergic neurones. There is probably at least one interneurone in the spinal cord between each afferent axon and the efferent cholinergic neurone. There are descending inhibitory pathways in the spinal cord whose activity increase the threshold stimulus needed to elicit the intestino-intestinal inhibitory reflex

withrawal of blood, or stimulation of somatic afferents was slow in onset and was eliminated by exclusion of the adrenal glands. This indicates that there was no cholinergic tone for increased adrenergic nerve activity to act against (see below). However, when the high threshold reflex was elicited, a prompt inhibition independent of central connexions but dependent on the integrity of mesenteric nerves was observed. This shows that adrenergic nerves were conducting impulses at a sufficient rate to act directly on the muscle. In contrast, the reflex elicited by JANSSON and MARTINSON (1966) was abolished by spinal anaesthesia and only caused prompt inhibition when there was some level of activity of enteric cholinergic neurones. This indicates that the discharge rate in the adrenergic neurones was too low for overflow of transmitter to the muscle to be significant.

2. Excitation of Afferent Nerves Outside the Intestine

Stimulation of somatic or autonomic afferent nerves by local mechanical disturbance, by electrical means, or during a general reflex commonly modifies intestinal movement. For example, when a depressor response is elicited by stimulation of aortic nerves or increased pressure in the caroid sinus, the movements of the alimentary canal are enhanced; the corresponding pressor reflex is associated with intestinal inhibition (BAYLISS, 1893; KISCH, 1926). Many other examples are given by ALVAREZ (1948) and YOUMANS (1968); these include mechanical stimuli applied to the urinary tract, peritoneum, skin, kidney, uterus, gallbladder and testes and appear to be associated with the stimulation of pain fibres. Unfortunately, analysis of most early experiments is very difficult, because the investigators did not appreciate that intestinal inhibition could be mediated by catecholamines released from the adrenal medulla as well as through the activation of mesenteric efferents. Furthermore, adrenergic nerves to the gastro-intestinal tract exert a major part of their inhibitory action on cholinergic neurones of the enteric ganglia, so that some degree of cholinergic tone is usually necessary to demonstrate their effects.

The relative roles of circulating and locally released catecholamines in inhibiting gastrointestinal motility have been reviewed recently (FURNESS and BURNSTOCK, 1974). It was concluded that, in the absence of activity of enteric cholinergic neurones, adrenergic nerves do not significantly inhibit motility but that the long-latency, sluggish relaxation which is observed arises from the release of catecholamines from the adrenal glands. This conclusion applies to reflex inhibition induced by systemic hypotension (YOUMANS et al., 1940; Hamilton et al., 1944; KOCK, 1959), stimulation of somatic afferents (KOCK, 1959; JANSSON, 1969), or stimulation of the hypothalamic defence area (JANSSON et al., 1969). In the presence of gastric contractions arising from cholinergic nerve activity, it has been shown that stimulation of somatic pressor afferents activates adrenergic fibres supplying the cat stomach and that stimulation of the corresponding depressor afferents suppresses the activity of adrenergic fibres induced by intestinal distension (JANSSON, 1969). The antagonism of the intestino-gastric inhibitory reflex is unaffected by section of the cervical cord, showing that supraspinal pathways are not involved. Although various results suggest that adrenergic inhibitory fibres might be activated during periods of hypotension, this has not been demonstrated directly. JANSSON et al. (1969) located hypothalamic areas whose stimulation increased and other areas whose stimulation decreased the activity of adrenergic fibres to the stomach.

The reflex enhancement and supression of activity in adrenergic fibres controlling gastrointestinal motility needs further investigation in carefully controlled experiments in which adrenal secretion is taken into account and the level of activity of enteric cholinergic neurones is known. It has yet to be determined to what extent parts of the gastrointestinal tract are differently affected although it is already known that adrenergic nerves in cat antagonize cholinergic contractions more readily in the distal ileum than in the jejunum (KEWENTER, 1965) and that the intestino-intestinal inhibitory reflex causes a greater effect in the proximal than in the distal colon (HULTÉN, 1969). It is not known what species differencies occur, although some might be expected (see Section II, C and E). The effects of a number of reflexes which influence sympatho-adrenal discharge (e.g. those triggered by haemorrhage, asphyxia or thermal changes) require further investigation in the light of modern knowledge.

Stimulation of afferent nerves or severe irritation of internal structures such as the kidney, bladder or peritoneum also causes contraction of sphincter muscle, probably mediated through adrenergic nerves. CARLSON et al. (1922) found that stimulation of the central ends of divided splanchnic nerves produced reflex spasm of the lower oesophageal sphincter mediated through splanchnic efferent nerves. SMETS (1936b) elicited a similar spasm in the ileo-colic sphincter and showed that a spinal pathway connecting with mesenteric (adrenergic) efferent nerves was involved. The ileo-colic sphincter also contracts in acute peritonitis and following irritation of the pleural cavity or the peritoneum (ELLIOTT, 1904a; HERTZ, 1913).

Changes in the discharge of adrenergic nerves to the gastrointestinal blood vessels occur in response to changes in the balance of regional demands on the circulation. In most circumstances a level of adrenergic tone is maintained in these vessels, permitting both increases and decreases and in local perfusion. The manner in which these nerves react to regional needs is well demonstrated by responses to heating and cooling (KULLMAN et al., 1970; WALTHER et al., 1970). When body temperature is raised, skin blood flow increases to dissipate excess heat and this change is accompanied by an increase in the frequency of action-potential firing in nerves supplying the abdominal viscera and a decrease in blood flow through the superior mesenteric artery. Conversely, lowering the body temperature decreases skin blood flow, decreases the activity of nerves supplying the splanchnic regions and increases intestinal blood flow. Similar reciprocal changes occur during exercise, when the needs of skeletal muscle place greater demands on the circulation. In these circumstances, the resistance of the intestinal vasculature is increased by adrenergic nerve activity and a smaller proportion of the cardiac output goes to the splanchnic region, although gastrointestinal flow probably does not change very much, the increase in resistance being matched by the increase in cardiac output (RUSHMER et al., 1961; BURNS and SCHENK, 1969; VATNER et al., 1971). The literature concerning increased activity of adrenergic constrictor fibres to the intestinal blood vessels during hypotension has been reviewed by FINE (1965), who has also discussed the possible contribution of sustained intestinal vasoconstriction to mucosal breakdown and the development of irreversible shock. The supposition that abnormally frequent or prolonged increases in adrenergic nerve activity could facilitate mucosal ulceration deserves further investigation.

From a teleological standpoint, blood flow to the gastrointestinal tract might be expected to be greater during digestion than during fasting, and from a physiological point of view, this might be expected to result from a diversion of blood from other regions, in part achieved by a decrease in the activity of adrenergic nerves which supply the vasculature of the gut. Most investigators agree that an increase in intestinal blood flow occurs, but disagree on the mechanism of the increase. GROLLMAN (1929) found that cardiac output increased in man after a meal and HERRICK et al. (1934) confirmed this result in dogs fed after an 18 h fast, finding that femoral, carotid and mesenteric artery flows all increased, suggesting no selective diversion of blood to the alimentary canal. In contrast, BURNS and SCHENK (1969), who also used conscious dogs previously fasted for 18h, found that mesenteric artery blood flow increased by an average of 71% with no increase in cardiac output. The percentage of the cardiac output taken by the superior mesenteric artery during fast was 7.5 and that during digestion was 14.6. GRIM and LINDSETH (1958) found that blood flow to the small intestine increased by 31% during digestion in the dog. Flow to the muscle was increased by 25% (jejunum) and 60% (ileum) and that to the submucosa by 25% (jejunum) and 20% (ileum), but there was little change in flow to the mucosa (15% increase and 2% decrease respectively). Blood flow to the mesenteric fat decreased by 60% (jejunum) and 30% (ileum). The gastrointestinal tract has a greater metabolic requirement during digestion, as demonstrated by the increased arterial-portal vein oxygen difference (SMYTHE et al., 1951), despite an increased oxygen intake and decreased systemic arterial-venous oxygen difference (GROLLMAN, 1929). The difference between the results of HERRICK et al. (1934) and BURNS and SCHENK (1969) might stem from differences in the levels of excitement of the dogs when given food. Some of the dogs used by the former authors were apparently highly excited, judging by the fact that they vomited their food without digesting it. On the other hand, the animals used by BURNS and SCHENK were placid, frequently falling asleep after feeding. BARCROFT and FLOREY (1929) have pointed out that excitement in dogs reduces intestinal blood flow and may mask the influence of other stimuli.

3. Adynamic Ileus

Adynamic ileus (often misleadingly referred to as paralytic ileus) occurs when the activation of adrenergic nerves leads to a long-lasting suspension of peristalsis, usually throughout the intestinal tract. Propulsion of the contents ceases and the intestine gradually becomes distended with gas (TUMEN, 1964). Because the adrenergic nerves act on the enteric ganglia, the reactivity of the muscle itself is not impaired (HOTZ, 1909). A number of stimuli can induce increased activity in adrenergic nerves supplying the intestine and this is reflected in the multitude of causes of adynamic ileus that have been reported (OCHSNER and GAGE, 1933; WANGENSTEEN, 1955). Common causes of ileus are exposure and handling of the intestine and peritonitis.

Reflexes through spinal pathways and peripheral reflexes through prevertebral ganglia can both mediate adynamic ileus and it can also arise when efferent pathways are irritated by injury to the spinal cord or to the region of the splanchnic nerves. It is likely that spinal pathways are mainly involved in mild ileus, as the condition usually can be relieved by spinal or splanchnic anaesthesia or, in experimentally induced ileus of animals, by splanchnic nerve section (CANNON and MURPHY, 1907; WAGNER, 1919; ARAI, 1922; MARKOWITZ and CAMPBELL, 1927; OCHSNER, GAGE and CUTTING, 1928; DOUGLAS and MANN, 1941). DAVID and LORING (1930) have pointed out that spinal anaesthesia is less effective when adynamic ileus results from severe and widespread peritonitis. This probably means that severe peritonitis stimulates entirely peripheral, in addition to spinal, pathways. CANNON and MURPHY (1907) and OLIVECRONA (1927) produced adynamic ileus in cats by rough handling of the exposed intestines under aseptic conditions. Both groups then closed the wound and used an X-ray technique to examine gastrointestinal movements. Peristaltic activity in the stomach and intestines ceased for several hours as a results of this treatment, but rhythmic contractions, which OLIVECRONA took to be myogenic, continued. The occurrence of inhibition after this trauma was consistently observed and was not affected by section of the splanchnic nerves, but was almost completely absent in cats from which the greater part of the coeliaco-mesenteric plexus had been removed 2–3 weeks previously. This clearly shows that reflexes leading to adynamic ileus can involve entirely peripheral, as well as spinal, pathways. In addition to the obvious suspension of peristalsis and subsequent intestinal distension, contraction of sphincter muscle occurs in adynamic ileus; for example, MacLean (1932) reported that chronic appendicitis causes stasis in the small intestine, accompanied by constriction of the distal ileum and ileo-colic sphincter. Earlier, Hertz (1913) had reported that peritonitis occurring in cases of chronic appendicitis results in contraction of the ileo-colic sphincter.

The rational treatment of adynamic ileus, especially when the basic irritation leading to increased adrenergic discharge cannot be dealt with immediately, is to use drugs which block the release of noradrenaline, such as guanethidine, or drugs which block the effect of adrenergic nerves on the enteric ganglia, such as phentolamine. This is the procedure advocated by Neely and Catchpole (1971) and Petri et al. (1971), who have applied it successfully in clinical situations. The therapy may be improved in some cases by combining blockade of adrenergic transmission with the administration of drugs which stimulate intestinal motility. Beta-receptor blockade is not worthwhile because the adrenergic nerves act on the ganglia via the α-receptors. In cases of mild post-operative ileus examined by Heimbach and Crout (1971) treatment with the adrenergic neurone-blocking drug bethanidine followed by the anticholinesterase neostigmine was no more effective than treatment with neostigmine alone.

VI. The Functions of the Adrenergic Nerves which Supply the Gastrointestinal Tract

We know that the adrenergic nerves controlling gastrointestinal movement are quite dispensible in animals exposed to a laboratory environment, and that when the nerves do act they suppress digestive action, by inhibiting the firing of cholinergic nerves, thereby paralyzing the peristaltic mechanism, by contracting the sphincters, and, to a slight extent, by relaxing the muscle of the non-sphincter regions of the gut. Thus the churning of the contents necessary for adequate absorption is inhibited and the movement of the digesta along the alimentary canal is suspended. There is some contribution to these effects, particularly direct inhibitory actions on the muscle, by catecholamines which are often released into the circulation from the adrenals on reflex activation of the adrenergic nerves to the gut.

The low-threshold spinal intestino-intestinal inhibitory reflex is observed in animals in which there is a minimum of interference and can be elicited by intraluminal pressures within the expected physiological range. On the other hand, the peripheral reflex in response to high intraluminal pressure is almost certainly initiated only in extreme conditions, when the spinal reflex mechanism has been unable to prevent the development of a crisis. The low threshold reflex probably acts as a buffer to prevent the development of excessive stress on the gut wall. If an obstruction develops in the intestine, the accumulation of the contents above this point would tend to activate excitatory cholinergic nerves through intrinsic reflex pathways (see KOSTERLITZ, 1968). The greater the build-up of matter in the lumen, the greater is the stimulus for the intestine to contract around it. This may aggravate the obstruction and hence the damage it causes. The intestino-intestinal inhibitory reflex can provide protection against this positive feedback and prevent the development of excessive pressures in the gastrointestinal tract, so that any obstruction is slowly and gently cleared. At the same time, the sphincters of the gastrointestinal tract are constricted, part because descending pressure waves which normally evoke their relaxation are inhibited and part because of the activity of their adrenergic nerve supply. The closure of the sphincters may help to prevent reflux or premature passage of the contents from one section of the gastrointestinal tract to another. When there is no threat of obstruction, the intestino-intestinal inhibitory reflex does not interfere with the normal peristalsis of the intestine, which is initiated by pressures below 5 cm H_2O; it only comes into play with the occurrence of higher pressures (greater than about 20 cm H_2O if there is no central suppression of the reflex).

The enterogastric reflex is also protective, limiting the development of acidity or hypertonicity of the contents of the upper small intestine by delaying gastric emptying. If the contents became too acid, ulceration could occur. However, the most important regulation of duodenal acidity seems to be mediated by hormones which are released in response to lowered duodenal pH and which inhibit gastric acid secretion and stimulate the flow of alkaline pancreatic juice (see ANDERSSON, 1973). Hypertonic conditions lead to the dumping syndrome, in which fluid is drawn from the plasma into the jejunum causing hypovolaemia and, if sufficiently intense, shock.

The precise function of inhibition of the intestine in response to noxious stimuli such as laparotomy, irritation of the peritoneum, handling of the viscera, or stimulation of pain fibres is not known. The reaction may have evolved as a preparation for evasive action, in which the digestive process is suppressed and a greater percentage of the cardiac output can be diverted to other regions, such as skeletal and cardiac muscle, where an increased blood supply might be demanded.

During more general reflexes such as occur in the defence reaction, both blood flow and motility are reduced. These two responses can be experimentally dissociated and therefore involve separate, though probably interconnected, efferent pathways.

Neurones are more sensitive than is smooth muscle to lack of an adequate blood supply, so that it is significant that adrenergic nerves supply the ganglion cells rather than the muscle of the gastrointestinal tract. This means that when the blood supply is reduced, the activity and hence the metabolic requirements of the enteric neurones also fall. In fact, it is possible that in some circumstances the function of the adrenergic nerves is to protect the ganglion cells and that suppression of peristalsis is incidental, although the relaxation of the intestine and the inhibition of cholinergic nerves may also lessen absorption and secretion through the mucosa.

The adrenergic nerves supplying the gastrointestinal blood vessels are tonically active. In keeping with mechanisms controlling blood flow in other regions, their activity is increased to decrease the proportion of the cardiac output feeding into the splanchnic vessels and to decrease the blood volume held in this region; conversely, their activity is decreased following demand in other regions to allow digestion to continue. After a meal, activity in splanchnic nerves is decreased and visceral blood flow is increased. The drowsiness which accompanies this change reduces the chance of increased requirements in other vascular beds usurping the needs of the digestive organs.

In essence, the role of the adrenergic nerves controlling motility is a protective one, allaying the damaging effects of excessive tension in the gut wall, inhibiting the development of hyperacidity and hypertonicity in the upper small intestine, and reducing the metabolic requirements of the intrinsic neurones (and secondarily of the muscle and mucosal cells) when a reduction in blood supply is imminent or present. The vasomotor nerves participate in cardiovascular and metabolic homeostasis, helping to ensure that the circulation is distributed according to the needs of the moment.

VII. Summary

Adrenergic nerves run from cell bodies in prevertebral autonomic ganglia to the blood vessels and intrinsic ganglia of the gastrointestinal tract. Only a few fibres supply the muscle of the non-sphincter regions and adrenergic innervation of other gastrointestinal tissue (e.g. glands, epithelia, lymphatics) is sparse or absent. Most sphincter muscle is supplied by a dense plexus of adrenergic nerves.

The adrenergic nerves reduce gastrointestinal motility by inhibiting the release of acetylcholine from excitatory cholinergic neurones and by contracting the muscle of the sphincters, both effects being mediated through α-adrenoceptors. The nerves have a slight direct inhibitory action, principally during discharge at high rates, on non-sphincter muscle. Inhibitory β-receptors are located on both non-sphincter and sphincter muscle and, depending on the area of the tract and the species, α-receptors on the muscle may be either excitatory or inhibitory. An analysis of the receptor mechanisms involved in the actions of adrenergic nerves on motility is presented in section IIF.

Noradrenaline released from vasomotor nerves constricts the arteries and veins which supply and drain the gut through an action on α-receptors. The restriction of blood flow through the muscle, submucosa and lamina propria of the mucosa is not maintained, whereas the reduction in blood flow through the rest of the mucosa and the constriction of the larger veins persists as long as adrenergic nerve activity continues. The decrease in mucosal blood flow supresses secretion and, in extreme cases, may reduce absorption.

The adrenergic nerves which run to the intrinsic ganglia of the gastrointestinal tract usually are inactive in the resting individual; their discharge is evoked through reflex pathways originating both within and outside the alimentary tract. Adrenergic reflexes arising from within the intestine act as buffers to regulate gastric emptying and intestinal peristalsis. If the contents of the upper intestine become hyperacidic or hypertonic, then an entero-gastric reflex mediated through adrenergic nerves reduces the rate of gastric emptying. Similarly, if regions of unusually high pressure develop within the intestines, an adrenergic intestino-intestinal inhibitory reflex supresses the activity in distension-sensitive cholinergic neurones whose continued excitation could aggravate an obstruction. Noxious stimuli, such as rough handling of the viscera or peritoneal infection, cause the reflex firing of adrenergic inhibitory nerves and if this reflex inhibits the peristaltic mechanism for an extended time the pathological condition, adynamic ileus, may develop. The adrenergic nerves inhibit the activity of enteric neurones and hence the motility and nutritional demands of the gastrointestinal tract when the splanchnic blood supply is reduced because of demands elsewhere in the body. The adrenergic nerves to the gastrointestinal vasculature are tonically active and they participate in cardiovascular homeostasis by adjusting the resistance and capacity of the splanchnic vascular bed to regional requirements.

Acknowledgements. Almost all of this article was written while J. B. F. held the John Halliday Fellowship of the Life Insurance Medical Research Fund of Australia and New Zealand. The experimental work was supported by grants from the National Heart Foundation of Australia, the Australian Research Grants Committee and the Life Insurance Medical Research Fund of Australia and New Zealand.

We thank Mrs. NANCY MITCHELL, formerly of the Barnes Library, University of Birmingham, for her invaluable assistance in obtaining reference material.

Professor G. BURNSTOCK, Dr. G. CAMPBELL and Dr. A. R. TUFFERY provided very helpful criticisms of the manuscript.

References

ÅBERG, G., ERÄNKÖ, O.: Localization of noradrenaline and acetylcholinesterase in the taenia of the guinea-pig caecum. Acta physiol. scand. **69,** 383–384 (1967).

ABRAHAMSSON, H.: Vago-vagal gastro-gastric relaxation. Rendic. Roman. Gastro-enterol. **3,** 114–115 (1971).

AHLQUIST, R. P.: A study of the adrenotropic receptors. Amer. J. Physiol. **153,** 586–600 (1948).

ALVAREZ, W. C.: An introduction to gastro-enterology, 4th ed. London: Heinemann 1948.

AMBACHE, N., FREEMAN, M. A.: Atropine-resistant longitudinal muscle spasms due to excitation of non-cholinergic neurones in Auerbach's plexus. J. Physiol. (Lond.) **199,** 705–727 (1968).

ANDERSSON, R., MOHME-LUNDHOLM, E.: Studies on relaxing actions mediated by stimulation of adrenergic α- and β-receptors in taenia coli of the rabbit and guinea-pig. Acta physiol. scand. **77,** 372–384 (1969).

ANDERSSON, R., MOHME-LUNDHOLM, E.: Metabolic actions in intestinal smooth muscle associated with relaxation mediated by adrenergic α- and β-receptors. Acta physiol. scand. **79,** 244–261 (1970).

ANDERSSON, S.: Secretion of gastrointestinal hormones. Ann. Rev. Physiol. **35,** 431–452 (1973).

ARAI, K.: Experimentelle Untersuchungen über die Magen-Darmbewegungen bei akuter Peritonitis. Naunyn-Schmiedebergs Arch. exp. Path. Pharmak. **94,** 149–189 (1922).

ÅSTRÖM, A.: Anti-sympathetic action of sympathomimetic amines. Acta physiol. scand. **18,** 295–307 (1949).

BABKIN, B. P.: Secretory mechanism of the digestive glands, 2nd ed. New York: Hoeber, 1950.

BAKER, R., MENDEL, D.: Some observations on 'autoregulatory escape' in cat intestine. J. Physiol. (Lond.) **190,** 229–240 (1967).

BARCROFT, J., FLOREY, H.: The effects of exercise on the vascular conditions in the spleen and the colon. J. Physiol. (Lond.) **68,** 181–189 (1929).

BASCH, S. von: Die Hemmung der Darmbewegung durch den Nervus Splanchnicus. Sitz. Akad. Wiss. Wien, Abth III, **68,** 7–28 (1873).

BAUMANN, J. A.: Distribution du nerf vague dans l'intestin de l'embryon du Cobaye. Schweiz. med. Wschr. **79**, 529 (1949).

BAUMGARTEN, H. G.: Über die Verteilung von Catecholaminen im Darm des Menschen. Z. Zellforsch. **83**, 133–146 (1967).

BAUMGARTEN, H. G., HOLSTEIN, A.-F., OWMAN, CH.: Auerbach's plexus of mammals and man: electron microscopic identification of three different types of neuronal processes in myenteric ganglia of the large intestine from rhesus monkeys, guinea-pigs and man. Z. Zellforsch. **106**, 376–397 (1970).

BAUMGARTEN, H. G., LANGE, W.: Adrenergic innervation of the oesophagus in the cat (felis domestica) and the rhesus monkey (macacus rhesus). Z. Zellforsch. **95**, 529–545 (1969a).

BAUMGARTEN, H. G., LANGE, W.: Extrinsic adrenergic innervation of the extrahepatic biliary duct system in guinea-pigs, cats and rhesus monkeys. Z. Zellforsch. **100**, 606–615 (1969b).

BAYLISS, W. M.: On the physiology of the depressor nerve. J. Physiol. (Lond.) **14**, 303–325 (1893).

BAYLISS, W. M., STARLING, E. H.: The movements and innervation of the small intestine. J. Physiol. (Lond.) **24**, 99–143 (1899).

BEANI, L., BIANCHI, C., CREMA, A.: The effect of catecholamines and sympathetic stimulation on the release of acetylcholine from the guinea-pig colon. Brit. J. Pharmacol. **36**, 1–17 (1969).

BENNETT, M. R., BURNSTOCK, G., HOLMAN, M. E.: Transmission from perivascular inhibitory nerves to the smooth muscle of the guinea-pig taenia coli. J. Physiol. (Lond.) **182**, 527–540 (1966a).

BENNETT, M. R., BURNSTOCK, G., HOLMAN, M. E.: Transmission from intramural inhibitory nerves to the smooth muscle of the guinea-pig taenia coli. J. Physiol. (Lond.) **182**, 541–558 (1966b).

BENNETT, M. R., ROGERS, D. C.: A study of the innervation of the taenia coli. J. Cell Biol. **33**, 573–596 (1967).

BENZI, G., BERTE, F., CREMA, A., FRIGO, G. M.: Actions of sympathomimetic drugs on the smooth muscle at the junction of the bile duct and duodenum studied in situ. Brit. J. Pharmacol. **23**, 101–114 (1964).

BORTOFF, A.: Digestion: motility. Ann. Rev. Physiol. **34**, 261–290 (1972).

BOTHA, G. S. M.: The gastro-oesophageal junction. London: Churchill 1962.

BOULLIN, D. J., COSTA, E., BRODIE, B. B.: Evidence that blockade of adrenergic receptors causes overflow of norepinephrine in cat's colon after nerve stimulation. J. Pharmacol. exp. Ther. **157**, 125–134 (1967).

BOWMAN, W. C., HALL, M. T.: Inhibition of rabbit intestine mediated by α- and β-adrenoceptors. Brit. J. Pharmacol. **38**, 399–415 (1970).

BRAAM-HOUCKGEEST, J. P. van: Zweite Mitteilung über Magen- und Darmperistaltik. Pflügers Arch. ges. Physiol. **8**, 163–171 (1874).

BRADLEY, S. E.: The hepatic circulation. In: Handbook of physiology sect. 2, vol. 2. Washington: American Physiological Society 1963.

BRODIE, B. B., BOGDANSKI, D. F., BONOMI, L.: Formation, storage and metabolism of serotonin (5-hydroxytryptamine) and catecholamines in lower vertebrates. In: Comparative neurochemistry, ed. D. Richter. Oxford: Pergamon 1964.

BRODY, M. J., DIAMOND, J.: Blockade of the biochemical correlates of contraction and relaxation in uterine and intestinal smooth muscle. Ann. N. Y. Acad. Sci. **139**, 772–780 (1967).

BRÜCKE, F. T. von., STERN, F.: Pharmakologische Untersuchungen über die Innervation des Mageneinganges. Naunyn-Schmiedebergs Arch. exp. Path. Pharmak. **189**, 311–326 (1938).

BUEDING, E., BÜLBRING, E., GERCKEN, G., HAWKINS, J. T., KURIYAMA, H.: The effect of adrenaline on the adenosine triphosphate and creatine phosphate content of intestinal smooth muscle. J. Physiol. (Lond.) **193**, 187–212 (1967).

BUEDING, E., BÜLBRING, E., KURIYAMA, H., GERCKEN, G.: Lack of activation of phosphorylase by adrenaline during its physiological effect on intestinal smooth muscle. Nature (Lond.) **196**, 944–946 (1962).

BUEDING, E., BUTCHER, R. W., HAWKINS, J., TIMMS, A. R., SUTHERLAND, E. W.: Effect of epinephrine on cyclic adenosine 3′, 5′-phosphate and hexose phosphates in intestinal smooth muscle. Biochim. biophys. Acta (Amst.) **115**, 173–178 (1966).

BÜLBRING, E.: Recent observations on smooth muscle. Acta biol. jugoslav. **3**, 239–248 (1967).

BÜLBRING, E., BURN, J. H. Sympathetic vasodilation in skin and intestine of the dog. J. Physiol. (Lond.) **87**, 254–274 (1936).

BÜLBRING, E., CASTEELS, R., KURIYAMA, H.: Membrane potential and ion content in cat and guinea-pig myometrium and the response to adrenaline and noradrenaline. Brit. J. Pharmacol. **34**, 388–407 (1968).

BÜLBRING, E., TOMITA, T.: Properties of the inhibitory potential of smooth muscle as observed in the response to field stimulation of the guinea-pig taenia coli. J. Physiol. (Lond.) **189**, 299–315, (1967).

BÜLBRING, E., TOMITA, T.: Increase of membrane conductance by adrenaline in the smooth muscle of guinea-pig taenia coli. Proc. roy. Soc. B **172**, 89–102 (1969a).

BÜLBRING, E., TOMITA, T.: Suppression of spontaneous spike generation by catecholamines in the smooth muscle of the guinea-pig taenia coli. Proc. roy. Soc. B **172**, 103–119 (1969b).

BÜLBRING, E., TOMITA, T.: Effect of calcium, barium and manganese on the action of adrenaline in the smooth muscle of the guinea-pig taenia coli. Proc. roy. Soc. B **172**, 121–136 (1969c).

BUNCH, J. L.: On the origin, course and cell-connections of the viscero-motor nerves of the small intestine. J. Physiol. (Lond.) **22**, 357–379 (1898).

BUNCH, J. L.: On the vasomotor nerves of the small intestine. J. Physiol. (Lond.) **24**, 72–98 (1899).

BURNS, G. P., SCHENK, W. G.: Effect of digestion and exercise on intestinal blood flow and cardiac output. An experimental study in the conscious dog. Arch. Surg. **98**, 790–794 (1969).

BURNSTOCK, G.: Purinergic nerves. Pharmacol. Rev. **24**, 509–581 (1972).

BURNSTOCK, G., CAMPBELL, G., RAND, M. J.: The inhibitory innervation of the taenia of the guinea-pig caecum. J. Physiol. (Lond.) **182**, 504–526 (1966).

BURNSTOCK, G., ROBINSON, P. M.: Localization of catecholamines and acetylcholinesterase in autonomic nerves. Circulat. Res. **20–21**, Suppl. 3, 43–55 (1967).

CAJAL, S. RAMON y: Sur les ganglions et plexus neureux de l'intestin. C. R. Soc. Biol. (Paris) **5**, 217–223 (1893).

CAJAL, S. RAMON y: Histologie du système nerveux de l'homme et des vertébrés, Vol II. Paris: Malione 1911.

CAMPBELL, G.: The inhibitory nerve fibres in the vagal supply to the guinea-pig stomach. J. Physiol., (Lond.) **185**, 600–612 (1966).

CANNON, W. B.: The motor activities of the stomach and small intestine after splanchnic and vagus section. Amer. J. Physiol. **17**, 429–442 (1906).

CANNON, W. B., MURPHY, F. T.: The movements of the stomach and intestine in some surgical conditions. Ann. Surg. **43**, 512–536 (1906).

CANNON, W. B., MURPHY, F. T.: Physiologic observations on experimentally produced ileus. J. Amer. med. Ass. **49**, 840–843 (1907).

CANNON, W. B., NEWTON, H. F., BRIGHT, E. M., MENKIN, V., MOORE, R. M.: Some aspects of the physiology of animals surviving complete exclusion of sympathetic nerve impulses. Amer. J. Physiol. **89**, 84–107 (1929).

CANNON, W. B., ROSENBLUETH, A.: Autonomic neuroeffector systems. New York: MacMillian 1937.

CAPURSO, L., FRIEDMANN, C. A., PARKS, A. G.: Adrenergic fibres in human intestine. Gut, **9**, 678–682 (1968).

CARLSON, A. J., BOYD, T. E., PEARCY, J. F.: The innervation of the cardia and the lower end of the esophagus in mammals. Amer. J. Physiol. **61**, 14–41 (1922).

CARPENTER, F. W.: A note on the connections, in the mammalian myenteric plexus, between the enteric neurones and the extrinsic nerve fibres. Anat. Rec. **28**, 149–156 (1924).

CASTEELS, R.: The action of ouabain on the smooth muscle cells of the guinea-pig's taenia coli. J. Physiol., (Lond.) **184**, 131–142 (1966).

CASTEELS, R., KURIYAMA, H.: Membrane potential and ion content in the smooth muscle of the guinea-pig's taenia coli at different external potassium concentrations. J. Physiol., (Lond.) **184**, 120–130 (1966).

CELANDER, O.: Are there any centrally controlled sympathetic inhibitory fibres to the musculature of the intestine? Acta physiol. scand. **47**, 299–309 (1959).

CELANDER, O., FOLKOW, B.: Are parasympathetic vasodilator fibres involved in depressor reflexes elicited from the baroreceptor regions? Acta physiol. scand. **23**, 64–72 (1951).

CHANG, P.-Y., HSU, F.-Y.: The localization of the intestinal inhibitory reflex arc. Quart. J. exp. Physiol. **31**, 311–318 (1942).

CLARK, G. A.: The vaso-dilator action of adrenaline. J. Physiol. (Lond.) **80**, 429–440 (1934).

COHEN, M. I., GOOTMAN, P. M.: Periodicities in efferent discharge of splanchnic nerve of the cat. Amer. J. Physiol. **218**, 1092–1101 (1970).

COSTA, M., FURNESS, J. B.: Storage, uptake and synthesis of catecholamines in the intrinsic adrenergic neurones in the proximal colon of the guinea-pig. Z. Zellforsch. **120**, 364–385 (1971).

COSTA, M., FURNESS, J. B.: The origins of the adrenergic fibres which innervate the internal anal sphincter, the rectum and other tissues of the pelvic region in the guinea-pig. Z. Anat. Entwickl.-Gesch. **140**, 129–142 (1973a).

COSTA, M., FURNESS, J. B.: The simultaneous demonstration of adrenergic fibres and enteric ganglion cells. Histochem. J. **5**, 343–349 (1973b).

COSTA, M., FURNESS, J. B., GABELLA, G.: Catecholamine containing nerve cells in the mammalian myenteric plexus. Histochemie **25**, 103–106 (1971).

COSTA, M., GABELLA, G.: Adrenergic innervation of the alimentary canal. Z. Zellforsch. **122**, 357–377 (1971).

COURTADE, D., GUYON, J. F.: Influence motrice du grand sympathique et du nerf érecteur sacré sur le gros intestine. Arch. Physiol. **9**, 880–890 (1897).

CREMA, A., FRIGO, G. M., LECCHINI, S.: A pharmacological analysis of the peristaltic reflex in the isolated colon of the guinea-pig or cat. Brit. J. Pharmacol. **39**, 334–345 (1970).

CROWCROFT, P. J., HOLMAN, M. E., SZURSZEWSKI, J. H.: Excitatory input from the distal colon to the inferior mesenteric ganglion in the guinea-pig. J. Physiol. (Lond.) **219**, 443–461 (1971).

DALE, H. H.: On the action of ergotoxine: with special reference to the existence of sympathetic vaso-dilators. J. Physiol. (Lond.) **46**, 291–300 (1913).

DALE, H. H.: Natural chemical stimulators. Edinb. med. J. **45**, 461–480 (1938).

DAVID, V. C., LORING, M.: Splanchnic anaesthesia in the treatment of paralytic ileus. Ann. Surg. **92**, 721–725 (1930).

DEAL, C. P., GREEN, H. D.: Comparison of changes in mesenteric resistance following splanchnic nerve stimulation with responses to epinephrine and nor-epinephrine. Circulat. Res. **4**, 38–44 (1956).

DERMIETZEL, R.: Elektronenmikroskopische Untersuchung über die Innervation der Pars pylorica des Mäusemagens. Z. mikr.-anat. Forsch. **84**, 225–255 (1971).

DEVINE, C. E., SIMPSON, F. O.: The fine structure of vascular sympathetic neuromuscular contacts in the rat. Amer. J. Anat. **121**, 153–174 (1967).

DIAMOND, J., BRODY, T. M.: Effect of catecholamines on smooth muscle motility and phosphorylase activity. J. Pharmacol. exp. Ther. **152**, 202–211 (1966).

DOGIEL, A. S.: Zur Frage über die Ganglion der Darmgeflechte bei den Säugetieren. Anat. Anz. **10**, 517–528 (1895).

DOUGLAS, D. M., MANN, F. C.: The effect of peritoneal irritation on the activity of the intestine. Brit. med. J. **1941 I**, 227–231.

DOWNMAN, C. B. B.: Distribution along the small intestine of afferent, vasoconstrictor and inhibitory fibres in the mesenteric nerve bundles. J. Physiol. (Lond.) **116**, 228–235 (1952).

DOYON, M.: De l'action exercée par le système nerveux sur l'appareil exréteur de la bile. Arch. Physiol. norm. path. **6**, 19–31 (1894).

DRESEL, P., FOLKOW, B., WALLENTIN, I.: Rubiduim^{86} clearance during neurogenic redistribution of intestinal blood flow. Acta physiol. scand. **67**, 173–184 (1966).

DRESEL, P., WALLENTIN, I.: Effects of sympathetic vasoconstrictor fibres, noradrenaline and vasopressin on the intestinal vascular resistance during constant blood flow or blood pressure. Acta physiol. scand. **66**, 427–436 (1966).

ECCLES, J. C.: Inhibitory pathways to motoneurons. In: Nervous inhibition, ed. E. Florey, p. 47–60. Oxford: Pergamon 1961.

ECCLES, J. C.: The physiology of synapses. Berlin-Göttingen-Heidelberg-New York: Springer 1964.

EDWARDS, D. A. W., ROWLANDS, E. N.: Physiology of the gastroduodenal junction. In: Handbook of physiology, sect. 6, vol. 4. Washington: American Physiological Sciety 1968.

ELLIOTT, T. R.: On the innervation of the ileo-colic sphincter. J. Physiol. (Lond.) **31**, 157–168 (1904a).

ELLIOTT, T. R.: On the action of adrenalin. J. Physiol. (Lond.) **31**, XX–XXI (1904b).

ELLIOTT, T. R.: The action of adrenalin. J. Physiol. (Lond.) **32**, 401–467 (1905).

EULER, U. S. von: The nature of adrenergic nerve mediators. Pharmacol. Rev. **3**, 247–277 (1951).

FARA, J. W.: Escape from tension induced by noradrenaline or electrical stimulation in isolated mesenteric arteries. Brit. J. Pharmacol. **43**, 865–867 (1971).

FARA, J. W., ROSS, G.: Escape from drug-induced constriction of isolated arterial segments from various beds. Angiologica **9**, 27–33 (1972).

FINE, J.: Shock and peripheral circulatory insufficiency. In: Handbook of physiology, Sect. 2, vol 3. American Physiological Society: Washington 1965.
FINKLEMAN, B.: On the nature of inhibition in the intestine. J. Physiol. (Lond.) **70**, 145-157 (1930).
FOERSTER, O., GAGEL, O., SHEEHAN, D.: Veränderung an den Endösen im Rückenmark des Affen nach Hinterwurzeldurchschneidung. Z. Anat. Entwickl.-Gesch. **101**, 553-565 (1933).
FOLKOW, B.: Nervous control of the blood vessels. Physiol. Rev. **35**, 629-663 (1955).
FOLKOW, B., FROST, J., UVNÄS, B.: Action of adrenaline, noradrenaline and some other sympathomimetic drugs on the muscular, cutaneous and splanchnic vessels of the cat. Acta physiol. scand. **15**, 412-420 (1948).
FOLKOW, B., LEWIS, D. H., LUNDGREN, O., MELLANDER, S., WALLENTIN, I.: The effect of graded vasoconstrictor fibre stimulation on the intestinal resistance and capacitance vessels. Acta physiol. scand. **61**, 445-457 (1964a).
FOLKOW, B., LEWIS, D. H., LUNDGREN, O., MELLANDER, S., WALLENTIN, I.: The effect of the sympathetic vasoconstrictor fibres on the distribution of capillary blood flow in the intestine. Acta physiol. scand. **61**, 458-466 (1964b).
FRANZ, D. N., EVANS, M. H., PERL, E. R.: Characteristics of viscerosympathetic reflexes in the spinal cat. Amer. J. Physiol. **211**, 1292-1298 (1966).
FRIEDMANN, C. A.: The action of nicotine and catecholamines on the human internal anal sphincter. Amer. J. dig. Dis. **13**, 428-431 (1968).
FURCHGOTT, R. F.: The classification of adrenoceptors (adrenergic receptors). An evaluation from the standpoint of receptor theory. In: Catecholamines (Handbook of experimental pharmacology, vol. 33), eds. H. Blaschko and E. Muscholl. Berlin-Heidelberg-New York: Springer 1972.
FURNESS, J. B.: The presence of inhibitory fibres in the colon after sympathetic denervation. Europ. J. Pharmacol. **6**, 349-352 (1969a).
FURNESS, J. B.: An electrophysiological study of the innervation of the smooth muscle of the colon. J. Physiol. (Lond.) **205**, 549-562 (1969b).
FURNESS, J. B.: The origin and distribution of adrenergic nerve fibres in the guinea-pig colon. Histochemie **21**, 295-306 (1970a).
FURNESS, J. B.: The excitatory input to a single smooth muscle cell. Pflügers Arch. **314**, 1-13 (1970b).
FURNESS, J. B.: The adrenergic innervation of the vessels supplying and draining the gastrointestinal tract. Z. Zellforsch. **113**, 67-82 (1971).
FURNESS, J. B., BURNSTOCK, G.: The role of circulating catecholamines in the gastrointestinal tract. In: Handbook of physiology, sect. 7. Washington: American Physiological Society 1974.
FURNESS, J. B., COSTA, M.: Morphology and distribution of intrinsic adrenergic neurones in the proximal colon of the guinea-pig. Z. Zellforsch. **120**, 346-363 (1971).
FURNESS, J. B., COSTA, M.: The nervous release and the action of substances which affect intestinal muscle through neither adrenoreceptors nor cholinoreceptors. Phil. Trans. B **265**, 123-133 (1973a).
FURNESS, J. B., COSTA, M.: The ramifications of adrenergic nerve terminals in the rectum, anal sphincter and anal accessory muscles of the guinea-pig. Z. Anat. Entwickl.-Gesch. **140**, 109-128 (1973b).
FURNESS, J. B., MARSHALL, J. M.: Constrictor responses and flow changes in the rat mesenteric microvasculature resulting from the stimulation of paravascular nerves. In: Proc. 7th Europ. Conf. Microcirc. Basel: Karger 1973.
GABELLA, G.: Electron microscopic observations on the innervation of the intestinal inner muscle layer. Experientia (Basel) **26**, 44-46 (1970).
GABELLA, G.: Synapses of adrenergic fibres. Experientia (Basel) **27**, 280-281 (1971).
GABELLA, G.: Fine structure of the myenteric plexus in the guinea-pig ileum. J. Anat. (Lond.) **111**, 69-97 (1972).
GABELLA, G., COSTA, M.: Le fibre adrenergiche nel canale alimentare. G. Accad. Med. Torino, **130**, 198-217 (1967).
GABELLA, G., COSTA, M.: Adrenergic fibres in the mucous membrane of guinea-pig alimentary tract. Experientia (Basel) **24**, 706-707 (1968).
GARRY, R. C.: The movements of the large intestine. Physiol. Rev. **14**, 103-132 (1934).
GERSHON, M. D.: Inhibition of gastrointestinal movement by sympathetic nerve stimulation: The site of action. J. Physiol. (Lond.) **189**, 317-327 (1967).
GILLESPIE, J. S.: Electrical responses of mammalian smooth muscle to stimulation of the extrinsic sympathetic and parasympathetic nerves. Physiologist **3**, 65 (1960).

GILLESPIE, J. S.: Spontaneous mechanical and electrical activity of stretched and unstretched intestinal smooth muscle cells and their response to sympathetic nerve stimulation. J. Physiol. (Lond.) **162**, 54–75 (1962).

GILLESPIE, J. S.: Electrical activity in the colon. In: Handbook of physiology, sect. 6, vol. 4. Washington: American Physiological Society 1968.

GILLESPIE, J. S., MACKENNA, B. R.: The inhibitory action of the sympathetic nerves on the smooth muscle of the rabbit gut, its reversal by reserpine and restoration by catecholamines and by dopa. J. Physiol. (Lond.) **156**, 17–34 (1961).

GILLESPIE, J. S., MAXWELL, J, D.: Adrenergic innervation of sphincteric and nonsphincteric smooth muscle in the rat intestine. J. Histochem. Cytochem. **19**, 676–681 (1971).

GOOTMAN, P. M., COHEN, M. I.: Efferent splanchnic activity and systemic arterial pressure. Amer. J. Physiol. **219**, 897–903 (1970).

GOVIER, W. C., SUGRUE, M. F., SHORE, P. A.: On the inability to produce supersensitivity to catecholamines in intestinal smooth muscle. J. Pharmacol. exp. Ther. **165**, 71–77 (1969).

GRAYSON, J., MENDEL, D.: Physiology of the splanchnic circulation. London: Arnold 1965.

GREEN, H. D., DEAL, C. P., BARDHANABAEDYA, S., DENISON, A. B.: Effects of adrenergic substances and ischemia on blood flow and peripheral resistance of canine mesenteric vascular bed before and during adrenergic blockade. J. Pharmacol. exp. Ther. **113**, 115–123 (1955).

GREEN, H. D., KEPCHAR, J. H.: Control of peripheral resistance in major systemic vascular beds. Physiol. Rev. **39**, 617–686 (1959).

GRILLO, M. A., PALAY, S. L.: Granule containing vesicles in the autonomic nervous system. In: Electron microscopy (Fifth International Congress for Electron Microscopy), ed. S. S. Barondes. New York-London: Academic Press 1962.

GRIM, E.: The flow of blood in the mesenteric vessels. In: Handbook of physiology, sect. 2, vol. 2. Washington: American Physiological Society 1963.

GRIM, E., LINDSETH, E. O.: Distribution of blood flow to the tissues of the small intestine of the dog. Univ. Minn. med. Bull. **30**, 138–145 (1958).

GROLLMAN, A.: Physiological variations in the cardiac output of man. III. The effect of the ingestion of food. Amer. J. Physiol. **89**, 366–370 (1929).

GUNN, M.: Histological and histochemical observations on the myenteric and submucous plexuses of mammals. J. Anat. (Lond.) **102**, 223–239 (1968).

HAGER, H., TAFURI, W. L.: Electronenoptischer Nachweis sog. neurosekretorischer Elementargranula in marklosen Nervernfasern des Plexus myentericus (Auerbach) des Meerschweinchens. Naturwissenschaften **46**, 332–333 (1959).

HÅKANSON, R., LILJA, B., OWMAN, C.: Cellular localization of histamine and monoamines in the gastric mucosa of man. Histochemie **18**, 74–86 (1969).

HAMBERGER, B., NORBERG, K. A.: Monoamines in sympathetic ganglia studied with fluorescence microscopy. Experientia (Basel) **19**, 580–581 (1963).

HAMILTON, A. S., COLLINS, D. A., OPPENHEIMER, M. J.: Effects of blood pressure on intestinal motility. Fed. Proc. **3**, 17 (1944).

HEIMBACH, D. M., CROUT, J. R.: Treatment of paralytic ileus with adrenergic neuronal blocking drugs. Surgery **69**, 582–587 (1971).

HENLE, J.: Handbuch der Nervenlehre des Menschen. Braunschweig: F. Vieweg u. Sohn 1879.

HENRICH, H., LUTZ, J.: Das vasculäre Escape-Phänomen am Intestinalkreislauf und seine Auslösung durch unterschiedliche vasoconstriktorische Substanzen. Pflügers Arch. **329**, 82–94 (1971).

HERRICK, J. H., ESSEX, H. E., MANN, F. C., BALDES, E. J.: The effect of digestion on the blood flow in certain blood vessels of the dog. Amer. J. Physiol. **108**, 621–628 (1934).

HERTZ, A. F.: The ileo-caecal sphincter. J. Physiol. (Lond.) **47**, 54–56 (1913).

HEYMANS, C., NEIL, E.: Reflexogenic areas of the cardiovascular system. London: Churchill 1958.

HINRICHSEN, J., IVY, A. C.: Studies on the ileo-caecal sphincter of the dog. Amer. J. Physiol. **96**, 494–507 (1931).

HÖKFELT, T.: *In vitro* studies on central and peripheral monoamine neurones at the ultrastructural level. Z. Zellforsch. **91**, 1–74 (1968).

HONJIN, R., TAKAHASHI, A., SHIMASAKI, S., MARUYAMA, H.: Two types of synaptic nerve processes in the ganglia of Auerbach's plexus of mice, as revealed by electronmicroscopy. J. electron Microscopy **14**, 43–49 (1965).

Hotz, G.: Beiträge zur Pathologie der Darmbewegungen. Mitt. Grenzgeb. Med. Chir. **20**, 257–318 (1909).

Howard, E. R., Garrett, J. R.: The intrinsic myenteric innervation of the hind-gut and accessory muscles of defaecation in the cat. Z. Zellforsch. **136**, 31–44 (1973).

Hukuhara, T., Nakayama, S., Nanba, R.: Locality of receptors concerned with the intestino-intestinal extrinsic and intestinal muscular intrinsic reflexes. Jap. J. Physiol. **10**, 414–419 (1960).

Hukuhara, T., Nakayama, S., Yamagami, M., Miyake, T.: On the intestinal extrinsic reflexes elicited from the small intestine. Acta med. Okayama. **13**, 113–121 (1959).

Hultén, L.: Reflex control of colonic motility and blood flow. Acta physiol. scand., Suppl. **335**, 77–93 (1969).

Hultén, L., Jodal, M.: Extrinsic nervous control of colonic motility. Acta physiol. scand., Suppl. **335**, 21–38 (1969).

Hultén, L., Jodal, M., Lundgren, O.: Local and nervous control of the consecutive vascular sections of the colon. Acta physiol. scand., Suppl **335**, 51–64 (1969).

Iggo, A., Vogt, M.: Preganglionic sympathetic activity in normal and in reserpine-treated cats. J. Physiol. (Lond.) **150**, 114–133 (1960).

Ingelfinger, F. J.: Esophageal motility. Physiol. Rev. **38**, 533–584 (1958).

Ivy, A. C., Grossman, M. I., Bachrach, W. H.: Peptic ulcer. London: Churchill 1950.

Izquierdo, J. J., Koch, E.: Über den Einfluß der Nervi splanchnici auf den arteriellen Blutdruck des Kaninchens. Z. Kreisl.-Forsch. **22**, 735–743 (1930).

Jacobowitz, D.: Histochemical studies of the autonomic innervation of the gut. J. Pharmacol. exp. Ther. **149**, 358–364 (1965).

Jacobson, E. D.: Secretion and blood flow in the gastrointestinal tract. In: Handbook of physiology, sect. 6, vol. 2. Washington: American Physiological Society 1967.

Jacobson, E. D.: Comparison of prostaglandin E_1 and norepinephrine on the gastric mucosal circulation. Proc. Soc. exp. Biol. (N. Y.) **133**, 516–519 (1970).

Jacobson, E. D., Linford, R. H., Grossman, M. I.: Gastric secretion in relation to mucosal blood flow studied by a clearance technique. J. clin. Invest. **45**, 1–13 (1966).

Jansson, G.: Vago-vagal reflex relaxation of the stomach in the cat. Acta physiol. scand. **75**, 245–252 (1969a).

Jansson, G.: Effects of reflexes of somatic afferents on the adrenergic outflow to the stomach in the cat. Acta physiol. scand. **77**, 17–22 (1969b).

Jansson, G., Lisander, B.: On adrenergic influence on gastric motility in chronically vagotomized cats. Acta physiol. scand. **76**, 463–471 (1969).

Jansson, G., Lisander, B., Martinson, J.: Hypothalamic control of adrenergic outflow to the stomach in the cat. Acta physiol. scand. **75**, 176–186 (1969).

Jansson, G., Martinson, J.: Studies on the ganglionic site of action of sympathetic outflow to the stomach. Acta physiol. scand. **68**, 184–192 (1966).

Jenkinson, D. H., Morton, I. K. M.: The role of α- and β-adrenergic receptors in some actions of catecholamines on intestinal smooth muscle. J. Physiol. (Lond.) **188**, 387–402 (1967).

Job, C., Lundberg, A.: Reflex excitation of cells in the inferior mesenteric ganglion on stimulation of the hypogastric nerve. Acta physiol. scand. **26**, 366–382 (1952).

Johansson, B., Jonsson, O., Ljung, B.: Supraspinal control of the intestino-intestinal inhibitory reflex. Acta physiol. scand. **63**, 442–449 (1965).

Johansson, B., Jonsson, O., Ljung, B.: Tonic supraspinal mechanisms influencing the intestino-intestinal inhibitory reflex. Acta physiol. scand. **72**, 200–204 (1968).

Johansson, B., Langston, J. B.: Reflex influence of mesenteric afferents on renal, intestinal and muscle blood flow and on intestinal motility. Acta physiol. scand. **61**, 400–412 (1964).

Katz, B.: Nerve, muscle, and synapse. New York: McGraw-Hill 1966.

Kewenter, J.: The vagal control of the jejunal and ileal motility and blood flow. Acta physiol. scand., Suppl. **257**, 1–68 (1965).

Kim, T. S., Shulman, J., Levine, R. A.: Relaxant effect of cyclic adenosine 3′,5′,-monophosphate on the isolated rabbit ileum. J. Pharmacol. exp. Ther. **163**, 36–42 (1968).

Kisch, B.: Die Irradiation autonomer Reflexe und ihre Beziehung zu gewissen pathologisch-physiologischen Erscheinungen. Z. ges. exp. Med. **52**, 499–507 (1926).

Knoll, J., Vizi, E. S.: Effect of frequency of stimulation on the inhibition by noradrenaline of the acetylcholine output from parasympathetic nerve terminals. Brit. J. Pharmacol. **42**, 263–272 (1971).

KOCK, N. G.: An experimental analysis of mechanisms engaged in reflex inhibition of intestinal motility. Acta physiol. scand. **47**, Suppl. 164, 1–54 (1959).

KOIZUMI, K., SUDA, I.: Induced modulations in autonomic efferent neuron activity. Amer. J. Physiol. **205**, 738–744 (1963).

KOSTERLITZ, H. W.: Intrinsic and extrinsic nervous control of motility of the stomach and the intestines. In: Handbook of physiology, sect. 6, vol 4. Washington: American Physiological Society 1968.

KREMER, M., WRIGHT, S.: The effects on blood pressure of section of the splanchnic nerves. Quart. J. exp. Physiol. **21**, 319–335 (1932).

KULLMANN, R., SCHÖNUNG, W., SIMON, E.: Antagonistic changes of blood flow and sympathetic activity in different vascular beds following central thermal stimulation. 1. Blood flow in skin, muscle and intestine during spinal cord heating and cooling in anesthetized dogs. Pflügers Arch. **319**, 146–161 (1970).

KUNTZ, A.: The structural organization of the inferior mesenteric ganglia. J. comp. Neurol. **72**, 371–382 (1940).

KUNTZ, A.: The autonomic nervous system. Philadelphia: Lea & Febiger 1953.

KUNTZ, A., BUSKIRK, C. VAN.: Reflex inhibition of bile flow and intestinal motility mediated through decentralized celiac plexus. Proc. Soc. exp. Biol. (N. Y.) **46**, 519–523 (1941).

KUNTZ, A., JACOBS, M. W.: Components of periarterial extensions of celiac and mesenteric plexuses. Anat. Rec. **123**, 509–520 (1955).

KUNTZ, A., SACCOMANNO, G.: Reflex inhibition of intestinal motility mediated through decentralized prevertebral ganglia. J. Neurophysiol. **7**, 163–170 (1944).

KURIYAMA, H., OSA, T., TOIDA, N.: Nervous factors influencing the membrane activity of intestinal smooth muscle. J. Physiol. (Lond.) **191**, 257–270 (1967).

LACROIX, E.: La circulation splanchnique. Acta gastro-ent. belg. **23**, 534–581 (1960).

LANGLEY, J. N.: Observations on the physiological action of extracts of the supra-renal bodies. J. Physiol. (Lond.) **27**, 237–256 (1901).

LANGLEY, J. N., ANDERSON, H. K.: On reflex action from sympathetic ganglia. J. Physiol. (Lond.) **16**, 410–440 (1894).

LANGLEY, J. N., ANDERSON, H. K.: The constituents of the hypogastric nerves. J. Physiol. (Lond.) **17**, 177–191 (1894–5).

LANGLEY, J. N., ANDERSON, H. K.: On the innervation of the pelvic and adjoining viscera. Part 1. The lower portion of the intestine. J. Physiol. (Lond.) **18**, 67–105 (1895a).

LANGLEY, J. N., ANDERSON, H. K.: The innervation of the pelvic and adjoining viscera. Part V. Position of the nerve cells on the course of the efferent nerve fibres. J. Physiol. (Lond.) **19**, 131–139 (1895b).

LANGLEY, J. N., DICKINSON, W. L.: On the local paralysis of the peripheral ganglia and on the connexion of different classes of nerve fibres with them. Proc. roy. Soc. **46**, 423–431 (1889).

LA VILLA, I.: Estructura de los ganglios intestinales. Rev. trim. micrograf. **3**, 1–13 (1898).

LEARMONTH, J. R., MARKOWITZ, J.: Studies on the function of the lumbar sympathetic outflow. I. The relation of the lumbar sympathetic outflow to the sphincter ani internus. Amer. J. Physiol. **89**, 686–691 (1929).

LEE, C. Y.: Adrenergic receptors in the intestine. In: Smooth muscle, eds. E. Bülbring, A. F. Brading, A. W. Jones, T. Tomita. London: Arnold 1970.

LI, P. L.: The intramural nervous system of the small intestine with special reference to the innervation of the inner subdivision of its circular muscle. J. Anat. (Lond.) **74**, 348–359 (1940).

LISTER, J.: Preliminary account of an inquiry into the functions of the visceral nerves, with special reference to the so-called "inhibitory system". Proc. roy. Soc. **9**, 367–380 (1858).

LLINAS, R.: A possible mecanism for presynaptic inhibition. In: Structure and function of inhibitory neuronal mechanisms, eds. C. v. Euler, S. Skoglund, U. Söderberg, p. 249–250. Oxford: Pergamon 1966.

LUDWIG, C.: Sympathischer Nerv. Lehrbuch der Physiologie des Menschen, Bd. 1, S. 175–185. Leipzig-Heidelberg: C. F. Winter 1852.

MACHADO, A. B. M.: Straight OsO_4 versus glutaraldehyde-OsO_4 in sequence as fixatives for the granular vesicles in sympathetic axons of the rat pineal body. Stain Technol. **42**, 293–300 (1967).

MACLEAN, A. B.: The gastro-ileal reflex in chronic appendicitis. Brit. med. J. **1932 II**, 1055–1056.

MALMFORS, T.: Studies on adrenergic nerves. Acta physiol. scand. **64**, Suppl. 248, 1–93 (1965).

MANN, M., WEST, G. B.: The nature of uterine and intestinal sympathin. Brit. J. Pharmacol. **6**, 79–82 (1951).
MARKOWITZ, J., CAMPBELL, W. R.: The relief of experimental ileus by spinal anesthesia. Amer. J. Physiol. **81**, 101–106 (1927).
MCCREA, E. D'A.: The nerves of the stomach and their relation to surgery. Brit. J. Surg. **13**, 621–648 (1926).
MCDOWALL, R. J. S.: The nervous control of the blood vessels. Physiol. Rev. **15**, 98–174 (1935).
MCLENNAN, H., PASCOE, J. E.: The origin of certain nonmedullated nerve fibres which form synapses in the inferior mesenteric ganglion of the rabbit. J. Physiol. (Lond.) **124**, 145–156 (1954).
MCSWINEY, B. A.: Innervation of the stomach. Physiol. Rev. **11**, 478–514 (1931).
M'FADDEN, G. D. F., LOUGHRIDGE, J. S., MILROY, T. H.: The nerve control of the distal colon. Quart. J. exp. Physiol. **25**, 315–327 (1935).
MIRKIN, B. L., BONNYCASTLE, D. D.: A pharmacological and chemical study of humoral mediators in the sympathetic nervous system. Amer. J. Physiol. **178**, 529–534 (1954).
MITCHELL, G. A. G.: Anatomy of the autonomic nervous system. Edinburgh-London: Livingstone 1953.
MONNIER, M.: Functions of the nervous system, vol. 1, General physiology, Autonomic functions. Amsterdam–London–New York: Elsevier 1968.
MOORE, R. M.: The dispensability of the sympathetic nervous system. In: The vegetative nervous system, eds. W. Timme, T. K. Davis, H. A. Riley. Baltimore: Williams & Wilkins 1930.
MORI, J., AZUMA, H., FUJIWARA, M.: Adrenergic innervation and receptors in the sphincter of Oddi. Europ. J. Pharmacol. **14**, 365–373 (1971).
MORIN, G., VIAL, J.: Sur les voies et les centres du réflexe inhibiteur intestino-intestinal. C. R. Soc. Biol. (Paris) **116**, 536–538 (1934).
MÜLLER, L. R.: Das vegetative Nervensystem. Berlin: Springer 1920.
MÜLLER, L. R.: Die Lebensnerven. Berlin: Springer 1924.
MUNRO, A. F.: Effect of autonomic drugs on the response of isolated preparations from the guinea-pig intestine to electrical stimulation. J. Physiol. (Lond.) **120**, 41–52 (1953).
MURYOBAYASHI, T., FUJIWARA, M., SHIMAMOTO, K.: Fluorescence histochemical findings of the stomach walls in response to ulcerogenic stimuli in rats. Jap. J. Pharmacol. **18**, 299–311 (1968).
NAGASAWA, J., MITO, S.: Electronmicroscopic observations on the innervation of the smooth muscle. Tohoku J. exp. Med. **91**, 277–293 (1967).
NEELY, J., CATCHPOLE, B.: Ileus: the restoration of alimentary tract motility by pharmacological means. Brit. J. Surg. **58**, 21–28 (1971).
NISIMARU, N.: Comparison of gastric and renal nerve activity. Amer. J. Physiol. **220**, 1303–1308 (1971).
NORBERG, K.-A.: Adrenergic innervation of the intestinal wall studied by fluorescence microscopy. Int. J. Neuropharmacol. **3**, 379–382 (1964).
NORBERG, K.-A., HAMBERGER, B.: The sympathetic adrenergic neuron. Some characteristics revealed by histochemical studies on the intraneuronal distribution of the transmitter. Acta physiol. scand. **63**, (Suppl. 238) 1–42 (1964).
OCHSNER, A., DEBAKEY, M.: Surgical consideration of achalasia. Arch. Surg. **41**, 1146–1183 (1940).
OCHSNER, A., GAGE, I. M.: Adynamic ileus. Amer. J. Surg. **20**, 378–404 (1933).
OCHSNER, A., GAGE, I. M., CUTTING, R. A.: Treatment of ileus by splanchnic anaesthesia. J. Amer. med. Ass. **90**, 1847–1853 (1928).
OHKAWA, H., PROSSER, C. L.: Electrical activity in myenteric and submucous plexuses of cat intestine. Amer. J. Physiol. **222**, 1412–1419 (1972a).
OHKAWA, H., PROSSER, C. L.: Functions of neurons in enteric plexuses of cat intestine. Amer. J. Physiol. **222**, 1420–1426 (1972b).
OLIVECRONA, H.: An experimental study of the post operative, so called paralytic ileus. Acta chir. scand. **61**, 485–534 (1927).
ONO, M.: Electron microscopic observations on the ganglia of Auerbach's plexus and autonomic nerve endings in muscularis externa of the mouse small intestine. Sapporo Med. J. (Sapporo Igaku Zasshi) **32**, 56–74 (1967).
OPENCHOWSKI, T.: Über die nervösen Vorrichtungen des Magens. Zbl. Physiol. **3**, 1–10 (1889).
PATON, W. D. M., VIZI, E. S.: The inhibitory action of noradrenaline and adrenaline on acetylcholine output by guinea-pig ileum longitudinal muscle strip. Brit. J. Pharmacol. **35**, 10–28 (1969).

PATON, W.D.M., VIZI, E.S., ZAR, M.A.: The mechanism of acetylcholine release from parasympathetic nerves. J. Physiol. (Lond.) **215**, 819–848 (1971).
PATON, W.D.M., ZAR, M.A.: The origin of acetylcholine released from guinea-pig intestine and longitudinal muscle strips. J. Physiol. (Lond.) **194**, 13–33 (1968).
PEARCY, J.F., LIERE, E.J. VAN: Studies on the visceral nervous system. XVII. Reflexes from the colon. 1. Reflexes to the stomach. Amer. J. Physiol. **78**, 64–73 (1926).
PERSSON, C.G.A.: Adrenoceptor functions in the cat choledochoduodenal junction in vitro. Brit. J. Pharmacol. **42**, 447–461 (1971).
PETRAS, J.M., CUMMING, J.F.: Anatomical basis of visceral reflexes in the spinal cord. Anat. Rec. **169**, 401 (1971).
PETRI, G., SZENOHRADSZKY, J., PÓRSZÁSZ-GIBISZER, K.: Sympatholytic treatment of "paralytic" ileus. Surgery **70**, 359–367 (1971).
PFLÜGER, E.: Über das Hemmungs-Nervensystem für die peristaltischen Bewegungen der Gedärme. Berlin: Hirschwald 1857.
PICK, J.: Fine structure of nerve terminals in the human gut. Anat. Rec. **159**, 131–146 (1967).
POMERANZ, B., WALL, P.D., WEBER, W.V.: Cord cells responding to fine myelinated afferents from viscera, muscle and skin. J. Physiol. (Lond.) **199**, 511–532 (1968).
RATZENHOFER, M., MÜLLER, O., BECKER, H.: Zur Innervation der Drüsen- und Stromazellen im Kaninchenmagen. Nach elektronen- und fluoreszenzmikroskopischen Untersuchungen. Mikroskopie **25**, 283–296 (1969).
READ, J.B., BURNSTOCK, G.: Adrenergic innervation of the gut musculature in vertebrates. Histochemie **17**, 263–272 (1969).
REED, J.D., SANDERS, D.J.: Pepsin secretion, gastric motility and mucosal blood flow in the anaesthetized cat. J. Physiol. (Lond.) **216**, 159–170 (1971).
REED, J.D., SANDERS, D.J., THORPE, V.: The effect of splanchnic nerve stimulation on gastric acid secretion and mucosal blood flow in the anaesthetized cat. J. Physiol. (Lond.) **214**, 1–13 (1971).
RICHARDSON, D.R., JOHNSON, P.C.: Changes in mesenteric capillary flow during norepinephrine infusion. Amer. J. Physiol. **219**, 1317–1323 (1970).
RICHARDSON, D.R., ZWEIFACH, B.W.: Pressure relationships in the macro and microcirculation of the mesentery. Microvasc. Res. **2**, 474–488 (1970).
RICHARDSON, K.C.: The fine structure of autonomic nerve endings in smooth muscle of the rat vas deferens. J. Anat. (Lond.) **96**, 427–442 (1962).
RINTOUL, J.R.: The comparative morphology of the enteric nerve plexuses. Ph. D. thesis, University of St. Andrews, 1960.
ROBISON, G.A., BUTCHER, R.W., SUTHERLAND, E.W.: Cyclic AMP. New York-London: Academic Press 1971.
ROSS, G.: Effects of epinephrine and norepinephrine on the mesenteric circulation of the cat. Amer. J. Physiol. **212**, 1037–1042 (1967).
ROSS, G.: Effects of catecholamines on splanchnic blood flow before and after beta adrenergic blockade. In: Cardiovascular beta adrenergic responses, eds. A.A. Kattus, G. Ross, V.E. Hall, p. 93–107. Los Angeles: University of California Press 1970.
ROSS, G.: Effects of norepinephrine infusion on mesenteric arterial blood flow and its tissue distribution. Proc. Soc. exp. Biol. (N. Y.) **137**, 921–924 (1971a).
ROSS, G.: The regional circulation. Ann. Rev. Physiol. **33**, 445–478 (1971b).
ROSS, G.: Escape of mesenteric vessels from adrenergic and noradrenergic vasoconstriction. Amer. J. Physiol. **221**, 1217–1222 (1971c).
ROSS, J.G.: On the presence of centripetal fibres in the superior mesenteric nerves of the rabbit. J. Anat. (Lond.) **92**, 189–197 (1958).
ROSS, L.L., GERSHON, M.D.: Adrenergic innervation of the myenteric plexus of the guinea-pig: ultrastructural, biochemical, and histofluorometric studies using 6-hydroxydopamine. Anat. Rec. **47**, 175 (1970a).
RUSHMER, R.F., FRANKLIN, D.L., CITTERS, R.L. VON, SMITH, O.A.: Changes in peripheral blood flow distribution in healthy dogs. Circulat. Res. **9**, 675–687 (1961).
SATO, A.: The relative involvement of different reflex pathways in somato-sympathetic reflexes analyzed in spontaneously active single preganglionic sympathetic units. Pflügers Arch. **333**, 70–81 (1972).
SCHABADASCH, A.: Intramurale Nervengeflechte des Darmrohrs. Z. Zellforsch. **10**, 320–385 (1930).

SCHANKER, L. S., SHORE, P. A., BRODIE, B. B., HOGBEN, C. A. M.: Absorption of drugs from the stomach. 1. The rat. J. Pharmacol. exp. Ther. **120**, 528–539 (1957).

SCHOFIELD, G. C.: Experimental studies on the innervation of the mucous membrane of the gut. Brain **83**, 490–514 (1960).

SCHOFIELD, G. C.: The enteric plexus of mammals. In: International review of general and experimental zoology, eds. W. J. L. Felts and R. J. Harrison. New York: Acad. Press 1968.

SEMBA, T.: Intestino-intestinal inhibitory reflexes. Jap. J. Physiol. **4**, 241–245 (1954a).

SEMBA, T.: Studies on the entero-gastric reflexes. Hiroshima J. med. Sci. **2**, 323–327 (1954b).

SEMBA, T., FUJII, Y.: Relationship between venous flow and colonic peristalsis. Jap. J. Physiol. **20**, 408–416 (1970).

SEMBA, T., SASAKI, H.: The effects of intestinal contraction on the portal venous pressure. Jap. J. Physiol. **3**, 18–24 (1953).

SHAPIRO, H., WOODWARD, E. R.: Inhibition of gastric motility by acid in the duodenum. J. appl. Physiol. **8**, 121–127 (1955).

SHAPIRO, H., WOODWARD, E. R.: Pathway of enterogastric reflex. Proc. Soc. exp. Biol. (N. Y.) **101**, 407–409 (1959).

SHEHADEH, Z., PRICE, W. E., JACOBSON, E. D.: Effects of vasoactive agents on intestinal blood flow and motility in the dog. Amer. J. Physiol. **216**, 386–392 (1969).

SHEPHERD, J. J., WRIGHT, P. G.: The response of the internal anal sphincter in man to stimulation of the presacral nerve. Amer. J. dig. Dis. **13**, 421–427 (1968).

SHORE, P. A.: A simple technique involving solvent extraction for the estimation of norepinephrine and epinephrine in tissues. Pharmacol. Rev. **11**, 276–277 (1959).

SIDKY, M., BEAN, J. W.: Influence of rhythmic and tonic contraction of intestinal muscle on blood flow and blood reservoir capacity in dog intestine. Amer. J. Physiol. **193**, 386–392 (1958).

SILVA, D. G., ROSS, G., OSBORNE, L. W.: Adrenergic innervation of the ileum of the cat. Amer. J. Physiol. **220**, 347–352 (1971).

SMETS, W.: La contraction de la valvule iléo-caecale. C. R. Soc. Biol. (Paris) **122**, 793–795 (1936a).

SMETS, W.: L'activité réflexe de la valvule iléo-caecale et du segment terminal de grêle. C. R. Soc. Biol. (Paris) **123**, 106–107 (1936b).

SMYTHE, C. MCC., FITZPATRICK, H. F., BLAKEMORE, A. H.: Studies of portal venous oxygen content in unanesthetized man. J. clin. Invest. **30**, 674 (1951).

STARLING, E. H.: Überblick über den gegenwärtigen Stand der Kenntnisse über die Bewegungen und die Innervation des Verdauungskanals. Ergebn. Physiol. **1**, 446–465 (1902).

STARLING, E. H.: Recent advances in the physiology of digestion. London: Constable 1906.

STÖHR, P.: Mikroskopische Studien zur Innervation des Magen-Darmkanales. Z. Zellforsch. **12**, 66–154 (1930).

TAFURI, W. L.: Ultrastructure of the vesicular component in the intramural nervous system of the guinea-pig's intestine. Z. Naturforsch. **19** B, 622–625 (1964).

TAXI, J.: Contribution à l'étude des connexions des neurones moteurs du système nerveux autonome. Ann. Sci. nat. Zool. **7**, 413–464 (1965).

TAXI, J.: Morphological and cytochemical studies on the synapses in the autonomic nervous system. Progr. Brain Res. **31**, 5–20 (1969).

TAXI, J., DROZ, B.: Localization d'amines biogènes dans le système neurovégétatif périphérique. Étude radioautographique en microscopie électronique après injection de noradrénaline-^{3}H et de 5-hydroxytryptophane-^{3}H. In: Neurosecretion, ed. F. Stutinsky. Berlin-Heidelberg-New York: Springer 1967.

TAXI, J., DROZ, B.: Radioautographic study of the accumulation of some biogenic amines in the autonomic nervous system. In: Cellular dynamics of the neuron, ed. S. H. Barondes, p. 175–190. New York: Academic Press 1969.

THOMPSON, J. E., VANE, J. R.: Gastric secretion induced by histamine and its relationship to the rate of blood flow. J. Physiol. (Lond.) **121**, 433–444 (1953).

TRANZER, J. P., THOENEN, H.: Significance of "empty vesicles" in post-ganglionic sympathetic nerve terminals. Experientia (Basel) **23**, 123–124 (1967).

TRENDELENBURG, U.: Some aspects of the pharmacology of autonomic ganglion cells. Ergebn. Physiol. **59**, 1–85 (1967).

TUMEN, H. J.: Intestinal obstruction. In: Gastroenterology, ed. H. L. Bockus, p. 343–403. Philadelphia: W. B. Saunders 1964.

UNGVÁRY, G., LÉRÁNTH, C.: Termination in the prevertebral abdominal sympathetic ganglia of axons arising from the local (terminal) vegetative plexus of visceral organs. Z. Zellforsch. **110**, 185–191 (1970).

VAN ORDEN, L. S., BLOOM, F. E., BARRNETT, R. J., GIARMAN, N. J.: Histochemical and functional relationships of catecholamines in adrenergic nerve endings. 1. Participation of granular vesicles. J. Pharmacol. exp. Ther. **154**, 185–199 (1966).

VAN ORDEN, L.S., SCHAEFER, J.-M., BURKE, J.P., LODOEN, F.V.: Differentiation of norepinephrine storage compartments in peripheral adrenergic nerves. J. Pharmacol. exp. Ther. **174**, 357–368 (1970).

VARRO, V., BLAHO, G., CSERNAY, L., JUNG, I., SZARVAS, F.: Effect of decreased local circulation on the absorptive capacity of a small intestine loop in the dog. Amer. J. dig. Dis. **10**, 170–177 (1965).

VATNER, S.F., HIGGINS, C.B., WHITE, S., PATRICK, T., FRANKLIN, D.: The peripheral vascular response to severe exercise in untethered dogs before and after complete heart block. J. clin. Invest. **50**, 1950–1960 (1971).

VIZI, E.S., KNOLL, J.: The effects of sympathetic nerve stimulation and guanethidine on parasympathetic neuroeffector transmission; the inhibition of acetylcholine release. J. Pharm. Pharmacol. **23**, 918–925 (1971).

WADDELL, M.C.: Anatomical evidence for existence of enteric reflex arcs following degeneration of extrinsic nerves. Proc. Soc. exp. Biol. (N. Y.) **26**, 867–869 (1929).

WAGNER, G.A.: Behandlung des paralytischen Ileus. Berl. klin. Wschr. **56**, 1221 (1919).

WALTHER, O.-E., IRIKI, M., SIMON, E.: Antagonistic changes of blood flow and sympathetic activity in different vascular beds following central thermal stimulation. II. Cutaneous and visceral sympathetic activity during spinal cord heating and cooling in anaesthetized rabbits and cats. Pflügers Arch. **319**, 162–184 (1970).

WANGENSTEEN, O. H.: Intestinal obstructions. Springfield: Thomas 1955.

WATT, A.J.: The inhibitory effect of perivascular stimulation on the response of the guinea-pig isolated ileum. J. Physiol. (Lond.) **219**, 38–39 P (1971).

WESTPHAL, K.: Muskelfunktion, Nervensystem und Pathologie der Gallenwege. Z. klin. Med. **96**, 22–150 (1923).

WILKENFELD, B.E., LEVY, B.: The effects of theophylline, diazoxide and imidazole on isoproterenol-induced inhibition of the rabbit ileum. J. Pharmacol. exp. Ther. **169**, 61–67 (1969).

WOLFE, D.E., POTTER, L.T., RICHARDSON, K.C., AXELROD, J.: Localizing tritiated norepinephrine in sympathetic axons by electron microscopic autoradiography. Science **138**, 440–442 (1962).

WOODS, G., NELSON, V.E., NELSON, E.E.: The effect of small amounts of ergotamine on the circulatory response to epinephrine. J. Pharmacol. exp. Ther. **45**, 403–418 (1932).

WYATT, A.P.: The relationship of the sphincter of Oddi to the stomach, duodenum and gall bladder. J. Physiol. (Lond.) **193**, 225–243 (1967).

YOUMANS, W.B.: Nervous and neurohumoral regulation of intestinal motility. New York: Interscience Publishers 1949.

YOUMANS, W.B.: Neural regulation of gastric and intestinal motility. Amer. J. Med. **13**, 209–226 (1952).

YOUMANS, W.B.: Innervation of the gastrointestinal tract. In: Handbook of physiology, sect. 6, vol. 4. Washington: American Physiological Society 1968.

YOUMANS, W.B., AUMANN, K.W., HANEY, H.F., WYNIA, F.: Reflex cardiac acceleration and liberation of sympathomimetic substances in unanaesthetized dogs during acetylcholine hypotension. Amer. J. Physiol. **128**, 467–474 (1940).

YOUMANS, W.B., KARSTENS, A.I., AUMANN, K.W.: Nervous pathways for the reflex regulation of intestinal pressure. Amer. J. Physiol. **135**, 619–627 (1942).

Regulation of Fatty-Acid Synthesis in Higher Animals

SHOSAKU NUMA *

Contents

I. Introduction . . . 53
II. Regulation at the Acetyl-CoA Carboxylase Step . . . 56
A. Generation of Cytoplasmic Acetyl-CoA . . . 56
1. Transport of Acetyl Group out of Mitochondria . . . 56
2. Control of Citrate Translocation and of Citrate-Cleavage Enzyme . . . 58
B. Control by Changes in the Quantity of Acetyl-CoA Carboxylase . . . 60
1. Tissue Content of Acetyl-CoA Carboxylase in Various Metabolic Conditions . . . 60
2. Synthesis and Degradation of Acetyl-CoA Carboxylase . . . 62
3. Loss of Control in Hepatomas . . . 64
C. Control by Changes in the Catalytic Efficiency of Acetyl-CoA Carboxylase . . . 65
1. Short-Term *vs.* Long-Term Control . . . 65
2. Reaction Mechanism . . . 66
3. Kinetic Effects of Allosteric Regulators . . . 68
4. Molecular Basis for Control of the Catalytic Efficiency . . . 72
5. Tissue Contents of Allosteric Regulators . . . 77
6. Hypolipidemic Agents . . . 78
7. Ratio of Holoenzyme to Apoenzyme . . . 78
III. Regulation at the Fatty-Acid Synthetase Step . . . 79
A. Generation of Cytoplasmic NADPH . . . 79
B. Control by Changes in the Quantity of Fatty-Acid Synthetase . . . 81
C. Control by Changes in the Catalytic Efficiency of Fatty-Acid Synthetase . . . 82
IV. Coordinate Response of Lipogenic Enzymes . . . 84
V. Concluding Remarks . . . 85
References . . . 85

I. Introduction

Fat is the most concentrated store of energy for the organism, yielding per gram over twice as many calories as carbohydrate or protein. The major homeostatic function of lipogenesis is to store as fat the chemical energy of foodstuffs ingested in excess of the immediate energy requirements of the organism. Therefore the lipogenic process must be precisely regulated in animals to respond to their fluctuating energy needs and to the quantity as well as the quality of foodstuffs ingested. For instance, lipogenesis is lowered in fasted or alloxan-diabetic animals and in animals fed a high-fat diet; in all these metabolic condi-

* Department of Medical Chemistry, Kyoto University Faculty of Medicine, Kyoto, Japan.

tions, carbohydrate utilization is restricted. On the other hand, when fasted animals are refed a fat-free high-carbohydrate diet, more fat is synthesized to replenish the fat store depleted during starvation.

What are the factors that determine the rate of lipogenesis? In higher animals, the endocrine and nervous systems obviously play essential roles in coordinating the metabolism of many substances, including lipid, in various tissues. In addition to the hormonal and nervous regulatory mechanisms, there are more basic regulatory mechanisms which operate at cellular level by controlling the quantity and the catalytic efficiency of the enzymes responsible for the metabolic pathway in question. Mechanisms of this type occur in all forms of life, including unicellular organisms that possess neither a hormonal nor a nervous system. Both sets of regulatory mechanisms function in higher animals: the messages provided by hormones and nerves are transmitted to the "basic" regulatory system, which then exhibits final modulating effects. Thus, control of lipogenesis in higher organisms is effected ultimately by changes in either the quantity or the catalytic efficiency of lipogenic enzymes.

The pathway for *de novo* synthesis of long-chain fatty acids proceeds via malonyl-CoA, as represented by Reactions 1 and 2 (WAKIL, 1958; LYNEN, 1959, 1961).

$$ATP + HCO_3^- + \text{acetyl-CoA} \overset{Mg^{2+}}{\rightleftharpoons} ADP + P_i + \text{malonyl-CoA} \quad (1)$$

$$\text{Acetyl-CoA} + 7\ \text{malonyl-CoA} + 14\ NADPH + 14\ H^+$$
$$\longrightarrow \text{palmitic acid} + 7\ CO_2 + 8\ COA + 14\ NADP^+ + 6\ H_2O \quad (2)$$

The intermediate, malonyl-CoA, is formed by carboxylation of acetyl-CoA catalyzed by acetyl-CoA carboxylase[1] (Reaction 1) and is converted to fatty acids through a series of reactions catalyzed by a multienzyme complex, fatty-acid synthetase (Reaction 2). Both enzymes are located in the cytoplasm. Fig. 1 illustrates, in addition to the pathway for fatty-acid synthesis, the supporting pathways which provide the carbon precursor, acetyl-CoA, and the reducing agent, NADPH, in the cytoplasm. The acetyl-CoA carboxylase reaction is the first step leading specifically to fatty-acid synthesis, since malonyl-CoA has no other apparent metabolic use. Hence, it would be of teleonomic significance to regulate fatty-acid synthesis at this carboxylase step. In fact, the tissue content of acetyl-CoA carboxylase changes in step with the rate of fatty-acid synthesis in a variety of metabolic conditions (see Section II, B, 1), and the catalytic efficiency of this enzyme is affected by a number of metabolites (see

[1] *Enzymes*. Acetyl-CoA carboxylase or acetyl-CoA:carbon-dioxide ligase (ADP) (EC 6.4.1.2); acetyl-CoA synthetase or acetate: CoA ligase (AMP) (EC 6.2.1.1); arginase or L-arginine amidinohydrolase (EC 3.5.3.1); carnitine acetyltransferase or acetyl-CoA:carnitine *O*-acetyltransferase (EC 2.3.1.7); citrate-cleavage enzyme or ATP:citrate oxaloacetate-lyase (CoA-acetylating and ATP-dephosphorylating) (EC 4.1.3.8); citrate synthase or citrate oxaloacetate-lyase (CoA-acetylating) (EC 4.1.3.7); glucose-6-phosphate dehydrogenase or D-glucose-6-phosphate: NADP oxidoreductase (EC 1.1.1.49); isocitrate dehydrogenase (NADP) or *threo*-D_s-isocitrate: NADP oxidoreductase (decarboxylating) (EC 1.1.1.42); malate dehydrogenase or L-malate: NAD oxidoreductase (EC 1.1.1.37); malic enzyme or L-malate: NADP oxidoreductase (decarboxylating) (EC 1.1.1.40); phosphogluconate dehydrogenase or 6-phospho-D-gluconate: NADP oxidoreductase (decarboxylating) (EC 1.1.1.44); pyruvate carboxylase or pyruvate: carbon-dioxide ligase (ADP) (EC 6.4.1.1); pyruvate dehydrogenase or pyruvate: lipoate oxidoreductase (acceptor-acetylating) (EC 1.2.4.1).

Section II, C, 1, 3 and 4). Although animal tissues contain a sufficient amount of acetyl-CoA carboxylase to match the catalytic capacity of fatty-acid synthetase when the carboxylase is fully activated, it is suggested that *in vivo* the carboxylase does not exhibit its maximal catalytic efficiency, and that the carboxylase step is therefore the most likely site of the regulation of fatty-acid synthesis (see Section II, C, 5).

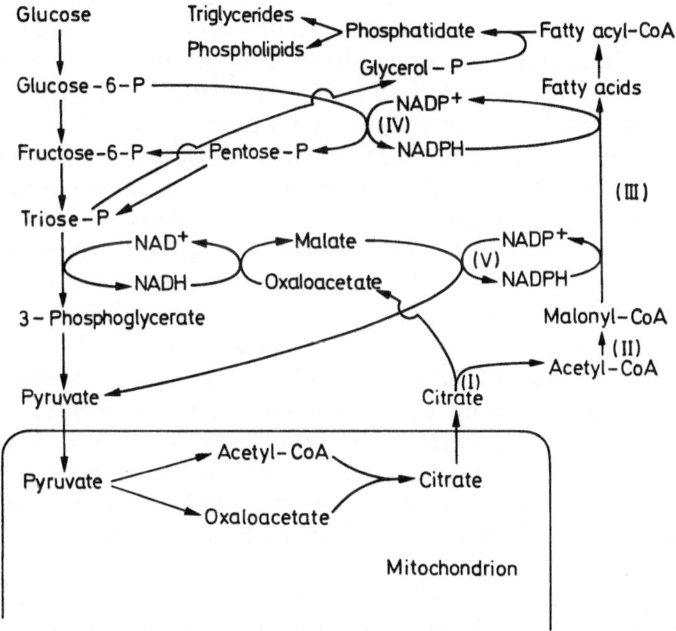

Fig. 1. Pathway for lipogenesis from glucose. The enzymes discussed in this review are as follows: citrate-cleavage enzyme (*I*); acetyl-CoA carboxylase (*II*); fatty-acid synthetase (*III*); hexose monophosphate shunt dehydrogenases (*IV*); malic enzyme (*V*)

The present review deals primarily with the "basic" regulatory mechanisms for fatty-acid synthesis in higher animals, namely, the control of lipogenic enzymes, and does not include studies on overall fatty-acid synthesis in whole animals, organs, and tissue slices. Theoretically, fatty-acid synthesis could be controlled at the steps of acetyl-CoA carboxylase and fatty-acid synthetase by changes in either substrate concentration or enzyme quantity or the catalytic efficiency of enzyme. In the following sections, these different modes of regulation and their relative importance in the control of fatty-acid synthesis are discussed. This review is restricted to *de novo* fatty-acid synthesis and does not cover transformation of preformed fatty acids, *e. g.* by chain elongation and desaturation. It is also beyond the scope of this review to discuss the regulation of glycolysis and of the oxidative decarboxylation of pyruvate, which constitute the pathway providing most of the carbon precursor for fatty-acid synthesis.

II. Regulation at the Acetyl-CoA Carboxylase Step

A. Generation of Cytoplasmic Acetyl-CoA

1. Transport of Acetyl Group out of Mitochondria

In the degradation of foodstuffs and cell constituents, two thirds of the carbon of carbohydrate, all carbon of common fatty acids and about one half of the carbon of amino acids are converted into acetyl-CoA (see KREBS and LOWENSTEIN, 1960). Most of the acetyl-CoA formed from these precursors arises in the mitochondria, since both pyruvate derived from carbohydrate and fatty acids derived from fat are converted into acetyl-CoA by intramitochondrial enzyme systems. That portion of intramitochondrial acetyl-CoA which is to be utilized for lipogenesis must be transported to the cytoplasm, where *de novo* synthesis of fatty acids occurs. Four alternative pathways have been considered for this translocation of acetyl-CoA (see LOWENSTEIN, 1963; SRERE, 1965a): 1. Incorporation of the acetyl group into citrate, followed by translocation in the form of citrate, or in a form biochemically related to citrate, with subsequent ATP-dependent cleavage giving acetyl-CoA and oxaloacetate (see Fig. 1 and Section III, A; SRERE and BHADURI, 1962; SPENCER and LOWENSTEIN, 1962); 2. Hydrolysis of the acetyl-CoA thioester, diffusion of acetate across the mitochondrial membrane, and its reactivation by acetyl-CoA synthetase (HOCHHEUSER et al., 1964; WIELAND and WEISS, 1966; KORNACKER and LOWENSTEIN, 1965a); 3. Transfer of the acetyl group in the form of acetyl carnitine (BREMER, 1962a, b; FRITZ, 1963; FRITZ and YUE, 1964); 4. Diffusion of acetyl-CoA itself out of mitochondria (LOWENSTEIN, 1963).

Pathway 1 is the generally accepted one for acetyl group translocation. Citrate is formed in the mitochondria by the transfer of the acetyl group from acetyl-CoA to oxaloacetate catalyzed by citrate synthase (Reaction 7 in Section III, A) and is transported out of the mitochondria. Acetyl-CoA is regenerated in the cytoplasm by citrate-cleavage enzyme (Reaction 8 in Section III, A; see SRERE, 1965a, 1972), which is located in the soluble fraction of the cell. This mechanism is supported by a variety of evidence. Citrate was found to be an efficient precursor of fatty-acid synthesis in extracts from avian liver (SRERE and BHADURI, 1962; FORMICA, 1962) and the mammary glands of lactating rats (SPENCER and LOWENSTEIN, 1962). Experiments with asymmetrically labeled citrate demonstrated good incorporation of the acetyl portion of citrate into fatty acids, whereas incorporation of the oxaloacetyl portion was poor (SPENCER and LOWENSTEIN, 1962; BHADURI and SRERE, 1963). The incorporation of radioactivity from [5-^{14}C]α-ketoglutarate into fatty acids in rat mammary gland slices (MADSEN et al., 1964) as well as in perfused rat liver (D'ADAMO and HAFT, 1965) is best accounted for by the conversion of α-ketoglutarate into citrate and its subsequent cleavage to acetyl-CoA and oxaloacetate. Experiments with rat mammary gland slices (BARTLEY et al., 1965) and adipose tissue (ROGNSTAD and KATZ, 1968) showed that tritiated acetyl-CoA generated in mitochondria loses a tritium atom during its incorporation into fatty acids. Such a loss would occur if acetyl-CoA condensed with oxaloacetate to form citrate, but would not occur in the other routes proposed for transport of the acetyl group out of mitochondria. Additional support for the citrate pathway is provided by the effect of inhibiting citrate-cleavage enzyme or citrate transport on fatty-acid synthesis or cytoplasmic acetyl-CoA generation. (−)-Hydroxycitrate, a potent inhibitor of citrate-cleavage enzyme (WATSON et al., 1969), strongly inhibits fatty-acid synthesis in rat liver *in vivo* (LOWENSTEIN, 1971) as well as in a system consisting of cytoplasm and mitochondria (WATSON and LOWENSTEIN, 1970). (+)-Hydroxycitrate, (−)-allohydroxycitrate and (+)-allohydroxycitrate exhibit practically no inhibitory effect either *in vivo* or *in vitro* (SULLIVAN et al., 1972). Fatty-acid synthesis *in vitro* is inhibited also by *n*-butylmalonate (WATSON and LOWENSTEIN, 1970), which is an inhibitor of malate permease (ROBINSON and CHAPPELL, 1967); this inhibition is reversed by malate. *n*-Butylmalonate appears to exert its effect by preventing the activation of citrate transport by malate. In a similar *in vitro* system from rat liver, antibody against citrate-cleavage enzyme reduces extramitochondrial acetylation of sulfanilamide or *p*-toluidine by acetyl-CoA generated from pyruvate to about 15 per cent of the control value (DAIKUHARA et al., 1968). Finally, parallel changes were observed for the level of citrate-cleavage enzyme activity and for fatty-acid synthesis in animals of different nutritional, hormonal, and genetic status, as discussed in more detail in Section II, A, 2. Some recent findings question whether the citrate pathway is the sole mechanism for acetyl group translocation. 2-Ethylcitrate and tricarballylate (propane 1,2,3-tricarboxylate), which are inhibitors of the citrate-transporting system, fail to inhibit fatty-acid synthesis from pyruvate in rat adipose tissue (ROBINSON and WILLIAMS, 1970). Furthermore, *in vivo* studies in mice showed that [2,4-$^{14}C_2$]citrate is incorporated into liver, carcass, and adipose tissue fatty acids to a much greater extent than [1,5-$^{14}C_2$]citrate (ROUS, 1971).

There has been little experimental evidence to support the mechanism involving translocation in the form of free acetate (Pathway 2). However, the recent finding that cytoplasmic acetyl-CoA synthetase in rat lipogenic tissues, in contrast to the enzyme in mitochondria, is subjected to diet-dependent changes (BARTH et al., 1971, 1972) suggests that this pathway needs reconsideration.

The organ distribution of carnitine acetyltransferase as well as its intracellular distribution is inconsistent with the involvement of this enzyme in the transfer of the acetyl group from the intra- to the extramitochondrial space in the cell (Pathway 3). For example, the level of activity of this enzyme in the rat is much higher in energy-utilizing tissues like heart and skeletal muscle than in liver, an organ with a high capacity for fatty-acid synthesis (BEENAKKERS and KLINGENBERG, 1964). Moreover, in rat heart muscle carnitine acetyltransferase is localized exclusively in the mitochondria (BEENAKKERS and KLINGENBERG, 1964). It has been suggested that this pathway is more likely to be involved in biological acetylation (BRESSLER and BRENDEL, 1966).

Direct translocation of acetyl-CoA across the mitochondrial membrane (Pathway 4) appears unlikely, since this process is too slow to account for the observed rates of fatty-acid synthesis (LOWENSTEIN, 1968).

Thus far, results obtained with nonruminant mammals and birds have been discussed. Tissues of ruminants appear to lack an active citrate-cleavage pathway (HANSON and BALLARD, 1967). In ruminants, dietary carbohydrate is fermented by the microbial flora during ruminal digestion mainly to acetate, propionate, and butyrate (ANNISON et al., 1957; ARMSTRONG and BLAXTER, 1957; REID et al., 1957). These short-chain acids, after being absorbed, serve as precursors for fatty-acid synthesis and gluconeogenesis. In fact, the level of activity of citrate-cleavage enzyme is low in ruminant tissues as compared with that in rat tissues (HANSON and BALLARD, 1967; BALLARD and HANSON, 1969).

2. Control of Citrate Translocation and of Citrate-Cleavage Enzyme

Cytoplasmic acetyl-CoA, which is a substrate of acetyl-CoA carboxylase, is derived mainly from carbohydrate via glycolysis, followed by oxidative decarboxylation of pyruvate and transport out of the mitochondria of the acetyl group in the form of citrate; control of fatty-acid synthesis could be achieved in any of these processes. (It is beyond the scope of this review to discuss the regulation of the whole pathway leading to the production of intramitochondrial citrate.) Intramitochondrial citrate can either be oxidized into CO_2 and H_2O through the tricarboxylic acid cycle, or be transported into the extramitochondrial space to serve as precursor for fatty-acid synthesis. The latter process could be controlled either at the step of citrate translocation (CHAPPELL, 1966; CHAPPELL and HAARHOFF, 1967; CHAPPELL and ROBINSON, 1968; MAX et al., 1970; KLINGENBERG, 1970; PALMIERI et al., 1972) or at that of citrate-cleavage enzyme.

Relatively little attention has been paid to the regulation of the system responsible for the transport of citrate out of mitochondria. Recent experiments with rat-liver mitochondria demonstrated that palmityl-CoA at low concentrations inhibits citrate transport (HALPERIN et al., 1972). This inhibition is overcome when the citrate concentration in the medium increases. This, together with the failure of palmityl-CoA to inhibit the dicarboxylate transport system at comparable concentrations, suggests that palmityl-CoA acts specifically on the tricarboxylate transport system.

The level of activity of citrate-cleavage enzyme varies with the nutritional state of the animal. The enzyme level in rat liver and adipose tissue is depressed by starvation and restored on refeeding fasted animals (KORNACKER and LOWENSTEIN, 1965a; KORNACKER and BALL, 1965). The increase in hepatic enzyme level depends on the type of diet used for refeeding; the largest increase occurs with a diet high in carbohydrate and low in fat, the smallest with a diet high in fat, and intermediate with a balanced diet. These fluctuations in the level of activity of hepatic citrate-cleavage enzyme reflect changes in the amount of the enzyme protein as measured immunochemically (SUZUKI et al., 1967). The adaptive increase in the enzyme level is blocked by injecting animals with puromycin at the onset of the refeeding period (GIBSON et al., 1966). This suggests that the response of citrate-cleavage enzyme to refeeding involves *de novo* synthesis of the enzyme. Recent results obtained with the combined use of immunochemical and isotopic techniques (for details, see Section II, B, 2) showed that hepatic citrate-cleavage enzyme is synthesized about three times more slowly than normal in rats starved for 48 hours and about six times faster than normal in rats refed a fat-free diet for 12 hours (GIBSON et al., 1972). The half-life for enzyme degradation is 27 hours in normally fed rats and 14 hours in refed rats. Therefore, the adaptive variations in the enzyme level after dietary manipulation are due to changes in the rate of enzyme synthesis. The activity level of citrate-cleavage enzyme in liver and adipose tissue is also lowered in alloxan-diabetic rats (KORNACKER and LOWENSTEIN, 1964b; BROWN and MCLEAN, 1965) but is restored on administration of insulin (KORNACKER and LOWENSTEIN, 1965b; SHRAGO and LARDY, 1966). Genetically obese hyperglycemic mice (C57BL/6J-*ob*), which carry a single recessive mutant gene for obesity (INGALLS et al., 1950), have an elevated level of hepatic citrate-cleavage enzyme (KORNACKER and LOWENSTEIN, 1964a; SPENCER and LOWENSTEIN, 1966). Administration of thyroid hormones results in an increased enzyme level in the liver (MURAD and FREEDLAND, 1965; YOUNG, 1968; LOCKWOOD et al., 1970; DIAMANT et al., 1972). The rate of enzyme synthesis, as measured by combined immunochemical and isotopic techniques, is elevated after hormone administration (GIBSON et al., 1972). Studies made during the development of rats between late fetal life and maturity showed that the level of hepatic citrate-cleavage enzyme falls after birth, is maintained at a low value until weaning, and rises again after weaning (HAHN and DRAHOTA, 1966; BALLARD and HANSON, 1967; TAYLOR et al., 1967). A similar rise in the hepatic enzyme level is observed also in mice upon weaning, and this rise is blocked by actinomycin D or puromycin (SMITH and ABRAHAM, 1970a). In chick liver, the enzyme level rises remarkably during the early part of the post-hatching period (FELICIOLI and GABRIELLI, 1967; GOODRIDGE, 1968). In rat mammary glands, there is a large increase in citrate-cleavage enzyme within two days of the onset of lactation with a rapid decline after weaning (HOWANITZ and LEVY, 1965; SPENCER and LOWENSTEIN, 1966; LOWENSTEIN, 1968). All these changes in the level of citrate-cleavage enzyme activity correlate well with changes in the rate of fatty-acid synthesis and support the view that this enzyme has a regulatory role in fatty-acid synthesis.

More recent work of SRERE's group (SRERE and FOSTER, 1967; FOSTER and SRERE, 1968) questions the role of citrate-cleavage enzyme in the control of fatty-acid synthesis. In recovery from fasting, fatty-acid synthesis in rat liver slices was seen to increase markedly without any change in the level of citrate-cleavage enzyme. Likewise, after administration of alloxan, hepatic fatty-acid synthesis decreases prior to any change in the level of citrate-cleavage enzyme. These studies were concerned with the relatively short-term control of fatty-acid synthesis, where regulation may be effected by changes in the catalytic efficiency of this enzyme rather than by quantitative changes; to vary the amount of the enzyme would require more time, since the half-life for its degradation is 14–27 hours (see above). In this regard, it is to be noted that citrate-cleavage enzyme is inhibited by ADP, which appears to compete with ATP (INOUE et al., 1966; ATKINSON and WALTON, 1967). ATKINSON and WALTON (1967) propose that the activity of this enzyme is a sensitive function of "energy charge", *i.e.* $ATP + \frac{1}{2}ADP)/(AMP + ADP + ATP)$. However, not all the available data on changes in this parameter upon refeeding seem to support this idea (TARNOWSKI and SEEMAN, 1967; GUYNN et al., 1972).

B. Control by Changes in the Quantity of Acetyl-CoA Carboxylase

1. Tissue Content of Acetyl-CoA Carboxylase in Various Metabolic Conditions

The level of acetyl-CoA carboxylase activity in animal tissues fluctuates as fatty-acid synthesis increases or decreases according to the alimentary, hormonal, developmental and genetic conditions. The carboxylase level in rat liver (NUMA et al., 1961; KORCHAK and MASORO, 1962; ALLMANN et al., 1965) and adipose tissue (SAGGERSON and GREENBAUM, 1970) is lowered in starvation and restored on refeeding fasted animals. The increase in the enzyme level after refeeding is more marked on a fat-free diet than on a balanced diet (ALLMANN et al., 1965). The addition to fat-free diet of methyl esters of polyunsaturated fatty acids, such as linoleic, linolenic, and arachidonic acid, tends to suppress the elevation of the carboxylase level observed when the fat-free diet alone is fed (ALLMANN et al., 1965; MUTO and GIBSON, 1970). Furthermore, enteral administration of fat to animals maintained on a fat-free diet lowers the hepatic enzyme level (BORTZ et al., 1963; PEARCE, 1968). Alloxan-diabetes is accompanied by a decrease in the carboxylase level in rat liver (WIELAND et al., 1963) and adipose tissue (SAGGERSON and GREENBAUM, 1970). On the other hand, genetically obese hyperglycemic mice (C57BL/6J-*ob*; see Section II, A, 2) exhibit an elevated hepatic enzyme level (CHANG et al., 1967; NAKANISHI and NUMA, 1971). In rats in the thyrotoxic state induced by triiodothyronine treatment, an increased carboxylase level in the liver and adipose tissue is observed (DIAMANT et al., 1972). The level of hepatic acetyl-CoA carboxylase varies remarkably during development of rats (LOCKWOOD et al., 1970) and mice (SMITH and ABRAHAM, 1970a). It is very low in fetal liver and after birth throughout the suckling period. There is a rapid increase in the enzyme level on weaning (21 days), and a maximum is reached between 40 and 50 days of age, after which the enzyme level declines to adult values. The developmental pattern of the carboxylase level parallels that of hepatic fatty-acid synthesis except in late fetal liver, in which lipogenesis is as high as in adult liver. The elevation in lipogenesis observed in chick liver after hatching is accompanied by a marked increase in the carboxylase level (ARINZE and MISTRY, 1970; RYDER, 1970). Glucose injection into embryonating eggs elevates the enzyme level in the liver (DONALDSON et al., 1971). In rat mammary glands, the carboxylase level rises rapidly during the first 4 days of lactation, remains at high levels and drops precipitously at the time of weaning (HOWANITZ and LEVY, 1965). All these findings demonstrate parallel changes in the level of acetyl-CoA carboxylase activity and fatty-acid synthesis under a variety of metabolic conditions.

The above studies relied exclusively on measurements of catalytic activity, which do not differentiate between changes in the number of enzyme molecules, *i.e.* enzyme quantity, and changes in catalytic efficiency per enzyme molecule. To determine whether the changes in the level of hepatic acetyl-CoA carboxylase activity observed in different nutritional, hormonal, and genetic states are actually due to changing quantities of enzyme protein, an immunochemical approach was adopted (NUMA et al., 1969, 1970a; MAJERUS and KILBURN, 1969; NAKANISHI and NUMA, 1970, 1971). For this purpose, antibody was prepared by injecting rabbits with homogeneous preparations of acetyl-CoA carboxylase from rat liver (NUMA et al., 1969; NAKANISHI and NUMA, 1970) or chicken liver (MAJERUS and KILBURN, 1969) (see Section II, C, 4). The antibody against rat-liver carboxylase cross-reacts with the mouse-liver enzyme, while that against chicken-liver carboxylase precipitates the rat-liver enzyme. Both the protomeric and polymeric forms of the carboxylase (see Section II, C, 4) are precipitated by antiacetyl-CoA carboxylase. The experimental animals used were rats subjected to different dietary conditions, including feeding of a balanced or high-fat diet, fasting, and fat-free refeeding (NUMA et al.,1969, 1970a; MAJERUS and KILBURN, 1969; NAKANISHI and NUMA, 1970), alloxan-diabetic rats (NAKANISHI and NUMA, 1970;

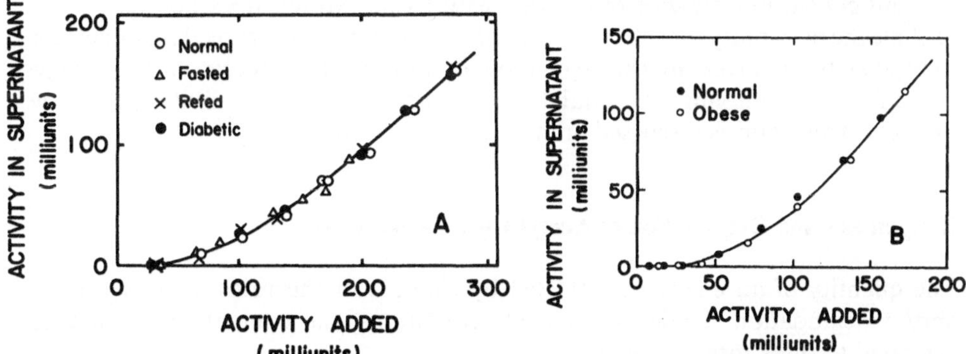

Fig. 2A and B. Immunochemical titration of hepatic acetyl-CoA carboxylase derived from: A, rats in different metabolic states; B, obese and nonobese mice. Increasing amounts of liver extracts containing the carboxylase activities indicated were added to a fixed amount of antiacetyl-CoA carboxylase. Following completion of precipitation, the supernatant fluids were assayed for carboxylase activity. Data taken from NUMA et al. (1969) and NAKANISHI and NUMA (1970, 1971)

NUMA et al., 1970a), and genetically obese hyperglycemic mice (NAKANISHI and NUMA, 1971). Fig. 2A shows the results of immunochemical titrations of liver extracts obtained from normally fed, fasted, fat-free refed, and alloxan-diabetic rats. Despite 9-fold fluctuations in the level of carboxylase activity derived from 1 g of liver, the equivalence point, *i.e.* the point at which enzyme activity first appeared in the supernatant fluid, was the same for all four types of liver extracts when based on the amount of enzyme activity added. Similar results were obtained independently by MAJERUS and KILBURN (1969), using rats under different dietary conditions. Fig. 2B represents analogous experiments with liver extracts derived from obese and nonobese mice, indicating the same equivalence point for both extracts. These results demonstrate that in all these conditions the level of carboxylase activity in liver extracts is proportional to the quantity of the enzyme protein as measured immunochemically with the use of specific antibody. In some instances, the proportionality between catalytic activity and immunochemically estimated enzyme protein was confirmed by another titration procedure, in which the amount of liver extracts was kept constant and increasing amounts of antibody were added. OUCHTERLONY double-diffusion analyses with enzyme preparations derived from animals in various metabolic states demonstrated the completeness of the precipitin bands, indicating that carboxylase molecules are immunologically similar in all the conditions. Furthermore, enzyme preparations derived from obese and nonobese mice exhibit no qualitative differences with regard to kinetic properties and heat stability (NAKANISHI and NUMA, 1971). It appears, therefore, that the mutation causing obesity does not affect the structure of the carboxylase, but alters the enzyme concentration in the liver. All these results indicate that catalytic efficiency per carboxylase molecule is the same in the various metabolic states associated with lowered or elevated lipogenesis; it is therefore concluded that the variations in the level of carboxylase activity in liver extracts are actually due to changes in the quantity of the enzyme.

This conclusion drawn from studies with tissue extracts does not necessarily exclude the possibility that changes in catalytic efficiency of carboxylase molecules may also be involved in the regulation of fatty-acid synthesis in cells, since the effectors modifying the catalytic efficiency are strongly diluted in tissue extracts. This point is discussed in more detail in Section II, C, 1.

2. Synthesis and Degradation of Acetyl-CoA Carboxylase

The quantity of an enzyme is affected by changes in the rates of its synthesis and/or degradation. Under steady-state conditions, the content of an enzyme is related to these rates as follows:

$$E = k_s / k_d \tag{3}$$

where E is the content of enzyme per mass, and k_s is a zero-order rate constant of synthesis per mass, and k_d is a first-order rate constant of degradation expressed as a reciprocal of time (see SCHIMKE and DOYLE, 1970).

In order to see whether the above-mentioned fluctuations in the quantity of hepatic acetyl-CoA carboxylase—which occur in various dietary, hormonal and genetic conditions (see Section II, B, 1)—are due to changes in the rate of enzyme synthesis or in that of enzyme degradation, combined immunochemical and isotopic studies were carried out with normally fed, fasted, and fat-free refed rats (MAJERUS and KILBURN, 1969; NAKANISHI and NUMA, 1970; NUMA et al., 1970a), alloxan-diabetic rats (NAKANISHI and NUMA, 1970; NUMA et al., 1970a) and genetically obese hyperglycemic mice (NAKANISHI and NUMA, 1971). The relative rate of synthesis of hepatic acetyl-CoA carboxylase was measured by injecting animals with a dose of [^{3}H]leucine and shortly thereafter determining the extent of isotope incorporation into the protein that is precipitated by antibody specific to the enzyme. The rate of carboxylase degradation was measured by following the loss of isotope from the prelabeled enzyme. The decay of radioactivity in the enzyme was shown to follow first-order kinetics. Table 1 summarizes the results of these measurements carried out in our laboratory. It is evident from this table that the rate constant of carboxylase degradation is essentially the same in normally fed, refed, and alloxan-diabetic rats ($t_{\frac{1}{2}} = 55$–59 hours), and that the increase or decrease in the carboxylase content in refed or alloxan-

Table 1. Synthesis and degradation of acetyl-CoA carboxylase in livers of animals in different metabolic conditions

Animals	Conditions	Enzyme content per liver (E)	Rate of synthesis per liver (k_s)	Rate constant of degradation (k_d)	k_s/k_d
Rats	Fed a balanced diet	100%	100%	100%	100%
	Fasted for 48 hours	28	59	190	31
	Fasted for 48 hours and subsequently refed a fat-free diet for 72 hours	376	405	107	378
	Alloxan-diabetic, fed a balanced diet	53	54	100	54
Mice	Nonobese, fed a balanced diet	100	100	100	100
	Obese, fed a balanced diet	1020	775	58	1340

Data taken from Nakanishi and Numa (1970, 1971).

diabetic rats can be attributed to accelerated or retarded synthesis of the enzyme. The elevated carboxylase content in genetically obese mice is due mainly to a faster rate of synthesis, and in a minor degree, to a lower degradation rate constant ($t_{\frac{1}{2}}=67$ and 115 hours in normal and obese mice, respectively). Normal and obese animals maintained on a balanced diet, rats refed a fat-free diet for more than 3 days, and rats suffering from prolonged diabetes can be assumed to be in a steady-state. The values for the enzyme content predicted from the ratio k_s/k_d according to Eq. (3) agree fairly well with the values actually found. Since the specific activity of homogeneous rat liver acetyl-CoA carboxylase is 15 units per milligram of protein at 37°C (NAKANISHI and NUMA, 1970), the rate of synthesis of hepatic acetyl-CoA carboxylase can be calculated by Eq. (3) to be 3.3 μg per hour per normally fed rat (mean body weight, 180 g), 13.2 μg per hour per refed rat (180 g), 1.7 μg per hour per diabetic rat (250 g), 0.25 μg per hour per nonobese mouse (24 g) and 1.47 μg per hour per obese mouse (40 g). On the other hand, fasted rats are not in a steady state, since both liver weight and enzyme content per liver decrease gradually during starvation. Although one cannot make a quantitative evaluation according to Eq. (3) in this instance, it can be concluded that the decrease in the carboxylase content in fasted rats is due both to diminished enzyme synthesis and to accelerated enzyme degradation ($t_{\frac{1}{2}}=31$ hours). The results obtained with normally fed, fasted, and fat-free refed rats are in general agreement with those of MAJERUS and KILBURN (1969), who made similar studies using antibody prepared against homogeneous chicken-liver acetyl-CoA carboxylase, although these investigators found slightly shorter $t_{\frac{1}{2}}$ values for carboxylase degradation (50, 48, and 18 hours for normally fed, refed, and fasted rats, respectively).

In experiments to determine the rate of enzyme degradation, reutilization of [^{3}H]leucine may lead to an overestimation of the half-life (KOCH, 1962). However, this factor does not appear to be of major significance in our experiments because the half-life for total soluble liver protein, estimated simultaneously by means of this isotope, was 3.8, 3.7, 3.4, and 2.9 days in normally fed, refed, diabetic, and fasted rats, respectively, while the half-life found by the use of guanido-labeled arginine, which is not reutilized, is 5.1 days in normally fed rats (ARIAS et al., 1969).

The fact that synthesis of hepatic acetyl-CoA carboxylase is accelerated in fat-free refed rats is supported also by the finding that actinomycin D or puromycin prevents the rise in the enzyme level observed upon refeeding (HICKS et al., 1965). Further evidence for adaptive enzyme synthesis is that actinomycin D blocks recovery of fatty-acid synthesis following treatment of alloxan-diabetic rats with insulin (GELLHORN and BENJAMIN, 1964, 1966). The increase in the hepatic carboxylase level observed in mice upon weaning is also blocked by actinomycin D or puromycin (SMITH and ABRAHAM, 1970a).

It is of interest that synthesis and degradation of acetyl-CoA carboxylase are controlled independently, and that under steady-state conditions the carboxylase content is regulated mainly by changes in the rate of synthesis, while under nonsteady-state conditions both the rate of synthesis and the rate of degradation are altered to adjust the carboxylase content. These findings suggest that the control of enzyme content by changes in the rate of degradation may play an important role only when the animal deviates from a steady state in adjusting to a new environment. This concept is supported also by the following findings of SCHIMKE (1964). Rats maintained on diets containing 8, 30, and 70 per cent casein show different steady-state arginase contents but essentially the same rate constants of arginase degradation. Upon switching rats from a diet containing 70 per cent protein to one containing 8 per cent protein, the rate constant of arginase degradation increases during the first 3 days. However, it then gradually approaches the steady-state value as the enzyme content attains a new steady-state level below the initial one. When rats are fasted, arginase degradation ceases and the enzyme content rises concomitantly. The mechanism responsible for the independent control of synthesis and degradation of acetyl-CoA carboxylase remains to be elucidated (see Section IV).

3. Loss of Control in Hepatomas

One of the striking features of minimum-deviation hepatomas is their failure to regulate fatty-acid synthesis upon dietary alteration (SABINE et al., 1968; ELWOOD and MORRIS, 1968). MAJERUS et al. (1968) showed that the levels of acetyl-CoA carboxylase and fatty-acid synthetase in hepatomas, in contrast to those in host livers, do not respond to fasting and fat-free refeeding of the tumor-bearing animal. The tumor carboxylase is essentially identical with the liver enzyme in terms of kinetic and other properties. Preliminary experiments indicate that the level of carboxylase activity in tumors parallels the content of immunochemically reactive protein, and that there is no change in the rate of enzyme synthesis in tumors after fat-free refeeding. These results suggest that the defect in the control of fatty-acid synthesis is due to the failure of tumors to regulate the amount of the carboxylase rather than to alteration of the enzyme structure. Using hepatic tissue transplanted to a subcutaneous environment and receiving no portal blood, BARTLEY and ABRAHAM (1972) showed that in autografts the levels of lipogenic enzymes, including acetyl-CoA carboxylase and fatty-acid synthetase, vary in response to fasting and refeeding in essentially the same manner as in normal liver. Since hepatic autotransplants with a blood supply comparable to that of hepatomas are able to respond to dietary manipulation, the site and blood supply of the tumor cannot explain the absence of dietary control in the neoplastic tissue.

C. Control by Changes in the Catalytic Efficiency of Acetyl-CoA Carboxylase

1. Short-Term *vs.* Long-Term Control

Evidence is accumulating to indicate that the rate of fatty acid synthesis is regulated not only by changes in the quantity of acetyl-CoA carboxylase but also by changes in the catalytic efficiency of the enzyme. KORCHAK and MASORO (1962) pointed out that at an early stage of fasting the capacity of liver slices to synthesize fatty acids is more depressed than can be accounted for by the level of acetyl-CoA carboxylase in liver extracts; after 24 hours' fasting, the carboxylase level falls by only 50 per cent, whereas a 99 per cent depression in fatty-acid synthesis is observed in liver slices. An analogous discrepancy was observed with alloxan-diabetic rats (WIELAND and NEUFELDT, 1963), as well as with rats fed a single dose of fat (BORTZ et al., 1963) at an acute or early stage. Even at later stages, the carboxylase level in liver extracts is generally less reduced than fatty acid synthesis in liver slices, although this discrepancy is not as marked as at early stages. These observations suggest that other factors, in addition to the carboxylase content, are involved in the regulation of fatty-acid synthesis in intact cells. In fact, the catalytic activity of acetyl-CoA carboxylase of animal origin is affected by various metabolites as positive and negative effectors. Tri- and dicarboxylic acids, most notably citrate and isocitrate, activate acetyl-CoA carboxylase from rat liver (MATSUHASHI et al., 1962; LYNEN et al., 1963; MATSUHASHI et al., 1964), chicken liver (WAITE, 1962; WAITE and WAKIL, 1962), rat adipose tissue (MARTIN and VAGELOS, 1962a, b, c), bovine adipose tissue (MOSS et al., 1969, 1972), and lactating rat mammary glands (KALLEN and LOWENSTEIN, 1962; MILLER and LEVY, 1969). In contrast, long-chain acyl-CoA thioesters (BORTZ and LYNEN, 1963a; NUMA et al., 1965a, b), malonyl-CoA (MATSUHASHI et al., 1964; GREGOLIN et al., 1966b; HASHIMOTO and NUMA, 1971; HASHIMOTO et al., 1971) and some metabolites of tryptophan including kynurenate and xanthurenate (HASHIMOTO et al., 1971) are inhibitors of the carboxylase from rat and chicken liver. In view of the fact that citrate is a precursor of acetyl-CoA, which is a substrate for this carboxylase, it is conceivable that citrate has a regulatory role as a positive feedforward activator. Inhibition by long-chain acyl-CoA can be regarded as a negative feedback mechanism due to end-product inhibition. Changes in the tissue contents of citrate and long-chain acyl-CoA thioesters in various metabolic states associated with elevated or lowered lipogenesis are generally consistent with the proposed regulatory roles of these allosteric effectors (see Section II, C, 5). Moreover, the tissue content of acetyl-CoA carboxylase cannot change rapidly, since the half-life for degradation of this enzyme ranges from 1 to 5 days, as described in Section II, B, 2; theoretically, the time required for the content of an enzyme to change to one half of the final change at a new steady state is equal to the half-life of the enzyme (see BERLIN and SCHIMKE, 1965). All these observations support the view that changes in the catalytic efficiency of acetyl-CoA carboxylase by activators or inhibitors are also involved in the regulation of fatty-acid synthesis. This mechanism must make a greater contribution to short-term control when the rate of lipogenesis changes promptly. In long-term control, on the other hand, the regulation of the enzyme content also plays an important role (see Section II, B, 1 and 2).

2. Reaction Mechanism

Knowledge of the enzymic reaction mechanism is necessary to understand the mode of action of allosteric effectors that modify the catalytic efficiency of enzyme molecules. Acetyl-CoA carboxylase is an enzyme containing biotin as the prosthetic group (WAKIL et al., 1958). A number of comprehensive reviews on biotin enzymes and their reaction mechanism are available (KAZIRO and OCHOA, 1964; LYNEN, 1967b; KNAPPE, 1970; MOSS and LANE, 1971), so only more recent findings concerning acetyl-CoA carboxylase are described in detail.

The acetyl-CoA carboxylase reaction proceeds in the following two steps:

$$\text{E-biotin} + \text{ATP} + \text{HCO}_3^- \overset{\text{Mg}^{2+}}{\rightleftharpoons} \text{E-biotin} \sim \text{CO}_2 + \text{ADP} + \text{P}_i \quad (4)$$

$$\text{E-biotin} \sim \text{CO}_2 + \text{acetyl-CoA} \rightleftharpoons \text{E-biotin} + \text{malonyl-CoA} \quad (5)$$

Overall:

$$\text{ATP} + \text{HCO}_3^- + \text{acetyl-CoA} \rightleftharpoons \text{ADP} + \text{P}_i + \text{malonyl-CoA} \quad (1)$$

where E-biotin denotes acetyl-CoA carboxylase.

There are three lines of evidence for this reaction mechanism. The first evidence was provided by isotope exchange studies (LYNEN et al., 1963; MATSUHASHI et al., 1964; GREGOLIN et al., 1968a; HASHIMOTO et al., 1971). ATP-$^{32}P_i$ exchange is demonstrable in the absence of acetyl-CoA and malonyl-CoA, indicating Reaction (4). Evidence for Reaction (5) is the occurrence of malonyl-CoA-[^{14}C]acetyl-CoA exchange without ATP, ADP,P_i, HCO_3^- and Mg^{2+}. Secondly, the carboxylated enzyme intermediate (E-biotin$\sim CO_2$) was isolated and shown to be active in transferring its carboxyl group to the carboxyl acceptor, acetyl-CoA (NUMA et al., 1964, 1965a). The active carboxyl is bound to the 1′-*N* atom of biotin, which is amide-linked to the ε-amino group of a lysyl residue in the enzyme protein. This structure is given in Fig. 3. Finally, detailed kinetic analysis demonstrated that the two-step mechanism in fact represents the principal pathway of the reaction (HASHIMOTO et al., 1970, 1971; NUMA et al., 1970a, 1972a, b; HASHIMOTO and NUMA, 1971). The results of initial velocity studies as well as product and dead-end inhibition studies lead to the conclusion that the acetyl-

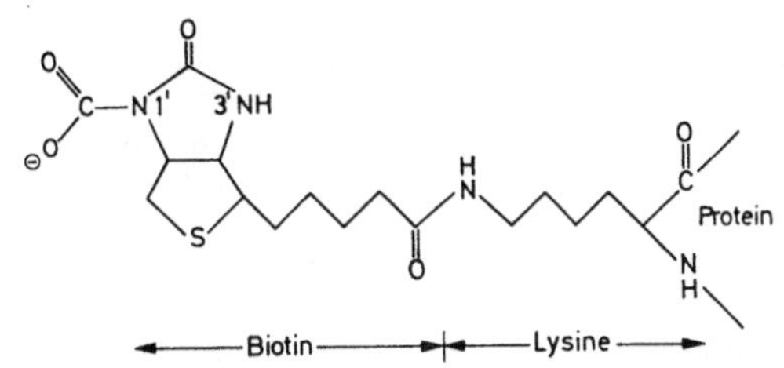

Fig. 3. Structure of the carboxylated active site of acetyl-CoA carboxylase. From NUMA et al. (1964)

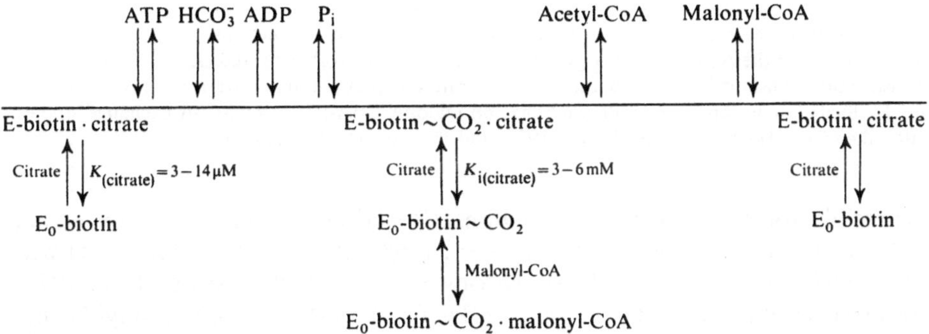

Fig. 4. Mechanism of the acetyl-CoA carboxylase reaction and mode of citrate activation. $K_{(citrate)}$ and $K_{i(citrate)}$ denote the dissociation constants of the uncarboxylated and carboxylated forms of the enzyme for citrate, respectively

CoA carboxylase reaction proceeds through an ordered bi-bi-uni-uni ping-pong mechanism; the order of addition of substrates to the enzyme is ATP, HCO_3^- and acetyl-CoA in the forward reaction, and malonyl-CoA, P_i and ADP in the reverse reaction, as shown in Fig. 4. Moreover, studies of malonyl-CoA-[^{14}C]acetyl-CoA exchange, in conjunction with the inhibition pattern produced by malonyl-CoA, revealed that malonyl-CoA forms a dead-end complex with the inactive (or less active) species of the carboxylated form of the enzyme (see Fig. 4). The kinetic constants for rat liver acetyl-CoA carboxylase are listed in Table 2.

On the basis of the isotope exchange and kinetic studies, and because convincing evidence to the contrary is lacking, the ATP-dependent carboxylation of biotin by HCO_3^- is generally thought to occur as a concerted mechanism (LYNEN et al., 1961; KAZIRO et al., 1962; see also the above-mentioned reviews). Recently, a stepwise chemical mechanism which involves carbonyl phosphate as a tightly bound enzyme intermediate has been suggested by POLAKIS et al. (1972). This mechanism is also consistent with earlier results and is supported by the finding that biotin carboxylase, a component of *Escherichia coli* acetyl-CoA carboxylase (see Section II, C, 4), catalyzes phosphoryl transfer from carbamyl phosphate, a closely related analog of carbonyl phosphate, to ADP to form ATP.

Increasing evidence indicates that biotin enzymes possess dual sites for the bicyclic ring of biotin, which are located adjacent to the two substrate sites

Table 2. Kinetic constants for acetyl-CoA carboxylase

Compound	Michaelis constant[a] (M)	Dissociation constant (M)
ATP	1.5×10^{-5}	
HCO_3^-	2.5×10^{-3}	
Acetyl-CoA	2.5×10^{-5}	5×10^{-6} [b]
Malonyl-CoA	1.6×10^{-5}	1×10^{-5} [b]
P_i	7.0×10^{-3}	
ADP	1.0×10^{-5}	
Citrate		$3\text{–}6 \times 10^{-3}$ [c]

[a] Estimated from the results of initial velocity studies on the overall reactions.
[b] Estimated from the results of malonyl-CoA-[^{14}C]acetyl-CoA exchange studies.
[c] Dissociation constant of the E-biotin ~ CO_2 · citrate complex estimated from the results of both initial velocity studies on the overall reactions and malonyl-CoA-[^{14}C]acetyl-CoA exchange studies. Data taken from HASHIMOTO and NUMA (1971) and HASHIMOTO et al. (1971).

responsible for the two partial reactions (Reactions 4 and 5) (LYNEN et al., 1961; GREEN, 1963; ALBERTS and VAGELOS, 1968; ALBERTS et al., 1969; GERWIN et al., 1969; DIMROTH et al., 1970; JACOBSON et al., 1970; AHMAD et al., 1970; GUCHHAIT et al., 1971; NERVI et al., 1971). The biocytin (ε-*N*-D-biotinyl-lysine) side chain, which anchors the functional, bicyclic ring of biotin to the apoprotein (see Fig. 3), can reach to 14 Å when fully extended (MILDVAN et al., 1966). Thus, the reaction mechanism of biotin enzymes appears to involve a shuttling of the biotin ring between the two separate sites. Consideration of this structure of biotin enzymes, together with the observed kinetics of the oxaloacetate transcarboxylase reaction, led NORTHROP (1969; NORTHROP and WOOD, 1969) to propose a novel mechanism for this enzymic reaction. This is designated hybrid ping-pong mechanism, to indicate that it involves intermediate formation of a substituted form of the enzyme and allows independent binding of substrates to two distinct sites on the enzyme. The pyruvate carboxylase reaction likewise proceeds through a similar two-site ping-pong mechanism (MCCLURE et al., 1971 a, b, c; BARDEN et al., 1972). There are several findings indicating the hybrid character of the mechanism of the acetyl-CoA carboxylase reaction (HASHIMOTO et al., 1971; NUMA et al., 1972b). These include a stimulatory effect of acetyl-CoA, a substrate for Reaction (5), on the rate of ATP-$^{32}P_i$ exchange (Reaction 4).

3. Kinetic Effects of Allosteric Regulators

Acetyl-CoA carboxylase from animal tissues exhibits nearly absolute requirement for tri- and dicarboxylate activators (for references, see Section II, C, 1). Studies on the activator specificity of rat-liver acetyl-CoA carboxylase (LYNEN et al., 1963; MATSUHASHI et al., 1964) showed that citrate, isocitrate and malonate are the most effective activators. Both enantiomers of isocitrate are equally effective, since *threo*-D_s-isocitrate (see PATTERSON et al., 1962) is as active as DL-isocitrate. Tricarballylate, a closely related tricarboxylate analog of citrate and isocitrate, has no activating effect. Monocarboxylic acids and amino acids are likewise inactive. In studies on the rat adipose-tissue enzyme, fluorocitrate, a potent inhibitor of the tricarboxylic acid cycle, was found to replace citrate effectively (VAGELOS et al., 1963). A similar activator specificity was shown also with the chicken-liver enzyme (GREGOLIN et al., 1968a). The stimulatory effects of malonate, L-isocitrate and fluorocitrate, together with other evidence, revealed that activation by citrate or isocitrate is not a consequence of their metabolic conversion, but is due to their direct action on the enzyme (MARTIN and VAGELOS, 1962b, c; LYNEN et al., 1963; MATSUHASHI et al., 1964).

Acetyl-CoA carboxylases from diverse animal sources differ in the conditions for their activation. The enzymes from rat liver (LYNEN et al., 1963; MATSUHASHI et al., 1964; NUMA and RINGELMANN, 1965) and rat adipose tissue (VAGELOS et al., 1963) require preliminary incubation with citrate at higher temperature (25–37° C) in order to exhibit full activity when subsequently assayed in the presence of citrate, and the degree of activation varies largely with the conditions of preincubation, such as enzyme concentration (VAGELOS et al., 1963; NUMA and RINGELMANN, 1965), temperature (NUMA and RINGELMANN, 1965) and the presence of Mg^{2+} (GREENSPAN and LOWENSTEIN, 1967, 1968). The rat-liver enzyme, previously activated by citrate at 25° C, is inactivated upon exposure to low temperature (0–7° C) (NUMA and RINGELMANN, 1965); this process is largely reversible. In contrast, the enzymes from chicken liver (NUMA et al., 1966, 1967; GREGOLIN et al., 1966a) and bovine adipose tissue (MOSS et al., 1969, 1972) are instantaneously activated without prior incubation and are not cold-labile. However, the assay temperature exerts a pronounced influence on the magnitude of isocitrate activation of the chicken-liver enzyme as well (GREGOLIN et al., 1968a); isocitrate causes very little activation at low temperature, and the transition temperature above which marked stimulation occurs is 20–22° C. These findings may imply the involvement of hydrophobic bonds in maintaining the active conformation of acetyl-CoA carboxylase, since theoretical studies with model systems suggest that, of the types of bonds postulated to maintain protein structure, only hydrophobic bonds have properties consistent with a decrease in stability at low temperature (KAUZMANN, 1959; NÉMETHY and SCHERAGA, 1966).

Both partial reactions (Reactions 4 and 5) involved in the overall carboxylation are dependent on a tricarboxylate activator. This was proven by studies of ATP-$^{32}P_i$ exchange (Reaction 4) and malonyl-CoA-[^{14}C]acetyl-CoA exchange (Reaction 5) (LYNEN et al., 1963; MATSUHASHI et al., 1964; GREGOLIN et al., 1968a), as well as by experiments with models of the partial reactions, *i.e.*, ATP-dependent carboxylation of free biotin (model of Reaction 4) and carboxyl transfer from carboxylated enzyme to acetyl pantetheine (model of Reaction 5) (STOLL et al., 1968).

On the basis of the reaction mechanism described in Section II, C, 2 (see Fig. 4), kinetic studies were carried out to answer the question, which of the enzyme forms involved is dependent on the allosteric activator, citrate (HASHIMOTO et al., 1970, 1971; NUMA et al., 1970a, 1972a, b; HASHIMOTO and NUMA, 1971). The results of these experiments are summarized in Table 3. In these studies, rat-liver enzyme preincubated with citrate for a sufficiently long time was used, so that only the citrate effect during the carboxylase catalysis was analyzed.

Table 3. Patterns of citrate activation of acetyl-CoA carboxylase

Reaction	Substrate varied	Citrate effect on double reciprocal plots
Forward	ATP	Intercept
	HCO_3^-	Intercept
	Acetyl-CoA	Slope
Reverse	Malonyl-CoA	Intercept
	P_i	Slope
	ADP	Slope and intercept
Malonyl-CoA-[^{14}C]acetyl-CoA exchange	Acetyl-CoA	Slope

Data taken from HASHIMOTO and NUMA (1971) and HASHIMOTO et al. (1971).

The results indicate that, of the obligatory enzyme forms, only the carboxylated form of the enzyme (E-biotin $\sim CO_2$) is dependent on the presence of citrate. The dissociation constant of the E-biotin $\sim CO_2 \cdot$ citrate complex is calculated from these data to be 3–6 mM (see Table 2).

On the other hand, it is well established that the effect of citrate is exerted also on the uncarboxylated form of the enzyme (E-biotin) to induce polymerization of protomeric carboxylase molecules (see Section II, C, 4). The apparent citrate-independence of the E-biotin form during the carboxylase catalysis—as found by the kinetic studies—suggests that this enzyme form is converted completely to the active polymeric form under the conditions employed for the kinetic experiments (citrate concentration varied from 1 to 5 mM). In fact, the dissociation constant of the E-biotin·citrate complex was estimated to be 3–14 μM in terms of [^{14}C]citrate binding (one tight citrate-binding site per protomer, see GREGOLIN et al., 1968b; HASHIMOTO et al., 1971), protection against avidin inactivation, protection against trypsin activation, and shift in the protomer-polymer equilibrium (HASHIMOTO et al., 1971; NUMA et al., 1972a, b). This value is about 10^3-fold smaller than the dissociation constant of the E-biotin $\sim CO_2 \cdot$ citrate complex (3–6 mM), so that the effect of citrate on the E-biotin form is not apparent upon kinetic analysis of the carboxylase-catalyzed reactions.

The above-mentioned findings indicate that carboxylation of the biotinyl prosthetic group gives rise to a conformational change at the citrate site, resulting in a marked decrease in the affinity of the enzyme for the allosteric activator. In addition, evidence was obtained which indicates that a conformational change is induced at the active site by the remote binding of a tricarboxylate activator. Studies on the inactivation of the chicken (RYDER et al., 1967; MOSS and LANE, 1972) and rat-liver enzymes (HASHIMOTO et al., 1971) by avidin, the specific biotin-binding protein from egg white, showed that in the citrate-activated enzyme the biotinyl prosthetic group is completely inaccessible to avidin, whereas in the absence of the activator the prosthetic group is readily accessible to this protein inhibitor. Apparently, the biotinyl prosthetic group becomes shielded by neighboring groups as a result of the conformational change at the active site induced by citrate. Furthermore, the reactivity of the carboxylated prosthetic group (1′-*N*-carboxybiotinyl group) (see Fig. 3) toward decarboxylation is enhanced by tricarboxylate activators, particularly in the presence of acetyl-CoA

(RYDER et al., 1967; OHTSU et al., 1968; LANE et al., 1970). All these findings suggest that there is a close conformational interaction between the catalytic and regulatory sites of the enzyme.

Long-chain acyl-CoA thioesters inhibit acetyl-CoA carboxylase from rat liver (BORTZ and LYNEN, 1963a; NUMA et al., 1965a, b) and rat mammary glands (MILLER et al., 1970) at extremely low concentrations. The inhibition is competitive with regard to the activator, citrate, but noncompetitive with regard to the substrates, acetyl-CoA, bicarbonate, and ATP (NUMA et al., 1965a, b). Kinetic analysis of the effects of these inhibitors, however, is complicated by several factors (NUMA et al., 1965b; HASHIMOTO et al., 1971). The maximal extent of inhibition is achieved only when the enzyme is incubated with the inhibitor prior to the assay. In addition, the degree of inhibition is influenced by the amount of protein present in the assay mixture, including the enzyme preparation itself and the serum albumin added to stabilize the enzyme. Nevertheless, it is clear that the inhibitory effect of acyl-CoA thioesters increases with increasing chain length, since the affinity of nonspecific protein for the thioesters can also be assumed to increase with increasing chain length, as shown by studies on the binding of free fatty acids (TERESI and LUCK, 1952; GOODMAN, 1958; SPECTOR et al., 1969) and alkyl sulfates (KARUSH and SONENBERG, 1949) to serum albumin. CoA thioesters of C_{16}-C_{18} fatty acids, which occur abundantly in animal tissues, are the most potent inhibitors, showing apparent inhibition constants of 0.3–2.5 μM, whereas shorter-chain acyl-CoA, such as hexanoyl-CoA, shows essentially no inhibitory effect (NUMA et al., 1965b). The inhibition can be partially reversed by the addition of serum albumin during the assay (NUMA et al., 1965b). Free fatty acids also inhibit acetyl-CoA carboxylase from rat mammary glands (LEVY, 1963; MILLER et al., 1970) and rat liver (YUGARI et al., 1964; KORCHAK and MASORO, 1964; Numa et al., 1965b), but much higher concentrations are required for the inhibition, as compared with their CoA derivatives.

It is known that long-chain acyl-CoA derivatives have strong detergent properties. The critical micellar concentrations of C_{16}–C_{18} acyl-CoA thioesters range from 2 to 6 μM (ZAHLER et al., 1968; BARDEN and CLELAND, 1969). Moreover, palmityl-CoA inhibits a number of enzymes of diverse metabolic function, some of which have no known relationship to this thioester (TAKETA and POGELL, 1966). For these reasons, the physiological role of fatty acyl-CoA derivatives as metabolic regulators was questioned (SRERE, 1965b; TAKETA and POGELL, 1966; DORSEY and PORTER, 1968). However, the competitive nature of the inhibition with respect to the activator, citrate, together with the reversibility of the inhibition and the extremely low inhibition constants, favors the regulatory role of these inhibitors. A difficulty in the interpretation of the physiological significance of long-chain acyl-CoA inhibition *in vivo* is that the effective concentration of these inhibitors at the site of enzyme reaction in the cell is unknown, owing to cell compartmentation and to interaction with intracellular proteins and lipid structures such as membranes. It is nevertheless an attractive hypothesis that this endproduct inhibition represents a negative feedback mechanism for fatty-acid synthesis.

(+)-Palmityl carnitine and other long-chain acyl carnitine derivatives stimulate fatty-acid synthesis at the step of the acetyl-CoA carboxylase reaction (FRITZ and HSU, 1966, 1967; GREENSPAN and LOWENSTEIN, 1968; MOSS et al., 1972). This activating effect is ascribed tentatively to the displacement by acyl carnitine of hydrophobic inhibitors, such as long-chain acyl-CoA thioesters and fatty acids, since the activation is observed with liver extracts and impure acetyl-CoA carboxylase preparations, but not with homogeneous enzyme preparations (LANE and MOSS, 1971; MOSS et al., 1972). Phospholipids were reported to activate crude acetyl-CoA carboxylase from rat liver (FOSTER and MCWHORTER, 1969) as well as purified rat mammary acetyl-CoA carboxylase (MILLER and LEVY, 1969).

4. Molecular Basis for Control of the Catalytic Efficiency

The citrate-induced activation of rat adipose tissue acetyl-CoA carboxylase was found to be accompanied by an increase in the sedimentation coefficient of the enzyme (VAGELOS et al., 1962, 1963). A similar finding was obtained also with the rat liver enzyme (LYNEN et al., 1963; MATSUHASHI et al., 1964; NUMA et al., 1965a, b; NUMA and RINGELMANN, 1965). The citrate-induced increase in the sedimentation coefficient of the rat liver enzyme is abolished either by an inhibitor, palmityl-CoA, or by exposure to cold, which annuls the activation by citrate (NUMA et al., 1965a, b; NUMA and RINGELMANN, 1965). Fig. 5 illustrates the sedimentation patterns of rat liver acetyl-CoA carboxylase in sucrose density

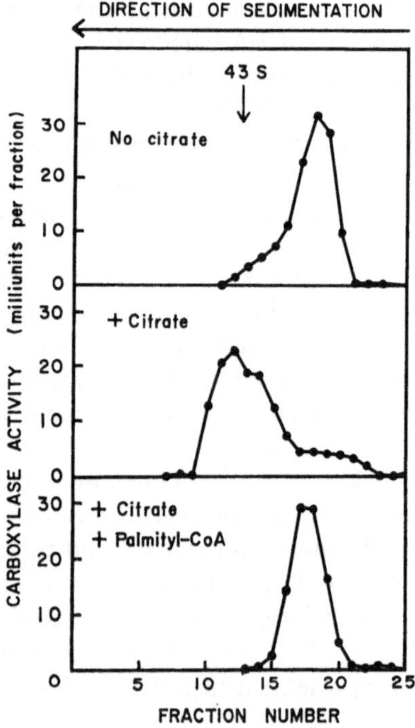

Fig. 5. Sedimentation patterns of acetyl-CoA carboxylase in sucrose density gradients in the presence and absence of its allosteric regulators. Data taken from NUMA et al. (1965b)

Table 4. Molecular properties of acetyl-CoA carboxylase of animal origin

Source	Molecular form	Sedimentation coefficient	Molecular weight	Reference
		S		
Chicken liver	Polymeric	40–60	$4–11 \times 10^6$	GREGOLIN et al. (1966a, b)
	Intermediate	27–33	–	NUMA et al. (1966)
	Protomeric	13–14	410000	HENNIGER (1969)
	Subunit	4.0	110000	HENNIGER and NUMA (1972)
Rat liver	Polymeric	40–60	–	NAKANISHI and NUMA (1970)
	Protomeric	16	~540000	NUMA et al. (1972)
	Subunit	3.3	110000	INOUE and LOWENSTEIN (1972)
Bovine adipose tissue	Polymeric	68	Several million	MOSS and LANE (1971)
				MOSS et al. (1972)
	Protomeric	15	560000	

gradients in the presence and absence of its allosteric regulators. Since in these experiments all fractions were assayed after prior incubation with citrate, the values given on the ordinate represent the amount of enzyme; both the activated enzyme and the enzyme that was not activated before and during centrifugation are included. It is evident from the data that the sedimentation velocity of the enzyme is increased by addition of citrate, while this citrate-induced increase is reversed by simultaneous addition of palmityl-CoA. These results suggested that activation and inhibition of the carboxylase are associated with polymerization and depolymerization of enzyme molecules, respectively.

Subsequent isolation of homogeneous enzyme preparations from chicken liver (NUMA et al., 1966; GREGOLIN et al., 1966a, 1968a; GOTO et al., 1967; NUMA, 1969), rat liver (NAKANISHI and NUMA, 1970) and bovine adipose tissue (MOSS et al., 1972) permitted more definite characterization of the molecular properties of acetyl-CoA carboxylase as summarized in Table 4. The protomeric form of the chicken (NUMA et al., 1966; GREGOLIN et al., 1966a) and rat liver enzymes (NAKANISHI and NUMA, 1970; OKAZAKI and NUMA, 1971; NUMA et al., 1972b) contains one molecule of covalently bound biotin, as determined microbiologically, and has one tight binding site for citrate and another for acetyl-CoA (GREGOLIN et al., 1968b; HASHIMOTO et al., 1971). Recent work reported that the rat liver protomer contains two molecules of biotin, as determined with the use of avidin (INOUE and LOWENSTEIN, 1972). The polymeric form, which consists of 10–27 protomers, exhibits a characteristic filamentous structure, as disclosed by electron microscopy of the chicken liver enzyme (GREGOLIN et al., 1966a; KLEINSCHMIDT et al., 1969). A similar structure was shown also with the enzymes from bovine adipose tissue (KLEINSCHMIDT et al., 1969; MOSS et al., 1972) and rat liver (NUMA et al., 1970b). Fig. 6 shows electron micrographs of the polymeric as well as protomeric molecules of chicken-liver acetyl-CoA carboxylase. The carboxylase filaments have widths of 70–100 Å and lengths of 500–5000 Å. Light-scattering measurements with chicken-liver enzyme at varying angles and enzyme concentrations revealed that the carboxylase polymer assumes

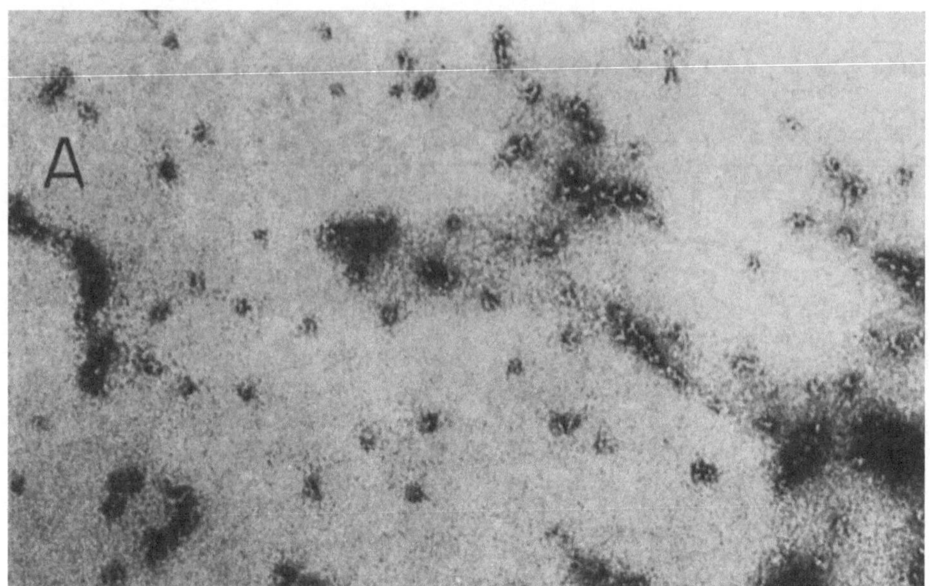

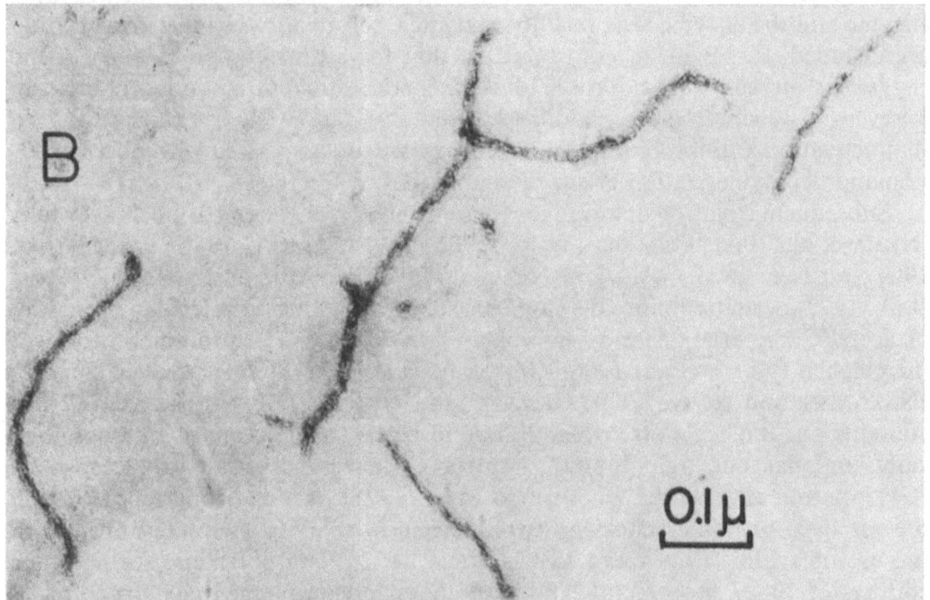

Fig. 6A and B. Electron micrographs of acetyl-CoA carboxylase. A, protomeric form; B, polymeric form. From GREGOLIN et al. (1966a)

a rodlike shape also in solution (HENNIGER, 1969; HENNIGER and NUMA, 1972). The length of the rod-shaped molecules was calculated to be approximately 3000 Å. This value is in fairly good agreement with the length found by electron microscopy. The highly asymmetric shape of the polymeric molecules was indicated also by the high intrinsic viscosity found for the polymeric form of the chicken liver enzyme (MOSS and LANE, 1972). In addition to the protomeric and polymeric forms, the intermediate form of the chicken-liver enzyme yielding a sedimentation coefficient of 27–33 S was observed under certain conditions (NUMA et al., 1966, 1967; GREGOLIN et al., 1968b). However, the molecular properties of the intermediate form have not been characterized, since no conditions have so far been found under which the enzyme is present exclusively in this molecular species in relatively high concentrations.

Extensive studies on the correlation between activity and structure of acetyl-CoA carboxylase under a variety of conditions demonstrated that the polymeric and intermediate forms represent the active conformation, whereas the protomeric form represents the inactive (or less active) conformation. The enzymes used in these studies were from rat adipose tissue (VAGELOS et al., 1962, 1963), rat liver (LYNEN et al., 1963; MATSUHASHI et al., 1964; NUMA et al., 1965a, b; NUMA and RINGELMANN, 1965; IRITANI et al., 1969), chicken liver (GREGOLIN et al., 1966a, b; NUMA et al., 1967) and bovine adipose tissue (MOSS et al., 1972). The three forms of the carboxylase are in an association-dissociation equilibrium. Tricarboxylate activators promote aggregation of protomers, while the inhibitor, palmityl-CoA, shifts the equilibrium toward the protomer (see Fig. 5). In addition, studies with the homogeneous chicken liver enzyme showed that a variety of substances and conditions (including substrates, buffer, pH, ionic strength, and protein concentration) influence the equilibrium (NUMA et al., 1966, 1967; GREGOLIN et al., 1966b, 1968b). ATP plus Mg^{2+} or malonyl-CoA tends to dissociate the enzyme. The dissociation by ATP plus Mg^{2+} is prevented by either acetyl-CoA or ADP. This effect of ADP is abolished by a large amount of bicarbonate. Since the medium contains some endogenous bicarbonate, the enzyme would exist in its carboxylated form (E-biotin$\sim CO_2$) in the presence of ATP plus Mg^{2+} or malonyl-CoA. Thus, the effects of substrates imply that the carboxylated form of the enzyme has a greater tendency to dissociate than the uncarboxylated form of the enzyme (E-biotin). In fact, carboxylation of the enzyme causes dissociation of the polymeric form (GREGOLIN et al., 1968b; OHTSU et al., 1968). Citrate and isocitrate, but not tricarballylate and P_i, are capable of promoting the transition of the carboxylated enzyme to the polymeric state. In accord with this is the observation that tricarballylate and P_i induce polymerization of the uncarboxylated enzyme in simple buffering media, whereas in the assay reaction mixture they, unlike tricarboxylate activators, fail to maintain the polymeric form. These results are of great interest in relation to the finding that the affinity of the carboxylase for citrate is 10^3-fold reduced upon carboxylation of the enzyme (see Section II, C, 3). The carboxylation at the 1′-*N* atom of the biotinyl prosthetic group induces a conformational strain at the regulatory site, so that only the specific allosteric activators can bind to the carboxylated enzyme and prevent dissociation. Kinetic analysis of the citrate effect revealed that the site of citrate action during the carboxylase-catalyzed reaction lies predominantly on the carboxylated enzyme (see Section II, C, 3). It is therefore concluded that the main role of tricarboxylate activators in the catalysis by acetyl-CoA carboxylase is to keep the carboxylated form of the enzyme in active conformation by shifting the equilibrium between the active and inactive (or less active) species of this enzyme form, as visualized in Fig. 4. (HASHIMOTO et al., 1970, 1971; NUMA et al., 1970a, 1972b; HASHIMOTO and NUMA, 1971).

Knowledge of the subunit structure of a regulatory enzyme is of great importance in order to understand its allosteric characters. Upon treatment with sodium dodecyl sulfate, protomeric acetyl-CoA carboxylase from chicken liver dissociates further into subunits with a molecular weight of 110000 (GREGOLIN et al., 1968b; see Table 4). The rat-liver enzyme is dissociated by 6.7 M guanidine-HCl into an apparently homogeneous species exhibiting an $s^0_{20,w}$ of 3.3 S (NUMA et al., 1972b) and a molecular weight of 110000 (INOUE and LOWENSTEIN, 1972) (see Table 4). Attempts to separate the subunit containing biotin from other component proteins have thus far been unsuccessful, although the presence of different subunits has recently been indicated by the results of sodium dodecyl sulfate gel electrophoresis (INOUE and LOWENSTEIN, 1972). In any case, the presence of nonidentical subunits is apparent from the fact that there is a single (or two) biotinyl prosthetic group and there are single binding sites for both citrate and acetyl-CoA per protomer, as described above. Recent studies on acetyl-CoA carboxylase from *Escherichia coli* have made an important contribution to the understanding of the subunit structure of the enzyme. The bacterial carboxylase is composed of three functionally dissimilar proteins, *i.e.* biotin carboxylase, carboxyl transferase, and biotin carboxyl carrier protein (BCCP) (ALBERTS and VAGELOS, 1968; ALBERTS et al., 1969; DIMROTH et al., 1970; GUCHHAIT et al., 1971; NERVI et al., 1971; ALBERTS et al., 1971). BCCP is a relatively low-molecular-weight protein to which biotin is covalently bound. Biotin carboxylase catalyzes ATP-dependent carboxylation of the biotin bound to BCCP (corresponding to Reaction 4). Carboxyl transferase is responsible for transfer of the carboxyl group from BCCP to acetyl-CoA to form malonyl-CoA (corresponding to Reaction 5). Thus BCCP in many respects resembles acyl carrier protein (ACP) (VAGELOS et al., 1966; MAJERUS and VAGELOS, 1967; VAGELOS, 1971). It is tempting to speculate that acetyl-CoA carboxylase from animal tissues is a complex of tightly bound subunits which correspond to the three proteins constituting the bacterial enzyme. Moreover, in view of the fact that the *E. coli* enzyme is not activated by citrate, the allosteric nature of the animal enzyme may be based on its complex character.

Another phenomenon of interest in relation to the allosteric nature of acetyl-CoA carboxylase is its modification by trypsin treatment. Using crude and partially purified rat-liver enzyme preparations, SWANSON et al., (1967) showed that the citrate effect seen during preincubation is replaced by trypsin treatment of the enzyme. Further studies with purified rat-liver enzyme (IRITANI et al., 1969) revealed that trypsin treatment modifies the enzyme in such a way that the carboxylase becomes less dependent on citrate for activation and less sensitive to inhibition by palmityl-CoA. Moreover, the treated enzyme, in contrast to the control enzyme, is in the polymeric state even in the absence of citrate. Palmityl-CoA or ATP plus Mg^{2+}, which is an inhibitor of the carboxylase and can reverse the tricarboxylate-induced polymerization of the control enzyme, fails to do so with the trypsin-treated enzyme. Thus the catalytic activity is well correlated with the aggregational state of the enzyme in these instances, too. These results, together with preliminary kinetic evidence (HASHIMOTO and NUMA, unpublished results), suggest that trypsin might modify the citrate site of the enzyme or a site closely related to it, so that the equilibrium between the active and inactive (or less active) conformations is shifted toward the active one.

5. Tissue Contents of Allosteric Regulators

It is of great importance in relation to the physiological regulation of fatty-acid synthesis to know to what extent acetyl-CoA carboxylase is activated *in vivo, i.e.* how much catalytic efficiency the enzyme molecules exhibit in cells. This depends naturally upon the intracellular concentrations of positive and negative allosteric effectors, such as citrate and long-chain acyl-CoA, and upon their interaction with the enzyme. In starvation and alloxan-diabetes, which are associated with depressed fatty-acid synthesis, the citrate content of rat liver (LYNEN et al., 1963; LYNEN, 1967a; START and NEWSHOLME, 1968; HERRERA and FREINKEL, 1968) and bovine liver (BALLARD et al., 1968) is lowered to one half to one sixth. Some conflicting results reported previously on the hepatic citrate content in starvation (ANGIELSKI and SZUTOWICZ, 1967; WILLIAMSON et al., 1966; SPENCER and LOWENSTEIN, 1967; TARNOWSKI and SEEMANN, 1967) and in alloxan-diabetes (PARMEGGIANI and BOWMAN, 1963; DIXIT et al., 1967) may be accounted for by unsatisfactory assays for citrate or by the possibility that the time at which food was removed from animals may not necessarily be the time at which they stopped eating (see START and NEWSHOLME, 1968). On the other hand, the long-chain acyl-CoA content of rat liver is elevated 2- to 4-fold in these conditions (BORTZ and LYNEN, 1963b; TUBBS and GARLAND, 1963, 1964). Refeeding fasted rats with sugar causes the level of hepatic long-chain acyl-CoA to fall to values below those for control animals fed a balanced diet. In contrast, refeeding with fat causes a further increase in the amount of long-chain acyl-CoA derivatives (TUBBS and GARLAND, 1963, 1964). Administration of a single dose of fat to rats fed a fat-free diet, which results in reduced hepatic fatty-acid synthesis, likewise leads to an increased long-chain acyl-CoA level in the liver (BORTZ, 1967). Thus, the observed variations in the citrate and long-chain acyl-CoA contents of the liver are generally consistent with the changes in hepatic fatty-acid synthesis in different metabolic states. However, the concentrations of these regulators were determined for the whole tissue and not for the cytoplasmic compartment where acetyl-CoA carboxylase is localized. No method so far used provides reliable data on the intracellular distribution of citrate or long-chain acyl-CoA, since these compounds are enzymatically too labile to permit conventional cell fractionation prior to their determination (see FAUPEL et al., 1972). If a method could be devised to measure the concentrations of metabolites in individual compartments of the cell, it would undoubtedly contribute a great deal to the understanding of the physiological role of the known effectors.

The citrate content of rat liver is 0.1–0.3 μmole per gram wet weight (LYNEN, 1967a; START and NEWSHOLME, 1968), while the long-chain acyl-CoA content is 0.01–0.14 μmole per gram wet weight (BORTZ and LYNEN, 1963b; TUBBS and GARLAND, 1963, 1964). These values correspond to intracellular concentrations of 0.2–0.6 mM and 0.02–0.28 mM, respectively (see FANG and LOWENSTEIN, 1967).

The activation constant of rat-liver acetyl-CoA carboxylase for citrate is 3–6 mM (NUMA et al., 1965b; HASHIMOTO and NUMA, 1971; HASHIMOTO et al., 1971; see Table 2), while the apparent inhibition constant for palmityl-CoA is 0.8–1.1 μM (NUMA et al., 1965b). Comparison of the intracellular effector concentrations with the activation and inhibition constants is complicated by the intracellular localization of the effectors and their interaction with various cellular proteins. Nevertheless, it appears reasonable to assume on the basis of the above-mentioned data that acetyl-CoA carboxylase is not fully activated in cells. This consideration supports the view that, although the catalytic capacity of hepatic acetyl-CoA carboxylase in the fully activated state nearly matches those of hepatic citrate-cleavage enzyme (KORNACKER and LOWENSTEIN, 1964a, 1965a; LOWENSTEIN, 1965a; SPENCER and LOWENSTEIN, 1966) and fatty-acid synthetase (CHANG et al., 1967; NUMA et al., 1970a), the carboxylase is not functioning at maximal capacity *in vivo*, and therefore represents the rate-limiting enzyme for fatty-acids synthesis, as suggested initially by GANGULY (1960) and NUMA et al. (1961).

6. Hypolipidemic Agents

The effects on acetyl-CoA carboxylase of some hypolipidemic agents, such as 2-methyl-2-[p-(1,2,3,4-tetrahydro-1-naphtyl)-phenoxy]-propionate and 2-(p-chlorophenoxy)-2-methyl-propionate, were investigated *in vitro* as well as *in vivo* (MARAGOUDAKIS, 1969, 1970a, b; MARAGOUDAKIS and HAUKIN, 1971; MARAGOUDAKIS et al., 1972). These drugs, which are known to reduce the level of plasma lipoproteins, are capable of inhibiting chicken- and rat-liver acetyl-CoA carboxylase. Fatty-acid synthetase is not inhibited at comparable concentrations of the drugs. Kinetic studies showed that the inhibition is competitive with regard to the activator, citrate or isocitrate. Furthermore, these agents reverse the activator-dependent polymerization of carboxylase protomers. Lipogenesis in cultured rat mammary cells is strongly reduced by the drugs (MARAGOUDAKIS, 1971). Exogenous citrate in the culture medium prevents or reverses the depression of lipogenesis caused by the agents. The level of measurable acetyl-CoA carboxylase activity is significantly lower in the livers of rats treated with a hypolipidemic agent than in those of control animals. This effect of the drug is annulled by dialysis of the enzyme preparation against a citrate-containing buffer. These results suggest that these compounds may act as lipid-lowering agents because acetyl-CoA carboxylase is inhibited *in vivo* due to competition with the tricarboxylate activator.

7. Ratio of Holoenzyme to Apoenzyme

In biotin-deficient rats, the level of acetyl-CoA carboxylase activity in the adipose tissue is decreased by a factor of 6, while that in the liver is reduced to only about 50 per cent of the control level (DAKSHINAMURTI and DESJARDINS, 1968, 1969). Immunochemical titration with the use of antiacetyl-CoA carboxylase indicated that the apoenzyme accumulates in the adipose tissue of biotin-deficient rats, whereas the liver contains little apoenzyme (JACOBS et al., 1970). Injection of biotin into deficient rats results in a rapid increase in adipose-tissue carboxylase activity. This, together with the demonstration of [^{3}H]biotin incorporation into the enzyme precipitated immunologically, indicates the conversion of the apoenzyme to the holoenzyme upon administration of biotin (JACOBS et al., 1970). Thus, the catalytic efficiency of acetyl-CoA carboxylase is regulated by the holoenzyme-to-apoenzyme ratio in biotin-deficient animals.

Recent results suggest that in chick liver acetyl-CoA carboxylase apoenzyme is synthesized in response to hatching regardless of feeding (RYDER, 1972). In this experiment, the apoenzyme was assayed by determining ATP-dependent incorporation of [^{14}C]biotin into crude preparations, but no test was made for concomitant increase in carboxylase activity.

III. Regulation at the Fatty-Acid Synthetase Step

A. Generation of Cytoplasmic NADPH

Fatty-acid synthetase requires NADPH as an efficient reducing agent (GIBSON et al., 1958). In the cytoplasm, NADPH can be generated by the hexose monophosphate shunt dehydrogenases, malic enzyme, and isocitrate dehydrogenase. However, current evidence indicates that hexose monophosphate shunt dehydrogenases and malic enzyme supply nearly all of the reducing equivalents required for fatty-acid synthesis, as discussed below. The involvement of the pentose phosphate cycle as a donor of NADPH was indicated by the finding that during stimulation by insulin of fatty-acid synthesis from glucose in rat adipose tissue, there occurs concomitantly a marked increase in the release of $^{14}CO_2$ from [1-^{14}C]glucose as compared with that from [6-^{14}C]glucose (WINEGRAD and RENOLD, 1958). However, balance studies of the reduced coenzymes produced and utilized in rat adipose tissue showed that, when lipogenesis is high, only 50–75 per cent of the reduced coenzymes required for fatty-acid synthesis are formed in the pentose phosphate cycle (FLATT and BALL, 1964; KATZ et al., 1966; KATZ and ROGNSTAD, 1966). The reducing equivalents needed to complete fatty-acid synthesis must therefore be furnished by the reduced coenzymes produced during the conversion of triose phosphates to acetyl-CoA. The utilization of NADH for fatty-acid synthesis was suggested by the transfer of tritium from [2-^{3}H]lactate or [2-^{3}H]glycerol to fatty-acids in rat-liver and mammary-gland slices (LOWENSTEIN, 1961; FOSTER and BLOOM, 1961). Furthermore, tritium was found to be incorporated from β-[^{3}H]NADH into fatty acids by particle-free supernatant fractions from rat mammary glands and by rat-liver homogenate preparations containing cytosol and microsomes (MATTHES et al., 1963). These findings suggested that a transhydrogenation mechanism is operating to effect the transfer of hydride ion from NADH to NADP^{+}. Evidence was subsequently provided to show the importance of the cytoplasmic transhydrogenation system involving coupled reactions between malate dehydrogenase and malic enzyme. The level of malic enzyme activity in rat liver and adipose tissue is altered by different nutritional and hormonal conditions in such a way as to suggest that this enzyme may play a role in fatty-acid synthesis by furnishing NADPH (for references, see below). The conversion of NADH to NADPH was also shown in rat liver and adipose tissue preparations containing malate dehydrogenase and malic enzyme (PANDE et al., 1964). The sequence of reactions involved in the transhydrogenation (see also Fig. 1) is as follows:

$$\text{Pyruvate}_m + CO_2 + \text{ATP} \xrightarrow{\text{pyruvate carboxylase}} \text{oxaloacetate}_m + \text{ADP} + \text{P}_i \quad (6)$$

$$\text{Oxaloacetate}_m + \text{acetyl-CoA}_m \xrightarrow{\text{citrate synthase}} \text{citrate}_m + \text{CoA} \quad (7)$$

$$\text{Citrate}_c + \text{CoA} + \text{ATP} \xrightarrow{\text{citrate-cleavage enzyme}} \text{oxaloacetate}_c + \text{acetyl-CoA}_c + \text{ADP} + \text{P}_i \quad (8)$$

$$\text{Oxaloacetate}_c + \text{NADH} + \text{H}^+ \xrightarrow{\text{malate dehydrogenase}} \text{malate}_c + \text{NAD}^+ \quad (9)$$

$$\text{Malate}_c + \text{NADP}^+ \xrightarrow{\text{malic enzyme}} \text{pyruvate}_c + CO_2 + \text{NADPH} + \text{H}^+ \quad (10)$$

Overall:

$$\text{NADH} + \text{NADP}^+ + 2\,\text{ATP} \rightarrow \text{NAD}^+ + \text{NADPH} + 2\,\text{ADP} + 2\,\text{P}_i \quad (11)$$

where m and c denote mitochondrial and cytoplasmic compartments, respectively. Reactions (6) and (7) take place in the mitochondria, while Reactions (8), (9) and (10) occur in the cytoplasm. Pyruvate enters the mitochondria, while citrate leaves them. It is evident that this process is of cyclic nature. The overall result of this citrate-malate cycle is an ATP-requiring transhydrogenation between NADH and NADP^+ to yield NADPH. The mechanism of acetyl-group transport via citrate involving Reactions (7) and (8) is discussed in Section II, A, 1. The citrate-cleavage enzyme generates one mole of acetyl-CoA and one mole of oxaloacetate. Therefore, the citrate-malate cycle can produce only one mole of NADPH for each acetyl unit used for fatty-acid synthesis. Since two moles of NADPH are required for the reduction of each acetyl unit, only half of the reduced coenzyme needed for fatty-acid synthesis is provided by this transhydrogenation mechanism. This deficit in reducing equivalents is supplied by the pentose phosphate cycle as described above. The involvement of NADP-linked isocitrate dehydrogenase in the generation of reducing equivalents used for fatty-acid synthesis is questioned, since the level of this enzyme in rat liver, in contrast to those of hexose monophosphate shunt dehydrogenases and malic enzyme, does not change in accord with the rate of fatty-acid synthesis (PANDE et al., 1964).

The levels of hexose monophosphate shunt dehydrogenases (TEPPERMAN and TEPPERMAN, 1958, 1964; FITCH and CHAIKOFF, 1960, 1962; WEBER, 1963; PANDE et al., 1964; MURAD and FREEDLAND, 1965; JOHNSON and SASSOON, 1967; YOUNG, 1968; SMITH and ABRAHAM, 1970a; LOCKWOOD et al., 1970; CHANG and SCHNEIDER, 1971; DIAMANT et al., 1972) and malic enzyme (FITCH and CHAIKOFF, 1960, 1962; SHRAGO et al., 1963; TEPPERMAN and TEPPERMAN, 1964; PANDE et al., 1964; WISE and BALL, 1964; YOUNG et al., 1964; MURAD and FREEDLAND, 1965;

YOUNG, 1968; SMITH and ABRAHAM, 1970a; LOCKWOOD et al., 1970; DIAMANT et al., 1972) in the liver and adipose tissue of rats and mice are subject to adaptive changes in different dietary, hormonal and developmental conditions that are accompanied by increased or decreased fatty-acid synthesis. The changes in the levels of rat-liver hexose monophosphate shunt dehydrogenases (RUDACK et al., 1971a,b) and malic enzyme (GIBSON et al., 1972) in response to dietary alteration were shown, by analyzing the time course of changes in the enzyme level (see SCHIMKE and DOYLE, 1970) or by combined immunochemical and isotopic techniques, to be due primarily to variations in the rate of enzyme synthesis. The rate of malic-enzyme synthesis in rat liver, as estimated by measuring incorporation of radioactive amino acid into the enzyme isolated by immunoprecipitation, is decreased by a factor of 2.3 after 2 days' starvation and increased about 5-fold after 12 hours' fat-free refeeding, while the half-life for degradation of the labeled enzyme is 63 hours in normal and 34 hours in refed rats (GIBSON et al., 1972). It was also shown by similar methods that in 8–11-day-old chicks the rate of synthesis of hepatic malic enzyme is reduced to about one third after 2 days' fasting, while the half-life for degradation of the enzyme is 55 hours in fed and 28 hours in fasted chicks (SILPANANTA and GOODRIDGE, 1971). Furthermore, thyroid hormones, which are capable of increasing the level of rat liver malic enzyme, were shown to accelerate enzyme synthesis (GIBSON et al., 1972). The increase in the hexose monophosphate shunt dehydrogenase level in rat liver upon refeeding is blocked by puromycin (POTTER and ONO, 1961), actinomycin D (JOHNSON and SASSOON, 1967) or ethionine (WEBER et al., 1962; WEBER, 1963; TEPPERMAN and TEPPERMAN, 1963). Similarly, actinomycin D or puromycin prevents the rise in the malic-enzyme level in mouse liver normally observed upon weaning (SMITH and ABRAHAM, 1970a). These results also support the conclusion that the adaptive changes in the enzyme level are due principally to altered rates of enzyme synthesis.

Recent work with isolated fat cells (KATHER et al., 1972) showed that rotenone, which inhibits electron transport in the respiratory chain and consequently blocks the conversion of glucose to fatty acids at the site of pyruvate dehydrogenase, diminishes both fatty-acid synthesis and the pentose phosphate cycle activity. On the other hand, phenazine methosulfate stimulates the pentose phosphate cycle activity without having much effect on fatty-acid synthesis. Rotenone has no direct effect on lipogenesis and on the pentose phosphate cycle, since the influence of this inhibitor on both pathways can be overcome by phenazine methosulfate. On the basis of these results, the rate of glucose catabolism via the pentose phosphate cycle in adipocytes appears to be controlled by the requirement of NADPH for lipogenesis. Evidence was also presented that indicates that this control is exerted at the step of glucose-6-phosphate dehydrogenase.

B. Control by Changes in the Quantity of Fatty-Acid Synthetase

The level of fatty-acid synthetase activity in rat liver (GIBSON and HUBBARD, 1960; HENNING et al., 1961; NUMA et al., 1961; ALLMANN et al., 1965; DAHLEN et al., 1968; CHANG et al., 1967; DIAMANT et al., 1972), mouse liver (ALLMANN and GIBSON, 1965; SMITH and ABRAHAM, 1970a), avian liver (BUTTERWORTH et al., 1966), and rat adipose tissue (SAGGERSON and GREENBAUM, 1970; DIAMANT et al., 1972) fluctuates in accord with the rate of fatty-acid synthesis under various nutritional, hormonal, developmental, and genetic conditions of the animal. Puromycin or actinomycin D injected into rats or mice blocks the anticipated rise in the hepatic enzyme level upon fat-free realimentation of fasted rats (HICKS et al., 1965; GIBSON et al., 1966) or upon weaning mice (SMITH and ABRAHAM, 1970a). This provides evidence for the involvement of *de novo* synthesis of the enzyme. It was also shown that the actual content of purified fatty acid synthetase in avian (BUTTERWORTH et al., 1966) and rat liver (BURTON et al., 1969b; CRAIG et al., 1972) varies in accord with the level of its activity, and that the increased level of hepatic fatty-acid synthetase during fat-free refeeding is due to adaptive enzyme synthesis, since [^{14}C]leucine is readily incorporated *in vivo* (BURTON et al., 1969b) or in isolated rat liver cells (BURTON et al., 1969a) into the purified fatty-acid synthetase formed during refeeding. Recent work reported that in starved rats, the rate of enzyme synthesis is reduced to 20 per cent of the control value (TWETO and LARRABEE, 1972; CRAIG et al., 1972), and that the apparent half-life for enzyme degradation is decreased from about 70 hours to about 20 hours (TWETO and LARRABEE, 1972). In rats refed a fat-free diet for 12–72 hours, on the other hand, the rate of enzyme synthesis is elevated about 4- to 14-fold as compared with that in normally fed rats (GIBSON et al., 1972; CRAIG et al., 1972), while the half-life for enzyme degradation is 69 hours both in refed and in control animals (CRAIG et al., 1972). In these studies, the isotopic method was used, except that the half-life for enzyme degradation in fasted rats was estimated from the time-course of decline of the enzyme level.

C. Control by Changes in the Catalytic Efficiency of Fatty-Acid Synthetase

Concerning the detailed mechanisms of the intermediate reactions catalyzed by the fatty-acid-synthetase multienzyme complex, a number of comprehensive reviews are available (LYNEN, 1961, 1967b,c; WAKIL, 1962; VAGELOS, 1964, 1971; VAGELOS et al., 1966; MAJERUS and VAGELOS, 1967; STUMPF, 1969; PORTER et al., 1971). Recent evidence indicated that in mammalian liver and mammary glands, as in pigeon liver and rat adipose tissue, the important primer in the fatty-acid synthetase-catalyzed reactions is butyryl-CoA rather than acetyl-CoA, and that the major pathway of fatty-acid synthesis in these tissues involves the initial formation of butyryl-CoA from acetyl-CoA via the reversal of β-oxidation catalyzed by the enzymes present in the cytoplasm (NANDEDKAR et al., 1969; NANDEDKAR and KUMAR, 1969; SMITH and ABRAHAM, 1971; LIN and KUMAR, 1971, 1972).

Fatty-acid synthetase from avian liver is activated 3- to 5-fold by various phosphorylated sugars, fructose 1,6-diphosphate being the most effective activator (WAKIL et al., 1966; PLATE et al., 1968). Fructose 1,6-diphosphate reverses substrate inhibition by malonyl-CoA, which competes with NADPH. The stimulatory effect of hexose diphosphate has not been confirmed with enzymes from avian and rat liver (PORTER et al., 1971) and rat mammary glands (SMITH and ABRAHAM,

1970b). Furthermore, malonyl-CoA was found to have no effect on the binding of NADPH by fatty-acid synthetase as measured by fluorescence emission spectroscopy (DUGAN and PORTER, 1970).

Fatty-acid synthetases from rat liver (TUBBS and GARLAND, 1964), avian liver (DORSEY and PORTER, 1968) and rat brain (ROBINSON et al., 1963) are inhibited by long-chain acyl-CoA. It is suggested that palmityl-CoA inhibits the enzyme by virtue of its detergent nature, since the inhibition is dependent on the presence of the critical micellar concentration of the thioester as well as on the molar ratio of the inhibitor to protein (DORSEY and PORTER, 1968). The physiological role of long-chain acyl-CoA derivatives as metabolic regulators is discussed in Section II, C, 3.

Cerulenin, an antibiotic of the structure (2S) (3R)2,3-epoxy-4-oxo-6,10-dodecadienoylamide, which inhibits the growth of a variety of yeasts, fungi, and bacteria, was found to be a potent inhibitor of fatty-acid synthetases from various microorganisms and from rat liver (VANCE et al., 1972). Cerulenin specifically blocks the activity of β-ketoacyl thioester synthetase (condensing enzyme). This effect may account for the inhibition of overall fatty-acid synthesis by the antibiotic.

The half-life for degradation of hepatic fatty-acid synthetase in normally fed rats is about 70 hours (see Section III, B). Recent work of TWETO et al. (1972) showed that the subunits of this multienzyme complex are degraded with half-lives which are related to their molecular weight in a manner similar to that found by SCHIMKE and his associates (DEHLINGER and SCHIMKE, 1970, 1971; DICE and SCHIMKE, 1972); subunits of greater molecular weights are degraded more rapidly. On the other hand, the exchange rate of the covalently bound prosthetic group, 4′-phosphopantetheine, with unbound pantothenate compounds is more than an order of magnitude greater than the rate of degradation of the enzyme complex (TWETO et al., 1971). This exchange of the prosthetic group is largely eliminated after fasting (TWETO and LARRABEE, 1972). On the basis of these results, a model is proposed in which the prosthetic group, or some small portion of the enzyme complex containing it, is periodically removed and replaced many times before an individual enzyme complex is catabolized. Such an exchange might represent a means of controlling fatty-acid synthetase activity.

An inhibitory effect of dibutyryl cyclic AMP on crude fatty-acid synthetase from rat brown adipose tissue was reported (GIACOBINE, 1971). In recent years, a large body of data has been accumulated, showing that cyclic AMP is an intracellular second messenger mediating many of the actions of a variety of different hormones (see ROBINSON et al., 1968). Recent results showed that cyclic AMP or its dibutyryl derivative depresses hepatic fatty-acid synthesis in tissue slices, as measured by incorporation of radioactive acetate (AKHTAR and BLOXHAM, 1970; BRICKER and LEVEY, 1972a; ALLRED and ROEHRIG, 1972). In contrast, rat hepatoma slices fail to respond to cyclic AMP (BRICKER and LEVEY, 1972b). Lipogenesis in isolated rat adipose tissue is stimulated by lower concentrations of dibutyryl cyclic AMP, but reduced by its higher concentrations (SKOSEY, 1970). These results appear to be concerned with short-term effects of the cyclic nucleotide on lipogenesis. The site and mode of these effects remain to be elucidated.

IV. Coordinate Response of Lipogenic Enzymes

The levels of acetyl-CoA carboxylase and fatty-acid synthetase as well as those of citrate-cleavage enzyme, hexose monophosphate shunt dehydrogenases and malic enzyme, which are involved in the generation of the carbon precursor and reducing agent required for fatty-acid synthesis, undergo coordinate adaptive changes when the rate of fatty-acid synthesis varies under a variety of metabolic conditions (see Sections II, A, 2; II, B, 1; III, A; III, B). The conditions include: (*a*) dietary alteration such as fasting, refeeding with various kinds of diets, and feeding of fat, fatty acids or carbohydrate (TEPPERMAN and TEPPERMAN, 1958, 1964; FITCH and CHAIKOFF, 1960; GIBSON and HUBBARD, 1960; HENNING et al., 1961; NUMA et al., 1961; POTTER and ONO, 1961; BORTZ et al., 1963; SHRAGO et al., 1963; WEBER, 1963; PANDE et al., 1964; WISE and BALL, 1964; YOUNG et al., 1964; ALLMANN and GIBSON, 1965; ALLMANN et al., 1965; KORNACKER and BALL, 1965; KORNACKER and LOWENSTEIN, 1965a; BUTTERWORTH et al., 1966; SHRAGO and LARDY, 1966; JOHNSON and SASSOON, 1967; PEARCE, 1968; BURTON et al., 1969b; DAKSHINAMURTI and DESJARDINS, 1969; MUTO and GIBSON, 1970; SAGGERSON and GREENBAUM, 1970; RUDACK et al., 1971a, b; SILPANANTA and GOODRIDGE, 1971); (*b*) hormonal alteration such as alloxan- or streptozotocin-diabetes and insulin administration (GIBSON and HUBBARD, 1960; HENNING et al., 1961; FITCH and CHAIKOFF, 1962; SHRAGO et al., 1963; WIELAND et al., 1963; KORNACKER and LOWENSTEIN, 1964b, 1965b; WISE and BALL, 1964; BROWN and MCLEAN, 1965; SHRAGO and LARDY, 1966; BURTON et al., 1969b; SAGGERSON and GREENBAUM, 1970; CHANG and SCHNEIDER, 1971; RUDACK et al., 1971a, b) and thyroid hormone administration (TEPPERMAN and TEPPERMAN, 1964; WISE and BALL, 1964; MURAD and FREEDLAND, 1965; YOUNG, 1968; LOCKWOOD et al., 1970; DIAMANT et al., 1972); (*c*) developmental alteration such as birth and growth (HAHN and DRAHOTA, 1966; BALLARD and HANSON, 1967; FELICIOLI and GABRIELLI, 1967; TAYLOR et al., 1967; GOODRIDGE, 1968; ARINZE and MISTRY, 1970; SMITH and ABRAHAM, 1970a; LOCKWOOD et al., 1970; RYDER, 1970) and lactation (HOWANITZ and LEVY, 1965; SPENCER and LOWENSTEIN, 1966; LOWENSTEIN, 1968); (*d*) genetic alteration such as obesity (KORNACKER and LOWENSTEIN, 1964a; SPENCER and LOWENSTEIN, 1966; CHANG et al., 1967; NAKANISHI and NUMA, 1971). As discussed above (see Section II, A, 2; II, B, 2; III, A; III, B), these changes in the levels of the lipogenic enzymes are due primarily to altered rates of enzyme synthesis except under nonsteady-state conditions like starvation.

The factors responsible for the coordinate synthesis of all these enzymes are unknown. MUTO and GIBSON (1970) showed that oral administration of methyl esters of polyunsaturated fatty acids, such as linoleic, linolenic, and arachidonic acid, to rats maintained on a fat-free high-carbohydrate diet brings about a progressive diminution of the levels of the above-mentioned lipogenic enzymes in the liver. Methyl esters of saturated and monounsaturated fatty acids exhibited little damping effect. The lipogenic enzymes behave as a constant-proportionality set, regardless of the kind of nutritional manipulation. The polyunsaturated-fatty-acid content of liver lipids and free fatty acids reflects the exogenous input of fatty acids. Thus there is a reciprocal relationship between the levels of the lipogenic enzymes and the polyunsaturated-fatty-acid content of the liver. This would support the concept that polyunsaturated fatty acids or substances metabolically related to them may be a factor responsible for the control of synthesis of these enzymes. In rat mammary glands and adipose tissue, likewise, the enzymes involved in lipogenesis fluctuate coordinately as a constant and specific proportion group upon change in lipogenesis (McLean et al., 1972). Further studies are required to understand the mechanism underlying the regulation of synthesis of the lipogenic enzymes.

V. Concluding Remarks

In the present review, the "basic" regulatory mechanisms for fatty-acid synthesis, *i.e.* the control mechanisms for lipogenic enzymes, are discussed, with special emphasis on acetyl-CoA carboxylase. In the last decade, a number of lipogenic enzymes have been purified to homogeneity, and much has been learned about their properties and reaction mechanisms. On the basis of these studies, great progress has been made in the field of the regulation of fatty-acid synthesis. Above all, the coordinate changes in the levels of lipogenic enzymes that occur in response to a variety of metabolic conditions associated with increased or decreased lipogenesis have been investigated in terms of enzyme synthesis and degradation. Extensive studies have also been made on the molecular basis for allosteric regulation of the catalytic efficiency of lipogenic enzymes. However, the factors that directly determine the rates of enzyme synthesis and degradation are still unknown, and it will be one of the most important aspects of research in the future to identify these factors. Another subject of interest will be to characterize the regulatory site of the enzymes materially and to elucidate its interaction with the catalytic site. It is hoped that research in this field will eventually contribute to the understanding and prevention of the derangement of lipid metabolism encountered in some pathological conditions.

Acknowledgements. The experimental work carried out in this laboratory was supported in part by research grants from the Ministry of Education of Japan, the Toray Science Foundation, the Japan Waksman Foundation, the Japanese Foundation of Metabolism and Diseases, the Naito Foundation, the Tanabe Amino Acid Research Foundation and the Japanese Medical Association.

References

Ahmad, F., Jacobson, B., Wood, H. G.: Transcarboxylase. X. Assembly of active transcarboxylase from its inactive subunits and incorporation of the biotin-carboxyl carrier protein. J. biol. Chem. **245**, 6486–6488 (1970).

Akhtar, M., Bloxham, D.P.: The co-ordinated inhibition of protein and lipid biosynthesis by adenosine 3′:5′-cyclic monophosphate. Biochem. J. **120**, 11P (1970).

Alberts, A. W., Gordon, S. G., Vagelos, P. R.: Acetyl CoA carboxylase: The purified transcarboxylase component. Proc. nat. Acad. Sci. (Wash.) **68**, 1259–1263 (1971).

Alberts, A. W., Nervi, A. M., Vagelos, P. R.: Acetyl CoA carboxylase. II. Demonstration of biotin-protein and biotin carboxylase subunits. Proc. nat. Acad. Sci. (Wash.) **63**, 1319–1326 (1969).

ALBERTS, A. W., VAGELOS, P. R.: Acetyl CoA carboxylase. I. Requirement for two protein fractions. Proc. nat. Acad. Sci. (Wash.) **59**, 561–568 (1968).

ALLMANN, D. W., GIBSON, D. M.: Fatty acid synthesis during early linoleic acid deficiency in the mouse. J. Lipid Res. **6**, 51–62 (1965).

ALLMANN, D. W., HUBBARD, D. D., GIBSON, D. M.: Fatty acid synthesis during fat-free refeeding of starved rats. J.Lipid Res. **6**, 63–74 (1965).

ALLRED, J. B., ROEHRIG, K. L.: Inhibition of hepatic lipogenesis by cyclic-3′,5′-nucleotide monophosphates. Biochem. biophys. Res. Commun. **46**, 1135–1139 (1972).

ANGIELSKI, S., SZUTOWICZ, A.: Tissue content of citrate and citrate-cleavage enzyme activity during starvation and refeeding. Nature (Lond.) **213**, 1252–1253 (1967).

ANNISON, E. F., HILL, K. J., LEWIS, D.: STUDIES Studies on the portal blood of sheep. 2. Absorption of volatile fatty acids from the rumen of the sheep. Biochem. J. **66**, 592–599 (1957).

ARIAS, I. M., DOYLE, D., SCHIMKE, R. T.: Studies on the synthesis and degradation of proteins of the endoplasmic reticulum of rat liver. J. biol. Chem. **244**, 3303–3315 (1969).

ARINZE, J. C., MISTRY, S. P.: Hepatic acetyl CoA carboxylase, propionyl CoA carboxylase and pyruvate carboxylase activities during embryonic development and growth in chickens. Proc. Soc. exp. Biol. (N. Y.). **135**, 553–556 (1970).

ARMSTRONG, D. G., BLAXTER, K. L.: The utilization of acetic, propionic, and butyric acids by fattening sheep. Brit. J. Nutr. **11**, 413–425 (1957).

ATKINSON, D. E., WALTON, G. M.: Adenosine triphosphate conservation in metabolic regulation. Rat liver citrate cleavage enzyme. J. biol. Chem. **242**, 3239–3241 (1967).

BALLARD, F. J., HANSON, R. W.: Changes in lipid synthesis in rat liver during development. Biochem. J. **102**, 952–958 (1967).

BALLARD, F. J., HANSON, R. W.: Gluconeogenesis and lipogenesis in tissue from ruminant and nonruminant animals. Fed. Proc. **28**, 218–231 (1969).

BALLARD, F. J., HANSON, R. W., KRONFELD, D. S., RAGGI, F.: Metabolic changes in liver associated with spontaneous ketosis and starvation in cows. J. Nutr. **95**, 160–172 (1968).

BARDEN, R. E., CLELAND, W. W.: 1-Acylglycerol 3-phosphate acyltransferase from rat liver. J. biol. Chem. **244**, 3677–3684 (1969).

BARDEN, R. E., FUNG, C. H., UTTER, M. F., SCRUTTON, M. C.: Pyruvate carboxylase from chicken liver. Steady state kinetic studies indicate a "two-site" ping-pong mechanism. J. biol. Chem. **247**, 1323–1333 (1972).

BARTH, C., SLADĘK, M., DECKER, K.: The subcellular distribution of short-chain fatty acyl-CoA synthetase activity in rat tissues. Biochim. biophys. Acta (Amst.) **248**, 24–33 (1971).

BARTH, C., SLADEK, M., DECKER, K.: Dietary changes of cytoplasmic acetyl-CoA synthetase in different rat tissues. Biochim. biophys. Acta (Amst.) **260**, 1–9 (1972).

BARTLEY, J. C., ABRAHAM, S.: Dietary regulation of fatty acid synthesis in rat liver and hepatic autotransplants. Biochim. biophys. Acta (Amst.) **260**, 169–177 (1972).

BARTLEY, J., ABRAHAM, S., CHAIKOFF, I. L.: Concerning the form in which acetyl units produced in mitochondria are transferred to the site of *de novo* fatty acid synthesis in the cell. Biochem. biophys. Res. Commun. **19**, 770–776 (1965).

BEENAKKERS, A. M. TH., KLINGENBERG, M.: Carnitine-coenzyme A transacetylase in mitochondria from variors organs. Biochim. biophys. Acta (Amst.) **84**, 205–207 (1964).

BERLIN, C. M., SCHIMKE, R. T.: Influence of turnover rates on the responses of enzymes to cortisone. Molec. Pharmacol. **1**, 149–156 (1965).

BHADURI, A., SRERE, P. A.: The incorporation of citrate carbon into fatty acids. Biochim. biophys. Acta (Amst.) **70**, 221–230 (1963).

BORTZ, W. M.: Fat feeding and cholesterol synthesis. Biochim. biophys. Acta (Amst.) **137**, 533–539 (1967).

BORTZ, W., ABRAHAM, S., CHAIKOFF, I. L.: Localization of the block in lipogenesis resulting from feeding fat. J. biol. Chem. **238**, 1266–1272 (1963).

BORTZ, W. M., LYNEN, F.: The inhibition of acetyl CoA carboxylase by long chain acyl CoA derivatives. Biochem. Z. **337**, 505–509 (1963a).

BORTZ, W. M., LYNEN, F.: Elevation of long chain acyl CoA derivatives in livers of fasted rats. Biochem. Z. **339**, 77–82 (1963b).

BREMER, J.: Carnitine in intermediary metabolism. Reversible acetylation of carnitine by mitochondria. J. biol. Chem. **237**, 2228–2331 (1962a).

BREMER, J.: Carnitine in intermediary metabolism. The metabolism of fatty acid esters of carnitine by mitochondria. J. biol. Chem. **237**, 3628–3632 (1962b).

BRESSLER, R., BRENDEL, K.: The role of carnitine and carnitine acyltransferase in biological acetylations and fatty acid synthesis. J. biol. Chem. **241**, 4092–4097 (1966).

BRICKER, L. A., LEVEY, G. S.: Evidence for regulation of cholesterol and fatty acid synthesis in liver by cyclic adenosine 3',5'-monophosphate. J. biol. Chem. **247**, 4914–4915 (1972a).

BRICKER, L. A., LEVEY, G.S.: Autonomous cholesterol and fatty acid synthesis in hepatomas: Deletion of the adenosine 3',5'-cyclic monophosphate control mechanism of normal liver. Biochem. biophys. Res. Commun. **48**, 362–365 (1972b).

BROWN, J., MCLEAN, P.: Effect of alloxan-diabetes on the activity of citrate cleavage enzyme in adipose tissue. Nature (Lond.) **207**, 407–408 (1965).

BURTON, D. N., COLLINS, J. M., KENNAN, A. L., PORTER, J. W.: The effects of nutritional and hormonal factors on the fatty acid synthetase level of rat liver. J. biol. Chem. **244**, 4510–4516 (1969b).

BURTON, D. N., COLLINS, J. M., PORTER, J. W.: Biosynthesis of the fatty acid synthetase by isolated rat lever cells. J. biol. Chem. **244**, 1076–1077 (1969a).

BUTTERWORTH, P. H. W., GUCHHAIT, R. B., BAUM, H., OLSON, E. B., MARGOLIS, S. A., PORTER, J. W.: Relationship between nutritional status and fatty acid synthesis by microsomal and soluble enzymes of pigeon liver. Arch. Biochem. Biophys. **116**, 453–457 (1966).

CHANG, A. Y., SCHNEIDER, D. I.: Hepatic enzyme activities in streptozotocin-diabetic rats before and after insulin treatment. Diabetes **20**, 71–77 (1971).

CHANG, H.-C., SEIDMAN, I., TEEBOR, G., LANE, M. D.: Liver acetyl CoA carboxylase and fatty acid synthetase: Relative activities in the normal state and in hereditary obesity. Biochem. biophys. Res. Commun. **28**, 682–686 (1967).

CHAPPELL, J. B.: The permeability properties of the mitochondrial membrane in relation to the distribution of enzymes in cells. Biochem. J. **100**, 43P (1966).

CHAPPELL, J. B., HAARHOFF, K. N.: The penetration of the mitochondrial membrane by anions and cations. In: E. C. SLATER, Z. KANIUGA and L. WOJTCZAK (eds.), Biochemistry of mitochondria, P. 75–91. London–New York: Academic Press 1967.

CHAPPELL, J. B., ROBINSON, B. H.: Penetration of the mitochondrial membrane by tricarboxylic acid anions. In: T. W. GOODWIN (ed.), Metabolic roles of citrate, P. 123–133. London-New York: Academic Press 1968.

CRAIG, M. C., NEPOKROEFF, C. M., LAKSHMANAN, M. R., PORTER, J. W.: Effect of dietary change on the rates of synthesis and degradation of rat liver fatty acid synthetase. Arch. Biochem. Biophys. **152**, 619–630 (1972).

D'ADAMO, A. F., HAFT, D. E.: An alternate pathway of α-ketoglutarate catabolism in the isolated perfused rat liver. J. biol. Chem. **240**, 613–617 (1965).

DAHLEN, J. V., KENNAN, A. L., PORTER, J. W.: Effect of alloxan and portacaval shunt on the synthesis of fatty acids and sterols by rat liver. Arch. Biochem. Biophys. **124**, 51–57 (1968).

DAIKUHARA, Y., TSUNEMI, T., TAKEDA, Y.: The role of ATP citrate lyase in the transfer of acetyl groups in rat liver. Biochim. biophys. Acta (Amst.) **158**, 51–61 (1968).

DAKSHINAMURTI, K., DESJARDINS, P.R.: Lipogenesis in biotin deficiency. Canad. J. Biochem. **46**, 1261–1267 (1968).

DAKSHINAMURTI, K., DESJARDINS, P. R.: Acetyl-CoA carboxylase from rat adipose tissue. Biochim. biophys. Acta (Amst.) **176**, 221–229 (1969).

DEHLINGER, P. J., SCHIMKE, R. T.: Effect of size on the relative rate of degradation of rat liver soluble proteins. Biochem. biophys. Res. Commun. **40**, 1473–1480 (1970).

DEHLINGER, P. J., SCHIMKE, R. T.: Size distribution of membrane proteins of rat liver and their relative rates of degradation. J. biol. Chem. **246**, 2574–2583 (1971).

DIAMANT, S., GORIN, E., SHAFRIR, E.: Enzyme activities related to fatty-acid synthesis in liver and adipose tissue of rats treated with triiodothyronine. Europ. J. Biochem. **26**, 553–559 (1972).

DICE, J. F., SCHIMKE, R. T.: Turnover and exchange of ribosomal proteins from rat liver. J. biol. Chem. **247**, 98–111 (1972).

DIMROTH, P., GUCHHAIT, R. B., STOLL, E., LANE, M. D.: Enzymatic carboxylation of biotin: Molecular and catalytic properties of a component enzyme of acetyl CoA carboxylase. Proc. nat. Acad. Sci. (Wash.) **67**, 1353–1360 (1970).

DIXIT, P. K., DEVILLIERS, D. C., LAZAROW, A.: Citrate metabolism in diabetes. II. Tissue citrate content, citrate synthase and oxidation in alloxan-diabetic rat. Metabolism **16**, 285–293 (1967).

DONALDSON, W. E., MUELLER, N. S., MASON, J. V.: Intracellular localization of fatty acid synthesis in chick-embryo liver and stimulation of synthesis by exogenous glucose. Biochim. biophys. Acta (Amst.) **248**, 34–40 (1971).

DORSEY, J. A., PORTER, J. W.: The effect of palmityl coenzyme A on pigeon liver fatty acid synthetase. J. biol. Chem. **243**, 3512–3516 (1968).

DUGAN, R. E., PORTER, J. W.: The binding of reduced nicotinamide adenine dinucleotide phosphate to mammalian and avian fatty acid synthetases. Number of binding sites and the effect of reagents and conditions on the binding of reduced nicotinamide adenine dinucleotide phosphate to enzyme. J. biol. Chem. **245**, 2051–2059 (1970).

ELWOOD, J. C., MORRIS, H. P.: Lack of adaptation in lipogenesis by hepatoma 9121. J. Lipid Res. **9**, 337–341 (1968).

FANG, M., LOWENSTEIN, J. M.: Citrate and the conversion of carbohydrate into fat. The regulation of fatty acid synthesis by rat liver extracts. Biochem. J. **105**, 803–811 (1967).

FAUPEL, R. P., SEITZ, H. J., TARNOWSKI, W.: The problem of tissue sampling from experimental animals with respect to freezing technique, anoxia, stress and narcosis. A new method for sampling rat liver tissue and the physiological values of glycolytic intermediates and related compounds. Arch. Biochem. Biophys. **148**, 509–522 (1972).

FELICIOLI, R. A., GABRIELLI, F.: The citrate cleavage enzyme activity in chick embryo and chicken liver during development. Experientia (Basel) **23**, 1000–1001 (1967).

FITCH, W. M., CHAIKOFF, I. L.: Extent and patterns of adaptation of enzyme activities in livers of normal rats fed diets high in glucose and fructose. J. biol. Chem. **235**, 554–557 (1960).

FITCH, W. M., CHAIKOFF, I. L.: Directions and patterns of adaptation induced in liver enzymes of the diabetic rat by the feeding of glucose and fructose. Biochim. biophys. Acta (Amst.) **57**, 588–595 (1962).

FLATT, J. P., BALL, E. G.: Studies on the metabolism of adipose tissue. XV. An evaluation of the major pathways of glucose catabolism as influenced by insulin and epinephrine. J. biol. Chem. **239**, 675–685 (1964).

FORMICA, J. V.: Utilization of citric acid carbon for the biosynthesis of long-chain fatty acids. Biochim. biophys. Acta (Amst.) **59**, 739–741 (1962).

FOSTER, D. W., BLOOM, B.: A comparative study of reduced di- and triphosphopyridine nucleotides in the intact cell. J. biol. Chem. **236**, 2548–2551 (1961).

FOSTER, D. W., MCWHORTER, W. P.: Microsomes, microsomal phospholipids, and fatty acid synthesis. J. biol. Chem. **244**, 260–267 (1969).

FOSTER, D. W., SRERE, P. A.: Citrate cleavage enzyme and fatty acid synthesis. J. biol. Chem. **243**, 1926–1930 (1968).

FRITZ, I. B.: Carnitine and its role of fatty acid metabolism. Advanc. Lipid Res. **1**, 285–334 (1963).

FRITZ, I. B., HSU, M. P.: The effects of palmitylcarnitine on hepatic fatty acid synthesis and on acetyl CoA carboxylase activity. Biochem. biophys. Res. Commun. **22**, 737–743 (1966).

FRITZ, I. B., HSU, M. P.: Studies on the control of fatty acid synthesis. I. Stimulation by (+)-palmitylcarnitine of fatty acid synthesis in liver preparations from fed and fasted rats. J. biol. Chem. **242**, 865–872 (1967).

FRITZ, I. B., YUE, K. T. N.: Effects of carnitine on acetyl-CoA oxidation by heart muscle mitochondria. Amer. J. Physiol. **206**, 531–535 (1964).

GANGULY, J.: Studies on the mechanism of fatty acid synthesis. VII. Biosynthesis of fatty acids from malonyl CoA. Biochim. biophys. Acta (Amst.) **40**, 110–118 (1960).

GELLHORN, A., BENJAMIN, W.: Insulin action in alloxan diabetes modified by actinomycin D. Science **146**, 1166–1168 (1964).

GELLHORN, A., BENJAMIN, W.: Fatty acid biosynthesis and RNA function in fasting, aging and diabetes. Advanc. Enzyme Regul. **4**, 19–41 (1966).

GERWIN, B. I., JACOBSON, B. E., WOOD, H. G.: Transcarboxylase, VIII. Isolation and properties of a biotin-carboxyl carrier protein. Proc. nat. Acad. Sci. (Wash.) **64**, 1315–1322 (1969).

GIACOBINO, J.-P.: Effects of cyclic-3′,5′-adenosine monophosphate and of dibutyryl-3′,5′-adenosine monophosphate on the fatty acid synthesizing system of rat brown adipose tissue. Life Sci., (Part II) **10**, 1089–1095 (1971).

GIBSON, D. M., HICKS, S. E., ALLMANN, D. W.: Adaptive enzyme formation during hyperlipogenesis. Advanc. Enzyme Regul. **4**, 239–246 (1966).

GIBSON, D. M., HUBBARD, D. D.: Incorporation of malonyl CoA into fatty acids by liver in starvation and alloxan-diabetes. Biochem. biophys. Res. Commun. **3**, 531–535 (1960).

GIBSON, D. M., LYONS, R. T., SCOTT, D. F., MUTO, Y.: Synthesis and degradation of the lipogenic enzymes of rat liver. Advanc. Enzyme Regul. **10**, 187–204 (1972).

GIBSON, D. M., TITCHENER, E. B., WAKIL, S. J.: Studies on the mechanism of fatty acid synthesis. V. Bicarbonate requirement for the synthesis of long-chain fatty acids. Biochim. biophys. Acta (Amst.) **30**, 376–383 (1958).

GOODMAN, D. S.: The interaction of human serum albumin with longchain fatty acid anions. J. Amer. chem. Soc. **80**, 3892–3898 (1958).

GOODRIDGE, A. G.: Citrate-cleavage enzyme, 'malic' enzyme and certain dehydrogenases in embryonic and growing chicks. Biochem. J. **108**, 663–666 (1968).

GOTO, T., RINGELMANN, E., RIEDEL, B., NUMA, S.: Purification and crystallization of acetyl CoA carboxylase. Life Sci. (Part II) **6**, 785–790 (1967).

GREEN, N. M.: Avidin. 1. The use of [^{14}C]biotin for kinetic studies and for assay. Biochem. J. **89**, 585–595 (1963).

GREENSPAN, M., LOWENSTEIN, J. M.: Effect of magnesium ions and adenosine triphosphate on the activity of acetyl coenzyme A carboxylase. Arch. Biochem. Biophys. **118**, 260–263 (1967).

GREENSPAN, M. D., LOWENSTEIN, J. M.: Effects of magnesium ions, adenosine triphosphate, palmitoylcarnitine, and palmitoyl coenzyme A on acetyl coenzyme A carboxylase. J. biol. Chem. **243**, 6273–6280 (1968).

GREGOLIN, C., RYDER, E., KLEINSCHMIDT, A. K., WARNER, R. C., LANE, M. D.: Molecular characteristics of liver acetyl CoA carboxylase. Proc. nat. Acad. Sci. (Wash.) **56**, 148–155 (1966a).

GREGOLIN, C., RYDER, E., LANE, M. D.: Liver acetyl coenzyme A carboxylase. I. Isolation and catalytic properties. J. biol. Chem. **243**, 4227–4235 (1968a).

GREGOLIN, C., RYDER, E., WARNER, R. C., KLEINSCHMIDT, A. K., CHANG, H.-C., LANE, M. D.: Liver acetyl coenzyme A carboxylase. II. Further molecular characterization. J. biol. Chem. **243**, 4236–4245 (1968b).

GREGOLIN, C., RYDER, E., WARNER, R. C., KLEINSCHMIDT, A. K., LANE, M. D.: Liver acetyl CoA carboxylase: The dissociation-reassociation process and its relation to catalytic activity. Proc. nat. Acad. Sci. (Wash.). **56**, 1751–1758 (1966b).

GUCHHAIT, R. B., MOSS, J., SOKOLSKI, W., LANE, M. D.: The carboxyl transferase component of acetyl CoA carboxylase: Structural evidence for intersubunit translocation of the biotin prosthetic group. Proc. nat. Acad. Sci. (Wash.) **68**, 653–657 (1971).

GUYNN, R. W., VELOSO, D., VEECH, R. I.: The concentration of malonyl-coenzyme A and the control of fatty acid synthesis *in vivo*. J. biol. Chem. **247**, 7325–7331 (1972).

HAHN, P., DRAHOTA, Z.: The activities of citrate cleavage enzyme, acetyl-CoA synthetase and lipoprotein lipase in white and brown adipose tissue and the liver of the rat during development. Experientia (Basel) **22**, 706–707 (1966).

HALPERIN, M. L., ROBINSON, B. H., FRITZ, I. B.: Effects of palmitoyl CoA on citrate and malate transport by rat liver mitochondria. Proc. nat. Acad. Sci. (Wash.) **69**, 1003–1007 (1972).

HANSON, R. W., BALLARD, F. J.: The relative significance of acetate and glucose as precursors for lipid synthesis in liver and adipose tissue from ruminants. Biochem. J. **105**, 529–536 (1967).

HASHIMOTO, T., IRITANI, N., NAKANISHI, S., NUMA, S.: Regulatory mechanisms for fatty acid synthesis. II. On the mode of action of allosteric effector on acetyl-CoA carboxylase. Proc. Jap. Conf. Biochem. Lipids **12**, 21–26 (1970).

HASHIMOTO, T., ISANO, H., IRITANI, N., NUMA, S.: Liver acetylcoenzyme-A carboxylase. Studies on kynurenate inhibition, isotope exchange and interaction of the uncarboxylated enzyme with citrate. Europ. J. Biochem. **24**, 128–139 (1971).

HASHIMOTO, T., NUMA, S.: Kinetic studies on the reaction mechanism and the citrate activation of liver acetyl coenzyme A carboxylase. Europ. J. Biochem. **18**, 319–331 (1971).

HENNIGER, G.: Untersuchungen zur Struktur der Acetyl-CoA-Carboxylase aus Hühnerleber. Dissertation, Universität München, 1969.

HENNIGER, G., NUMA, S.: Lichtstreuungsmessungen mit der polymeren Form der Acetyl-CoA-Carboxylase aus Hühnerleber. Hoppe-Seylers Z. physiol. Chem. **353**, 459–462 (1972).

HENNING, U., LYNEN, F., WIELAND, O., NEUFELDT, I.: Zur Störung der Fettsynthese beim Diabetes. 1. Europäisches Symposium für medizinische Enzymologie. Milano 1960, p. 526–531. Basel: S. Karger, 1961.

HERRERA, E., FREINKEL, N.: Interrelationships between liver composition, plasma glucose and ketones, and hepatic acetyl-CoA and citric acid during prolonged starvation in the male rat. Biochim. biophys. Acta (Amst.) **170**, 244–253 (1968).

HICKS, S.E., ALLMANN, D.W., GIBSON, D.M.: Inhibition of hyperlipogenesis with puromycin or actinomycin D. Biochim. biophys. Acta (Amst.) **106**, 441–444 (1965).

HOCHHEUSER, W., WEISS, H., WIELAND, O.: Über den Stoffwechsel der freien Essigsäure im Tierkörper und seine Beziehung zum Atherogeneseproblem. Z. klin. Chem. **2**, 175–181 (1964).

HOWANITZ, P.J., LEVY, H.R.: Acetyl-CoA carboxylase and citrate cleavage enzyme in the rat mammary gland. Biochim. biophys. Acta (Amst.) **106**, 430–433 (1965).

INGALLS, A.M., DICKIE, M.M., SNELL, G.D.: A new mutation in the house mouse. J. Hered. **41**, 317–318 (1950).

INOUE, H., LOWENSTEIN, J.M.: Acetyl coenzyme A carboxylase from rat liver. Purification and demonstration of different subunits. J. biol. Chem. **247**, 4825–4832 (1972).

INOUE, H., SUZUKI, F., FUKUNISHI, K., ADACHI, K., TAKEDA, Y.: Studies on ATP citrate lyase of rat liver. I. Purification and some properties. J. Biochem. **60**, 543–553 (1966).

IRITANI, N., NAKANISHI, S., NUMA, S.: Alterations in the sensitivity to effectors of rat liver acetyl-CoA carboxylase following trypsin treatment. Life Sci. (Part II) **8**, 1157– 1165 (1969).

JACOBS, R., KILBURN, E., MAJERUS, P.W.: Acetyl coenzyme A carboxylase. The effects of biotin deficiency on enzyme in rat liver and adipose tissue. J. biol. Chem. **245**, 6462–6467 (1970).

JACOBSON, B., GERWIN, B.I., AHMAD, F., WAEGELL, P., WOOD, H.G.: Transcarboxylase. IX. Parameters effecting dissociation and reassociation of the enzyme. J. biol. Chem. **245**, 6471–6483 (1970).

JOHNSON, B.C., SASSOON, H.F.: Studies on the induction of liver glucose-6-phosphate dehydrogenase in the rat. Advanc. Enzyme Regul. **5**, 93–106 (1967).

KALLEN, R.G., LOWENSTEIN, J.M.: The stimulation of fatty acid synthesis by isocitrate and malonate. Arch. Biochem. Biophys. **96**, 188–190 (1962).

KARUSH, F., SONENBERG, M.: Interaction of homologous alkyl sulfates with bovine serum albumin. J. Amer. chem. Soc. **71**, 1369–1376 (1949).

KATHER, H., RIVERA, M., BRAND, K.: Interrelationship and control of glucose metabolism and lipogenesis in isolated fat-cells. Biochem. J. **128**, 1097–1102 (1972).

KATZ, J., LANDAU, B.R., BARTSCH, G.E.: The pentose cycle, triose phosphate isomerization, and lipogenesis in rat adipose tissue. J. biol. Chem. **241**, 727–740 (1966).

KATZ, J., ROGNSTAD, R.: The metabolism of tritiated glucose by rat adipose tissue. J. biol. Chem. **241**, 3600–3610 (1966).

KAUZMANN, W.: Some factors in the interpretation of protein denaturation. Advanc. Protein Chem. **14**, 1–63 (1959).

KAZIRO, Y., HASS, L.F., BOYER, P.D., OCHOA, S.: Mechanism of the propionyl carboxylase reaction. II. Isotopic exchange and tracer experiments. J. biol. Chem. **237**, 1460–1468 (1962).

KAZIRO, Y., OCHOA, S.: The metabolism of propionic acid. Advanc. Enzymol. **26**, 283–378 (1964).

KLEINSCHMIDT, A.K., MOSS, J., LANE, M.D.: Acetyl coenzyme A carboxylase: Filamentous nature of the animal enzymes. Science **166**, 1276–1278 (1969).

KLINGENBERG, M.: Metabolite transport in mitochondria: Example for intracellular membrane function. Essays Biochem. **6**, 119–159 (1970).

KNAPPE, J.: Mechanism of biotin action. Ann. Rev. Biochem. **39**, 757–776 (1970).

KOCH, A.R.: The evaluation of the rates of biological-processes from tracer kinetic data. I. The influence of labile metabolic pools. J. theor. Biol. **3**, 283–303 (1962).

KORCHAK, H.M., MASORO, E.J.: Changes in the level of the fatty acid synthesizing enzymes during starvation. Biochim. biophys. Acta (Amst.) **58**, 354–356 (1962).

KORCHAK, H.M., MASORO, E.J.: Free fatty acids as lipogenic inhibitors. Biochim. biophys. Acta (Amst.) **84**, 750–753 (1964).

KORNACKER, M.S., BALL, E.G.: Citrate cleavage in adipose tissue. Proc. nat. Acad. Sci. (Wash.) **54**, 899–904 (1965).

KORNACKER, M.S., LOWENSTEIN, J.M.: Citrate cleavage enzyme in livers of obese and nonobese mice. Science **144**, 1027–1028 (1964a).

KORNACKER, M.S., LOWENSTEIN, J.M.: Citrate cleavage and acetate activation in livers of normal and diabetic rats. Biochim. biophys. Acta (Amst.) **84**, 490–492 (1964b).

KORNACKER, M.S., LOWENSTEIN, J.M.: Citrate and the conversion of carbohydrate into fat. The activities of citrate-cleavage enzyme and acetate thiokinase in livers of starved and re-fed rats. Biochem. J. **94**, 209–215 (1965a).

KORNACKER, M. S., LOWENSTEIN, J. M.: Citrate and the conversion of carbohydrate into fat. Activities of citrate-cleavage enzyme and acetate thiokinase in livers of normal and diabetic rats. Biochem. J. **95**, 832–837 (1965b).

KREBS, H. A., LOWENSTEIN, J. M.: The tricarboxylic acid cycle. In: D. M. GREENBERG (ed.), Metabolic pathways, 2nd ed., vol. 1, p. 129–203. New York: Academic Press 1960.

LANE, M. D., EDWARDS, J., STOLL, E., MOSS, J.: Tricarboxylic acid activator-induced changes at the active site of acetyl-CoA carboxylase. Vitam. and Horm. **28**, 345–363 (1970).

LANE, M. D., MOSS, J.: Regulation of fatty acid synthesis in animal tissues. In: H. J. VOGEL (ed.), Metabolic pathways, 3rd ed., vol. 5, p. 23 54. New York: Academic Press 1971.

LEVY, H. R.: Inhibition of mammary gland acetyl CoA carboxylase by fatty acids. Biochem. biophys. Res. Commun. **13**, 267–272 (1963).

LIN, C. Y., KUMAR, S.: Primer specificity of mammary fatty acid synthetase and the role of the soluble β-oxidative enzymes. J. biol. Chem. **246**, 3284 3290 (1971).

LIN, C. Y., KUMAR, S.: Pathway for the synthesis of fatty acids in mammalian tissues. J. biol. Chem. **247**, 604–606 (1972).

LOCKWOOD, E. A., BAILEY, E., TAYLOR, C. B.: Factors involved in changes in hepatic lipogenesis during development of the rat. Biochem. J. **118**, 155–162 (1970).

LOWENSTEIN, J. M.: The pathway of hydrogen in biosyntheses. I. Experiments with glucose-1-H^3 and lactate-2-H^3. J. biol. Chem. **236**, 1213–1216 (1961).

LOWENSTEIN, J. M.: The supply of precursors for the extramitochondrial synthesis of fatty acids. In: J. K. GRANT, (ed.), The control of lipid metabolism, p. 57–61. London-New York: Academic Press 1963.

LOWENSTEIN, J. M.: Adaptive behavior of citrate cleavage enzyme. In: B. Chance (ed.), Control of energy metabolism, p. 261–265. New York: Academic Press 1965.

LOWENSTEIN, J. M.: Citrate and the conversion of carbohydrate into fat. In: T. W. GOODWIN (ed.), The metabolic roles of citrate, p. 61–86. London-New York: Academic Press 1968.

LOWENSTEIN, J. M.: Effect of (–)-hydroxycitrate on fatty acid synthesis by rat liver *in vivo*. J. biol. Chem. **246**, 629–632 (1971).

LYNEN, F.: Participation of acyl-CoA in carbon chain biosynthesis. J. cell. comp. Physiol. **54**, Suppl. 1, 33–49 (1959).

LYNEN, F.: Biosynthesis of saturated fatty acids. Fed. Proc. **20**, 941–951 (1961).

LYNEN, F.: Mechanism and biological regulation of fatty acid synthesis. Progr. biochem. Pharmacol. **3**, 1–31 (1967a).

LYNEN, F.: The role of biotin-dependent carboxylations in biosynthetic reactions. Biochem. J. **102**, 381–400 (1967b).

LYNEN, F.: Multienzyme complex of fatty acid synthetase. In: H. J. VOGEL, J. O. LAMPEN and V. BRYSON (eds.), Organizational biosynthesis, p. 243–266. New York-London: Academic Press 1967c.

LYNEN, F., KNAPPE, J., LORCH, E., JÜTTING, G., RINGELMANN, E., LACHANCE, J.-P.: Zur biochemischen Funktion des Biotins. II. Reinigung und Wirkungsweise der β-Methyl-crotonyl-Carboxylase. Biochem. Z. **335**, 123–167 (1961).

LYNEN, F., MATSUHASHI, M., NUMA, S., SCHWEIZER, E.: The cellular control of fatty acid synthesis at the enzymatic level. In: J. K. GRANT (ed.), The control of lipid metabolism, p. 43–56. London-New York: Academic Press 1963.

MADSEN, J., ABRAHAM, S., CHAIKOFF, I. L.: The conversion of glutamate carbon to fatty acid carbon via citrate. J. biol. Chem. **239**, 1305–1309 (1964).

MAJERUS, P. W., JACOBS, R., SMITH, M. B.: The regulation of fatty acid biosynthesis in rat hepatomas. J. biol. Chem. **243**, 3588–3595 (1968).

MAJERUS, P. W., KILBURN, E.: Acetyl coenzyme A carboxylase. The roles of synthesis and degradation in regulation of enzyme levels in rat liver. J. biol. Chem. **244**, 6254–6262 (1969).

MAJERUS, P. W., VAGELOS, P. R.: Fatty acid biosynthesis and the role of the acyl carrier protein. Advanc. Lipid Res. **5**, 1–33 (1967).

MARAGOUDAKIS, M. E.: Inhibition of hepatic acetyl coenzyme A carboxylase by hypolipidemic agents. J. biol Chem. **244**, 5005–5013 (1969).

MARAGOUDAKIS, M. E.: On the mode of action of lipid-lowering agents. II. *In vitro* inhibition of acetyl coenzyme A carboxylase by a hypolipidemic drug. Biochemistry **9**, 413–417 (1970a).

MARAGOUDAKIS, M. E.: On the mode of action of Lipid-lowering agents. III. Kinetics of activation and inhibition of acetyl coenzyme A carboxylase. J. biol. Chem. **245**, 4136–4140 (1970b).

MARAGOUDAKIS, M. E.: On the mode of action of lipid-lowering agents. VI. Inhibition of lipogenesis in rat mammary gland cell culture. J. biol. Chem. **246**, 4046–4052 (1971).

MARAGOUDAKIS, M. E., HANKIN, H.: On the mode of action of lipid-lowering agents. V. Kinetics of the inhibition *In vitro* of rat acetyl coenzyme A carboxylase. J. biol. Chem. **246**, 348–358 (1971).

MARAGOUDAKIS, M. E., HANKIN, H., WASVARY, J. M.: On the mode of action of lipid-lowering agents. VII. *In vivo* inhibition and reversible binding of hepatic acetyl coenzyme A carboxylase by hypolipidemic drugs. J. biol. Chem. **247**, 342–347 (1972).

MARTIN, D. B., VAGELOS, P. R.: Tricarboxylic acid cycle regulation of fatty acid biosynthesis. Fed. Proc. **21**, 289 (1962a).

MARTIN, D. B., VAGELOS, P. R.: Mechanism of tricarboxylic acid cycle regulation of fatty acid synthesis. Biochem. biophs. Res. Commun. **7**, 101–106 (1962b).

MARTIN, D. B., VAGELOS, P. R.: The mechanism of tricarboxylic acid cycle regulation of fatty acid synthesis. J. biol. Chem. **237**, 1787–1792 (1962c).

MATSUHASHI, M., MATSUHASHI, S., LYNEN, F.: Zur Biosynthese der Fettsäuren. V. Die Acetyl-CoA Carboxylase aus Rattenleber und ihre Aktivierung durch Citronensäure. Biochem. Z. **340**, 263–289 (1964).

MATSUHASHI, M., MATSUHASHI, S., NUMA, S., LYNEN, F.: Citrate-dependent carboxylation of acetyl CoA in rat liver. Fed. Proc. **21**, 288 (1962).

MATTHES, K. J., ABRAHAM, S., CHAIKOFF, I. L.: Hydrogen transfer in fatty acid synthesis by rat liver and mammary-gland cell-free preparations studied with tritium-labeled pyridine nucleotides and glucose. Biochim. biophys. Acta (Amst.) **70**, 242–259 (1963).

MAX, S. R., SCORPIO, R. M., PURVIS, J. L.: Citrate and isocitrate transport in isolated rat liver mitochondria. J. biol. Chem. **245**, 4807–4813 (1970).

MCCLURE, W. R., LARDY, H. A., CLELAND, W. W.: Rat liver pyruvate carboxylase. III. Isotopic exchange studies of the first partial reaction. J. biol. Chem. **246**, 3584–3590 (1971c).

MCCLURE, W. R., LARDY, H. A., KNEIFEL, H. P.: Rat liver pyruvate carboxylase. I. Preparation, properties, and cation specificity. J. biol. Chem. **246**, 3569–3578 (1971a).

MCCLURE, W. R., LARDY, H. A., WAGNER, M., CLELAND, W. W.: Rat liver pyruvate carboxylase. II. Kinetic studies of the forward reaction. J. biol. Chem. **246**, 3579–3583 (1971b).

MCLEAN, P., GREENBAUM, A. L., GUMAA, K. A.: Constant and specific proportion groups of enzymes in rat mammary gland and adipose tissue in relation to lipogenesis. FEBS Letters **20**, 277–282 (1972).

MILDVAN, A. S., SCRUTTON, M. C., UTTER, M. F.: Pyruvate carboxylase. VII. A possible role for tightly bound manganese. J. biol. Chem. **241**, 3488–3498 (1966).

MILLER, A. L., GEROCH, M. E., LEVY, H. R.: Rat mammary-gland acetyl-coenzyme A carboxylase. Interaction with milk fatty acids. Biochem. J. **118**, 645–657 (1970).

MILLER, A. L., LEVY, H. R.: Rat mammary acetyl coenzyme A carboxylase. I. Isolation and characterization. J. biol. Chem. **244**, 2334–2342 (1969).

MOSS, J., LANE, M. D.: The biotin-dependent enzymes. Advanc. Enzymol. **35**, 321–442 (1971).

MOSS, J., LANE, M. D.: Acetyl coenzyme A carboxylase. III. Further studies on the relation of catalytic activity to polymeric state. J. biol. Chem. **247**, 4944–4951 (1972).

MOSS, J., YAMAGISHI, M., KLEINSCHMIDT, A. K., LANE, M. D.: Acetyl CoA carboxylase: Properties of the mammalian adipose tissue enzyme. Fed. Proc. **28**, 538 (1969).

MOSS, J., YAMAGISHI, M., KLEINSCHMIDT, A. K., LANE, M. D.: Acetyl coenzyme A carboxylase. Purification and properties of the bovine adipose tissue enzyme. Biochemistry **11**, 3779–3786 (1972).

MURAD, S., FREEDLAND, R. A.: The *in vivo* effect of thyroxine on citrate cleavage enzyme and NADP linked dehydrogenase activities. Life Sci. (Part II) **4**, 527–531 (1965).

MUTO, Y., GIBSON, D. M.: Selective dampening of lipogenic enzymes of liver by exogenous polyunsaturated fatty acids. Biochem. biophys. Res. Commun. **38**, 9–15 (1970).

NAKANISHI, S., NUMA, S.: Purification of rat liver acetyl coenzyme A carboxylase and immunochemical studies on its synthesis and degradation. Europ. J. Biochem. **16**, 161–173 (1970).

NAKANISHI, S., NUMA, S.: Synthesis and degradation of liver acetyl coenzyme A carboxylase in genetically obese mice. Proc nat. Acad. Sci. (Wash.) **68**, 2288–2292 (1971).

NANDEDKAR, A. K. N., KUMAR, S.: Biosynthesis of fatty acids in mammary tissue. II. Synthesis of butyrate in lactation rabbit mammary supernatant fraction by the reversal of β-oxidation. Arch. Biochem. Biophys. **134**, 563–571 (1969).

NANDEDKAR, A. K. N., SCHIRMER, E. W., PYNADATH, T. I., KUMAR, S.: Biosynthesis of fatty acids in mammary tissue. I. Purification and properties of fatty acid synthetase from lactating-goat mammary tissue. Arch. Biochem. Biophys. **134**, 554–562 (1969).

NÉMETHY, G., SCHERAGA, H. A.: The structure of water and hydrophobic bonding in proteins. III. The thermodynamic properties of hydrophobic bonds in proteins. J. Phys. Chem. **66**, 1733–1789 (1966).

NERVI, A. M., ALBERTS, A. W., VAGELOS, P. R.: Acetyl CoA carboxylase. III. Purification and properties of a biotin carboxyl carrier protein. Arch. Biochem. Biophys. **143**, 401–411 (1971).

NORTHROP, D. B.: Transcarboxylase. VI. Kinetic analysis of the reaction mechanism. J. biol. Chem. **244**, 5808–5819 (1969).

NORTHROP, D. B., WOOD, H. G.: Transcarboxylase. VII. Exchange reactions and kinetics of oxalate inhibition. J. biol. Chem. **244**, 5820–5827 (1969).

NUMA, S.: Acetyl-CoA carboxylase from chicken and rat liver. Methods Enzymol. **14**, 9–16 (1969).

NUMA, S., BORTZ, W. M., LYNEN, F.: Regulation of fatty acid synthesis at the acetyl-CoA carboxylation step. Advanc. Enzymol. Regul. **3**, 407–423 (1965a).

NUMA, S., GOTO, T., RINGELMANN, E., RIEDEL, B.: Studies on the relationship between activity and structure of liver acetyl-CoA carboxylase. Effects of magnesium ions, adenine nucleotides and acetyl-CoA. Europ. J. Biochem. **3**, 124–128 (1967).

NUMA, S., HASHIMOTO, T., IRITANI, N.: Modification of rat liver acetyl-CoA carboxylase by trypsin. Reported at the 8th Intern. Congr. Biochem., Lucerne, 1970b.

NUMA, S., HASHIMOTO, T., NAKANISHI, S., OKAZAKI, T.: Regulatory mechanisms for liver acetyl coenzyme A carboxylase. Biochem. J. **128**, 2P (1972a).

NUMA, S., HASHIMOTO, T., NAKANISHI, S., OKAZAKI, T.: Regulatory mechanisms for liver acetyl coenzyme A carboxylase. In: R. M. S. SMELLIE (ed.), Current trends in the biochemistry of lipids, p. 27–39. London-New York: Academic Press 1972b.

NUMA, S., MATSUHASHI, M., LYNEN, F.: Zur Störung der Fettsäuresynthese bei Hunger und Alloxan-Diabetes. I. Fettsäuresynthese in der Leber normaler und hungernder Ratten. Biochem. Z. **334**, 203–217 (1961).

NUMA, S., NAKANISHI, S., HASHIMOTO, T., IRITANI, N., OKAZAKI, T.: Role of acetyl coenzyme A carboxylase in the control of fatty acid synthesis. Vitam. and Horm. **28**, 213–243 (1970a).

NUMA, S., NAKANISHI, S., IRITANI, N.: Regulatory mechanisms for fatty acid biosynthesis at the acetyl-CoA carboxylase step. Proc. Jap. Conf. Biochem. Lipids **11**, 235–240 (1969).

NUMA, S., RINGELMANN, E.: Zur Aufhebung der Citrat-Aktivierung der Acetyl-CoA-Carboxylase durch Kälte. Biochem. Z. **343**, 258–268 (1965).

NUMA, S., RINGELMANN, E., LYNEN, F.: Zur biochemischen Funktion des Biotins. VIII. Die Bindung der Kohlensäure in der carboxylierten Acetyl-CoA-Carboxylase. Biochem. Z. **340**, 228–242 (1964).

NUMA, S., RINGELMANN, E., LYNEN, F.: Zur Hemmung der Acetyl-CoA- Carboxylase durch Fettsäure-Coenzym A-Verbindungen. Biochem. Z. **343**, 243–257 (1965b).

NUMA, S., RINGELMANN, E., RIEDEL, B.: Further evidence for polymeric structure of liver acetyl CoA carboxylase. Biochem. biophys. Res. Commun. **24**, 750–757 (1966).

OHTSU, E., AKAGAMI, H., OKAZAKI, T., NUMA, S.: On the mechanism for activation of acetyl CoA carboxylase by citrate. Seikagaku **40**, 698 (1968).

OKAZAKI, T., NUMA, S.: The biotin-containing protein component of hepatic acetyl CoA carboxylase. Proc. Jap. Conf. Biochem. Lipids **13**, 147–150 (1971).

PALMIERI, F., STIPANI, I., QUAGLIARIELLO, E.: Kinetic study of the tricarboxylate carrier in rat liver mitochondria. Europ. J. Biochem. **26**, 587–594 (1972).

PANDE, S. V., KHAN, R. P., VENKITASUBRAMANIAN, T. A.: Nicotinamide adenine dinucleotide phosphate-specific dehydrogenases in relation to lipogenesis. Biochim. biophys. Acta (Amst.) **84**, 239–250 (1964).

PARMEGGIANI, A., BOWMAN, R. H.: Regulation of phosphofructokinase activity by citrate in normal and diabetic muscle. Biochem. biophys. Res. Commun. **12**, 268–273 (1963).

PATTERSON, A. L., JOHNSON, C. K., HELM, D. van der, MINKIN, J. A.: The absolute configuration of naturally occurring isocitric acid. J. Amer. chem. Soc. **84**, 309–310 (1962).

PEARCE, J.: The effect of dietary fat on lipogenic enzymes in the liver of the domestic fowl. Biochem. J. **109**, 702–704 (1968).

PLATE, C. A., JOSHI, V. C., SEDGWICK, B., WAKIL, S. J.: Studies on the mechanism of fatty acid synthesis. XXI. The role of fructose 1,6-diphosphate in the stimulation of the fatty acid synthetase from pigeon liver. J. biol. Chem. **243**, 5439–5445 (1968).

POLAKIS, S. E., GUCHHAIT, R. B., LANE, M. D.: On the possible involvement of a carbonyl phosphate intermediate in the adenosine triphosphate-dependent carboxylation of biotin. J. biol. Chem. **247**, 1335–1337 (1972).

PORTER, J. W., KUMAR, S., DUGAN, R. E.: Synthesis of fatty acids by enzymes of avian and mammalian species. Progr. biochem. Pharmacol. **6**, 1–101 (1971).

POTTER, R., ONO, T.: Enzyme patterns in rat liver and morris hepatoma 5123 during metabolic transitions. Cold Spr. Harb. Symp. quant. Biol. **26**, 355–362 (1961).

REID, R. L., HOGAN, J. P., BRICKS, P. K.: The effect of diet on individual volatile fatty acids in the rumen of sheep with particular reference to the effect of low rumen pH and adaptation on high starch diets. Aust. J. agr. Res. **8**, 691–710 (1957).

ROBINSON, B. H., CHAPPELL, J. B.: The inhibition of malate, tricarboxylate and oxoglutarate entry into mitochondria by 2-n-butylmalonate. Biochem. biophys. Res. Commun. **28**, 249–255 (1967).

ROBINSON, B. H., WILLIAMS, G. R.: The effects of 2-ethylcitrate and tricarballylate on citrate transport in rat liver mitochondria and fatty acid synthesis in rat white adipose tissue. Europ. J. Biochem. **15**, 263–272 (1970).

ROBINSON, G. A., BUTCHER, R. W., SUTHERLAND, E. W.: Cyclic AMP. Ann. Rev. Biochem. **37**, 149–174 (1968).

ROBINSON, J. D., BRADY, R. O., BRADLEY, R. M.: Biosynthesis of fatty acids. IV. Studies with inhibitors. J. Lipid Res. **4**, 144–150 (1963).

ROGNSTAD, R., KATZ, J.: Acetyl group tranfer in lipogenesis II. Fatty acid synthesis from intra- and extramitochondrial acetyl CoA. Arch. Biochem. Biophys. **127**, 437–444 (1968).

ROUS, S.: Rôle modeste du citrate comme transporteur d'acétyl-CoA chez l'animal vivant. FEBS Letters **12**, 338–340 (1971).

RUDACK, D., CHISHOLM, E. M., HOLTEN, D.: Rat liver glucose 6-phosphate dehydrogenase. Regulation by carbohydrate diet and insulin. J. biol. Chem. **246**, 1249–1254 (1971 a).

RUDACK, D., GOZUKARA, E. M., CHISHOLM, E. M., HOLTEN, D.: The effect of dietary carbohydrate and fat on the synthesis of rat liver 6-phospogluconate dehydrogenase. Biochim. biophys. Acta (Amst.) **252**, 305–313 (1971 b).

RYDER, E.: Effect of development on chicken liver acetyl-coenzyme A carboxylase. Biochem. J. **119**, 929–930 (1970).

RYDER, E.: The presence of acetyl-coenzyme A carboxylase apoenzyme in the liver of newly hatched chicks. Biochem. J. **128**, 183–185 (1972).

RYDER, E., GREGOLIN, C., CHANG, H.-C., LANE, M. D.: Liver acetyl CoA carboxylase: Insight into the mechanism of activation by tricarboxylic acids and acetyl CoA. Proc. nat. Acad. Sci. (Wash.) **57**, 1455–1462 (1967).

SABINE, J. R., ABRAHAM, S., MORRIS, H. P.: Defective dietary control of fatty acid metabolism in four transplantable rat hepatomas: Numbers 5123 C, 7793, 7795, and 7800. Cancer Res. **28**, 46–51 (1968).

SAGGERSON, E. D., GREENBAUM, A. L.: The regulation of triglyceride synthesis and fatty acid synthesis in rat epididymal adipose tissue. Effects of altered dietary and hormonal conditions. Biochem. J. **119**, 221–242 (1970).

SCHIMKE, R. T.: The importance of both synthesis and degradation in the control of arginase levels in rat liver. J. biol. Chem. **239**, 3808–3817 (1964).

SCHIMKE, R. T., DOYLE, D.: Control of enzyme levels in animal tissues. Ann. Rev. Biochem. **39**, 929–976 (1970).

SHRAGO, E., LARDY, H. A.: Paths of carbon in gluconeogenesis and lipogenesis. II. Conversion of precursors to phosphoenolpyruvate in liver cytosol. J. biol. Chem. **241**, 663–668 (1966).

SHRAGO, E., LARDY, H. A., NORDLIE, R. C., FOSTER, D. O.: Metabolic and hormonal control of phosphoenolpyruvate carboxykinase and malic enzyme in rat liver. J. biol. Chem. **238**, 3188–3192 (1963).

SILPANANTA, P., GOODRIDGE, A. G.: Synthesis and degradation of malic enzyme in chick liver. J. biol. Chem. **246**, 5754–5761 (1971).

SKOSEY, J. L.: Adrenocorticotropin stimulation of 2-ketoglutarate oxidation by isolated rat epididymal adipose tissue. Mediation by cyclic adenosine 3′, 5′-monophosphate and relationship to lipid metabolism. J. biol. Chem. **245**, 510–518 (1970).

SMITH, S., ABRAHAM, S.: Fatty acid synthesis in developing mouse liver. Arch. Biochem. Biophys. **136**, 112–121 (1970a).

SMITH, S., ABRAHAM, S.: Fatty acid synthetase from lactating rat mammary gland I. Isolation and properties. J. biol. Chem. **245**, 3209–3217 (1970b).

SMITH, S., ABRAHAM, S.: Fatty acid synthetase from lactating rat mammary gland, II. Studies on the termination sequence. J. biol. Chem. **246**, 2537–2542 (1971).

SPECTOR, A. A., JOHN, K., FLETCHER, J. E.: Binding of long-chain fatty acids to bovine serum albumin. J. Lipid Res. **10**, 56–67 (1969).

SPENCER, A. F., LOWENSTEIN, J. M.: The supply of precursors for the synthesis of fatty acids. J. biol. Chem. **237**, 3640–3648 (1962).

SPENCER, A. F., LOWENSTEIN, J. M.: Citrate and the conversion of carbohydrate into fat. Citrate cleavage in obesity and lactation. Biochem. J. **99**, 760–765 (1966).

SPENCER, A. F., LOWENSTEIN, J. M.: Citrate content of liver and kidney of rat in various metabolic states and in flouroacetate poisoning. Biochem. J. **103**, 342–348 (1967).

SRERE, P. A.: The molecular physiology of citrate. Nature (Lond.) **205**, 766–770 (1965a).

SRERE, P. A.: Palmityl-coenzyme A inhibition of the citratecondensing enzyme. Biochim. biophys. Acta (Amst.) **106**, 445–455 (1965b).

SRERE, P. A.: The citrate enzymes: Their structures, mechanisms, and biological functions. In: B. L. HORECKER and E. R. STADTMAN (eds.), Current topics in cellular regulation, vol. 5, p. 229–283. New York: Academic Press 1972.

SRERE, P. A., BHADURI, A.: Incorporation of radioactive citrate into fatty acids. Biochim. biophys. Acta (Amst.) **59**, 487–489 (1962).

SRERE, P. A., FOSTER, D. W.: On the proposed relation of citrate enzymes to fatty acid synthesis and ketosis in starvation. Biochem. biophys. Res. Commun. **26**, 556–561 (1967).

START, C., NEWSHOLME, E. A.: The effects of starvation and alloxandiabetes on the contents of citrate and other metabolic intermediates in rat liver. Biochem. J. **107**, 411–415 (1968).

STOLL, E., RYDER, E., EDWARDS, J. B., LANE, M. D.: Liver acetyl coenzyme A carboxylase: Activation of model partial reactions by tricarboxylic acids. Proc. nat. Acad. Sci. (Wash.) **60**, 986–981 (1968).

STUMPF, P. K.: Metabolism of fatty acids. Ann. Rev. Biochem. **38**, 159–212 (1969).

SULLIVAN, A. C., HAMILTON, J. G., MILLER, O.N., WHEATLEY, V. R.: Inhibition of lipogenesis in rat liver by (−)-hydroxycitrate. Arch Biochem. Biophys. **150**, 183–190 (1972).

SUZUKI, F., FUKUNISHI, K., DAIKUHARA, Y., TAKEDA, Y.: Studies on ATP citrate lyase of rat liver. II. Immunochemical properties. J. Biochem. **62**, 170–178 (1967).

SWANSON, R. F., CURRY, W. M., ANKER, H. S.: The activation of rat liver acetyl-CoA carboxylase by trypsin. Proc. nat. Acad. Sci. (Wash.) **58**, 1243–1248 (1967).

TAKETA, K., POGELL, B. M.: The effect of palmityl coenzyme A on glucose 6-phosphate dehydrogenase and other enzymes. J. biol. Chem. **241**, 720–726 (1966).

TARNOWSKI, W., SEEMANN, M.: Konzentrationsänderungen von Effektoren gluconeogenetischer Schlüsselenzyme in der Rattenleber unter gluconeogenetischen Bedingungen. Hoppe-Sylers Z. physiol. Chem. **348**, 829–838 (1967).

TAYLOR, C. B., BAILEY, E., BARTLEY, W.: Changes in hepatic lipogenesis during development of the rat. Biochem. J. **105**, 717–722 (1967).

TEPPERMAN, H. M., TEPPERMAN, J.: On the response of hepatic glucose-6-phosphate dehydrogenase activity to changes in diet composition and food intake pattern. Advanc. Enzyme Regul. **1**, 121–136 (1963).

TEPPERMAN, H. M., TEPPERMAN, J.: Patterns of dietary and hormonal induction of certain NADP-linked liver enzymes. Amer. J. Physiol. **206**, 357–361 (1964).

TEPPERMAN, J., TEPPERMAN, H. M.: Effects of antecedent food intake pattern on hepatic lipogenesis. Amer. J. Physiol. **193**, 55–64 (1958).

TERESI, J. D., LUCK, J.M.: The combination of organic anions with serum albumin. VIII. Fatty acid salts. J. biol. Chem. **194**, 823–834 (1952).

TUBBS, P. K., GARLAND, P. B.: Fatty acyl thio esters of coenzyme A: Inhibition of fatty acid synthesis in vitro and determination of levels in liver in normal, fasted, and fat- or sugar-fed rats. Biochem. J. **89**, 25P (1963).

TUBBS, P. K., GARLAND, P. B.: Variations in tissue contents of coenzyme A thio esters and possible metabolic implications. Biochem. J. **93**, 550–557 (1964).

TWETO, J., DEHLINGER, P., LARRABEE, A. R.: Relative turnover rates of subunits of rat liver fatty acid synthetase. Biochem. biophys. Res. Commun. **48**, 1371–1377 (1972).

TWETO, J., LARRABEE, A. R.: The effect of fasting on synthesis and 4'-phosphopantetheine exchange in rat liver fatty acid synthetase. J. biol. Chem. **247**, 4900–4904 (1972).

TWETO, J., LIBERATI, M., LARRABEE, A. R.: Protein turnover and 4'-phosphopantetheine exchange in rat liver fatty acid synthetase. J. biol. Chem. **246**, 2468–2471 (1971).

VAGELOS, P. R.: Lipid metabolism. Ann. Rev. Biochem. **33**, 139–172 (1964).

VAGELOS, P. R.: Regulation of fatty acid biosynthesis. In: B. L. HORECKER and E. R. STADTMAN (eds.), Current topics in cellular regulation, vol. 4, p. 119–166. New York: Academic Press 1971.

VAGELOS, P. R., ALBERTS, A. W., MARTIN, D. B.: Activation of acetyl-CoA carboxylase and associated alteration of sedimentation characteristics of the enzyme. Biochem. biophys. Res. Commun. **8**, 4–8 (1962).

VAGELOS, P. R., ALBERTS, A. W., MARTIN, D. B.: Studies on the mechanism of activation acetyl coenzyme A carboxylase by citrate. J. biol. Chem. **238**, 533–540 (1963).

VAGELOS, P. R., MAJERUS, P. W., ALBERTS, A. W., LARRABEE, A. R., AILHAUD, G. P.: Structure and function of the acyl carrier protein. Fed. Proc. **25**, 1485–1494 (1966).

VANCE, D., GOLDBERG, I., MITSUHASHI, O., BLOCH, K., OMURA, S., NOMURA, S.: Inhibition of fatty acid syntetases by the antibiotic cerulenin. Biochem. biophys. Res. Commun. **48**, 649–656 (1972).

WAITE, M.: The carboxylation of acetyl CoA by acetyl carboxylase. Fed. Proc. **21**, 287 (1962).

WAITE, M., WAKIL, S. J.: Studies on the mechanism of fatty acid synthesis. XII. Acetyl conezyme A carboxylase. J. biol. Chem. **237**, 2750–2757 (1962).

WAKIL, S. J.: A malonic acid derivative as an intermediate in fatty acid synthesis. J. Amer. chem. Soc. **80**, 6465 (1958).

WAKIL, S. J.: Lipid metabolism. Ann. Rev. Biochem. **31**, 369–406 (1962).

WAKIL, S. J., GOLDMAN, J. K., WILLIAMSON, I. P, TOOMEY, R. E.: Stimulation of fatty acid biosynthesis by phosphorylated sugars. Proc. nat. Acad Sci. (Wash.) **55**, 880–887 (1966).

WAKIL, S. J., TITCHENER, E. B., GIBSON, D. M.: Evidence for the participation of biotin in the enzymic synthesis of fatty acids. Biochim. biophys. Acta (Amst.) **29**, 225–226 (1958).

WATSON, J. A., FANG, M., LOWENSTEIN, J. M.: Tricarballylate and hydroxycitrate: Substrate and inhibitor of ATP: citrate oxaloactate lyase. Arch. Biochem. Biophys. **135**, 209–217 (1969).

WATSON, J. A., LOWENSTEIN, J. M.: Citrate and the conversion of carbohydrate into fat. Fatty acid synthesis by a combination of cytoplasm and mitochondria. J. biol. Chem. **245**, 5993–6002 (1970).

WEBER, G.: Study and evaluation of regulation of enzyme activity and synthesis in mammalian liver. Advanc. Enzyme Regul. **1**, 1–35 (1963).

WEBER, G., BANERJEE, G., BRONSTEIN, S. B.: Selective induction and suppression of liver enzyme synthesis. Amer. J. Physiol. **202**, 137–144 (1962).

WIELAND, O., NEUFELDT, I. E.: Zur Hemmung der Lipoidsynthese der Leber beim Diabetes. Biochem. Z. **337**, 349–359 (1963).

WIELAND, O., NEUFELDT, I., NUMA, S., LYNEN, F.: Zur Störung der Fettsäuresynthese bei Hunger und Alloxandiabetes. II. Fettsäuresynthese in der Leber alloxandiabetischer Ratten. Biochem. Z. **336**, 455–459 (1963).

WIELAND, O., WEISS, H.: Sur le métabolisme de l'acide acétique libre chez les animaux. Bull. Soc. Chim. biol. **46**, 223 (1964).

WILLIAMSON, J. R., HERCZEG, B., COLES, H., DANISH, R.: Studies on the ketogenic effect of glucagen in intact rat liver. Biochem. biophys. Res. Commun. **24**, 437–442 (1966).

WINEGRAD, A. I., RENOLD, A. E.: Studies on rat adipose tissue in vitro. II. Effects of insulin on the metabolism of specifically labeled glucose. J. biol. Chem. **233**, 273–276 (1958).

WISE, E. M., BALL, E. G.: Malic enzyme and lipogenesis. Proc. nat. Acad. Sci. (Wash.) **52**, 1255–1263 (1964).

YOUNG, J. W.: Effects of D- and L-thyroxine on enzymes in liver and adipose tissue of rats. Amer. J. Physiol. **214**, 378–383 (1968).

YOUNG, J. W., SHRAGO, E., LARDY, H. A.: Metabolic control of enzymes involved in lipogenesis and gluconeogenesis. Biochemistry **3**, 1687–1692 (1964).

YUGARI, Y., MATSUDA, T., SUDA, M.: Control of fatty acid synthesis by long-chain fatty acids in rat liver. Abstr. 6th Intern. Congr. Biochem., p. 602, New York, 1964.

ZAHLER, W. L., BARDEN, R. E., CLELAND, W. W.: Some physical properties of palmityl-coenzyme A micelles. Biochim. biophys. Acta (Amst.) **164**, 1–11 (1968).

Amino Acid Transmitters in the Mammalian Central Nervous System

DAVID R. CURTIS and GRAHAM A. R. JOHNSTON *

Content

1. Introduction and Scope of this Review 98
2. Evidence for Amino Acids as Transmitters 99
   2.1. Synthesis and Storage 100
   2.2. Synaptic Release 102
   2.3. Postsynaptic Action 102
   2.4. Postsynaptic Antagonists 103
   2.5. Inactivation and Removal 104
3. Neurochemistry and Neuropharmacology of Amino Acid Transmitters 106
   3.1. Glycine 107
   3.2. GABA 110
   3.3. Taurine 113
   3.4. Structurally Related Depressant Amino Acids 114
   3.5. Aspartate 116
   3.6. Glutamate 117
   3.7. Structurally Related Excitant Amino Acids 119
4. Anatomical Organization of Amino Acid Transmitters 120
   4.1. Cerebral Cortex 120
      4.1.1. Excitation 120
      4.1.2. Inhibition 122
   4.2. Cerebellum 127
      4.2.1. Excitation 127
      4.2.2. Inhibition 127
   4.3. Rhinencephalon 130
      4.3.1. Excitation 130
      4.3.2. Inhibition 130
   4.4. Retina 131
   4.5. Olfactory Bulb 132
   4.6. Thalamus 134
      4.6.1. Neurochemistry 134
      4.6.2. Excitation 135
      4.6.3. Inhibition 136
   4.7. Hypothalamus 137
   4.8. Basal Ganglia and Red Nucleus 137
      4.8.1. Substantia Nigra 137
      4.8.2. Caudate and Other Nuclei 139
      4.8.3. Red Nucleus 140
   4.9. Vestibular Nuclei 140
      4.9.1. Neurochemistry 140
      4.9.2. Inhibition 141
   4.10. Brain Stem 142
      4.10.1. Cranial Motor Nuclei 142

* Department of Pharmacology, John Curtin School of Medical Research, Australian National University, Canberra, Australia

4.10.2. Reticular Formation 144
4.11. Dorsal Column Nuclei 145
4.11.1. Excitation 145
4.11.2. Inhibition 146
4.12. Spinal Cord 147
4.12.1. Neurochemistry 147
4.12.1.1. Aspartate 150
4.12.1.2. Glutamate 150
4.12.1.3. Glycine 150
4.12.1.4. GABA 152
4.12.1.5. Other Amino Acids 153
4.12.2. Excitation 153
4.12.3. Inhibition 155
4.12.3.1. Glycine 155
4.12.3.2. GABA 159
4.13. Non-Central Neurones 161
4.13.1. Dorsal Root Ganglia 161
4.13.2. Autonomic Ganglia 162
5. Summarizing and Concluding Remarks 163
5.1. Glycine 163
5.2. GABA 164
5.3. Aspartate and Glutamate 164
5.4. Other Amino Acids 164
5.5. Concluding Remarks 164
References 165

1. Introduction and Scope of this Review

Two decades ago insufficient information was available for amino acids to be seriously considered as synaptic transmitters in the mammalian central nervous system (CNS). Even as recently as ten years ago amino acids were regarded as mainly concerned in the brain with intermediary metabolism (ELLIOTT, PAGE, and QUASTEL, 1962), although considerable significance had already been attached to γ-aminobutyric acid (GABA) as a possible vertebrate and invertebrate inhibitory transmitter, a subject to which a whole symposium was devoted in 1959 (ROBERTS, 1960).

Largely as a consequence of increasing understanding of neurochemistry, neuroanatomy, neurophysiology and neuropharmacology, and advances in techniques at both the multi- and single neurone level, a stage has now been reached where the metabolism of glycine, GABA, aspartate and glutamate within nervous tissue might even be considered primarily of importance in relation to their synthesis and inactivation as transmitters. At the present time there may be, in fact, sufficient evidence to justify the use of terms such as glycinergic, gabergic, aspartergic and glutamergic to indicate neurones and terminals which release these amino acids, and synaptic processes in which they participate. These terms, or perhaps more euphonious versions of them, are required for the same reasons proposed 40 years ago by DALE when introducing "cholinergic" and "adrenergic", some 29 years after the concept of chemical transmission was put forward by ELLIOTT: "Such a usage would assist clear thinking without committing us to precise chemical identification, which may be long in coming" (DALE, 1933). There may even be a need for less specific terms than those above which indicate the participation of particular amino acids, since groups of closely related amino acids may activate the same "receptors" of a number of membrane and enzyme processes, and the onus of proof for transmitter function becomes eventually a chemical problem requiring techniques applicable at the synaptic level.

Aspects of central amino acid transmission have been reviewed previously from both the neurochemical and neuropharmacological point of view (ROBERTS, 1956, 1960, 1968, 1971; HAYASHI, 1959, 1966; ELLIOTT and JASPER, 1959; ROBERTS and EIDELBERG, 1960; GRUNDFEST, 1960; FLOREY, 1964; CURTIS and WATKINS, 1965; CURTIS and CRAWFORD, 1969; CURTIS, 1970; CURTIS and JOHNSTON, 1970; APRISON and WERMAN, 1968; KRNJEVIĆ, 1970; BLOOM, 1972; HAMMERSCHLAG and WEINREICH, 1972; IVERSEN, 1972; JOHNSON, 1972a; OBATA, 1972b; OTSUKA, 1972; SNYDER, LOGAN, BENNETT, and ARREGUI, 1972; SYTINSKY, 1972; WERMAN, 1972). The purpose of this present review is to consider the types of evidence upon which claims can be based for suggesting an amino acid as the transmitter for a certain type of synapse (Section 2), to discuss the general aspects of the metabolism and neuropharmacology of particular amino acid transmitters (Section 3), and to provide an analysis of those pathways and neurones for which there is evidence that amino acids could be transmitters (Section 4).

Drug induced alterations in the activity of the CNS associated with changed levels of certain amino acids have not been considered in detail. In view of the complex metabolic inter-relationships between amino acid transmitters (WATKINS, 1972), and the impossibility of determining the distribution of any one between various metabolic and anatomical compartments, it is difficult to ascertain whether alterations in the level of an amino acid cause or result from abnormal neuronal activity. Nevertheless the importance of both inhibitory and excitatory amino acids in these abnormal states, and in a number of neurological disorders, must not be underestimated.

Much of the subsequent text is concerned with the central nervous system of the cat, since most neurophysiological and neuropharmacological investigations, and an increasing number of neurochemical studies concerned with amino acid transmitters, have been carried out in this animal. There are, however, no major reasons for suspecting that conclusions made about the cat nervous system are not directly applicable to other mammals, including man. Reference will be made to other species when information of the particular type is either not available or reinforces findings made in the cat.

2. Evidence for Amino Acids as Transmitters

The evidence for certain amino acids as synaptic transmitters includes information about the synthesis and storage within presynaptic nerve cells, the release from presynaptic terminals, the interaction with receptors on the postsynaptic neurones and the consequent transient alterations in ionic permeability of the subsynaptic membrane, the ability of certain substances to antagonise both amino acid- and synaptically-induced postsynaptic effects, and the processes associated with the inactivation and removal of amino acids from the synaptic environment. Experimental difficulties abound in all aspects of these investigations and, as discussed below, necessitate compromises regarding the rigid application of lists of criteria for transmitter identification which were drawn up largely on the basis of experience regarding the transmitter function of acetylcholine at vertebrate neuromuscular and ganglionic synapses.

2.1. Synthesis and Storage

Any transmitter must occur in the appropriate presynaptic terminals. It may be synthesised in, or accumulated by, these terminals or transported to the terminals from cell bodies. An appreciable presynaptic store of transmitter may not be required if the transmitter can be synthesised rapidly and efficiently on demand. In addition, since most amino acids have a variety of intracellular functions, a localised presynaptic store of "transmitter" amino acid may not be readily apparent. It may be possible, however, to establish an association between an amino acid, and/or its synthesising enzyme, with the fibres and terminals of a particular pathway. Measurement of turnover rates may provide more useful information than that of steady state levels of amino acids.

Amino acid levels are usually determined by conventional chemical and biochemical procedures (TALLAN, MOORE and STEIN, 1954; SHAW and HEINE, 1965; BAXTER, 1972), but enzyme recycling (OTSUKA and MIYATA, 1972), gas-liquid chromatography (SHIMADA, KURIMOTO, WADA, HINO, SUGINOSHITA, and KIHARA, 1971), and methods involving the formation of radioactive derivatives (e.g. dansyl derivatives, BRIEL and NEUHOFF, 1972; SNODGRASS and IVERSEN, 1973a) are finding increasing application to smaller pieces of tissue. Postmortem changes have been observed in the levels of most amino acids (TEWS, CARTER, ROA, and STONE, 1963; SHANK and APRISON, 1971; PERRY, HANSEN, BERRY, MOK, and LESK, 1971b; PERRY, SANDERS, HANSEN, LESK, KLOSTER, and GRAVLIN, 1972), and the figures usually reported are probably overestimates of those *in vivo*. When available biopsy figures are included in the Tables.

Regional Studies. Regional variations in the levels of amino acids and the activities of related enzymes and transport systems may provide indications of transmitter function: in fact the unusually high levels of glycine in the spinal grey matter prompted investigation of this amino acid as a spinal transmitter (APRISON, DAVIDOFF, and WERMAN, 1970). Amino acid analyses have been performed on single neurones of a particular class isolated from nervous tissue, or on small blocks of nervous tissue (OTSUKA and MIYATA, 1972). Determination of the distribution of amino acid synthesising enzymes may offer certain advantages (FONNUM, 1972), moreover histochemical methods may lead to direct visualization of enzyme location (HÖKFELT and LJUNGDAHL, 1972c). The localization and even the concentration of amino acids and enzymes within nerve terminals can be assessed by comparing tissue levels and morphology before and after section of the appropriate nervous pathway (see FONNUM and WALBERG, 1973).

Autoradiographic procedures, at the light and electron microscope level, have been used to study the distribution of amino acid transport systems, and hence that of certain endogenous amino acid compartments, under both *in vitro* and *in vivo* conditions. Such methods are of considerable value, especially when the radiolabel is confined to terminals or neurones of one particular type (MATUS and DENNISON, 1972; HÖKFELT and LJUNGDAHL, 1972a, b, c). Labelled amino acids have also been used to study dendritic and axonal transport (GLOBUS, LUX, and SCHUBERT, 1968; SCHUBERT, KREUTZBERG, and LUX, 1972), the latter being useful in analyses of fiber projections.

Subcellular Studies. Considerable use has been made of the preparation of pinched-off nerve endings, "synaptosomes", from CNS tissue using carefully controlled homogenization and centrifugation techniques (DE ROBERTIS, 1968; WHITTAKER, 1968; WHITTAKER and BARKER, 1972; COTMAN, 1972). Synaptosomes cannot be isolated in pure form, being contaminated in particular with glial fragments (COTMAN, HERSCHMAN, and TAYLOR, 1971), and the endings of a particular pathway cannot be readily identified. Furthermore, during the preparative stage there may be a redistribution of soluble components between intra- and extra-cellular phases (WHITTAKER, 1965; NEAL and IVERSEN, 1969; TACHIKI, DE FEUDIS, and APRISON, 1972). Nevertheless these subcellular particles have been used to study the synthesis, storage, uptake and release of amino acid transmitters, synaptosomes presumably associated with different amino acids being separable on the basis of density differences (SNYDER, LOGAN, BENNETT, and ARREGUI, 1972).

Studies on such contaminated and frequently morphologically unidentified "synaptosomal preparations" have yielded some supportive information relevant to amino acid transmitter function: the synthesising enzymes for GABA and taurine appear to be associated with these preparations as do active transport systems for L-glutamate/L-aspartate, glycine and GABA. It should be emphasised, however, that these studies do not usually distinguish between neuronal and glial elements, and that the results obtained should be compared with those obtained by other methods, such as autoradiography at the electron microscope level, preferably using more intact tissue preparations.

Inhibitors. Evidence for the synthesis and/or storage of a compound at a particular synapse may be provided by the use of inhibitors of transmitter synthesis and/or storage in a relatively specific manner, for example by limiting the access of a precursor to the site of synthesis, or by inhibiting an essential enzyme. With respect to amino acid transmitters evidence of this type so far has been obtained only for inhibitors of glutamic acid decarboxylase (GAD), relating the fall in GABA levels to changes in the operation of GABA-releasing synaptic pathways, although the inhibitors used are far from selective for this particular enzyme.

Studies of this type have not been carried out on single neurones using electrophoretic administration of an amino acid synthesis antagonist. Since there are probably substantial transmitter stores within presynaptic terminals, high frequency activation of the pathway of interest for prolonged periods may be required to exhaust the store. Furthermore it is possible that enzyme inhibitors active in homogenates may not readily gain access to enzyme systems contained within presynaptic terminals *in vivo*.

2.2. Synaptic Release

Largely because of difficulties in the collection of measurable amounts the investigation of the synaptic release of amino acids *in vivo* has not proved easy. Collection can only be remote from sites of release within a complex tissue, either from the surface of the brain or cord, from the ventricles or spinal central canal, from the venous effluent or from artificial tissue spaces created at the tip of a "push-pull" cannula. Since the extracellular concentrations of amino acid transmitters must normally be maintained by tissue mechanisms at a low level, the released amino acid requires "protection" during its diffusion through nervous tissue. However the major method of transmitter inactivation appears to be cellular uptake (Section 2.5), and no specific antagonists have been found of the uptake of any of the putative amino acid transmitters which do not have either excitant or depressant effects on neurones. Hence success in demonstrating the release of a particular amino acid by impulses in a certain pathway seems difficult.

Some progress has been made in this direction using a relatively non-specific antagonist of amino acid uptake (JORDAN and WEBSTER, 1971), inhibitors of amino acid metabolising enzymes which may also indirectly modify uptake (OBATA and TAKEDA, 1969; IVERSEN, MITCHELL, and SRINIVASAN, 1971), or relatively non-physiological stimuli which possibly "overwhelm" transmitter inactivating mechanisms. Use has been made of the measurement of the efflux of substances other than transmitters preloaded into the tissue to ascertain whether enhanced release is not merely a consequence of increased metabolic activity of stimulated structures. Additionally, diminished release of a possible transmitter in the absence of calcium ions in the perfusing medium is usually advanced as evidence of synaptic release (RUBIN, 1970), neglecting the complex effects of such a solution on the activity of synaptically interconnected neurones.

Results from these investigations provide, at the most, indirect evidence of synaptic release of amino acid by impulses in particular pathways. The *in vivo* release of endogenous GABA, taurine, L-glutamate and L-aspartate has been observed to be, at least qualitatively, correlated with stimulation of certain pathways. The evoked efflux of preloaded labelled (i. e. exogenous) GABA and glycine has also been studied *in vivo*, but equating such release with that of the endogenous material is complicated by metabolic compartmentation (BERL and CLARKE, 1969; BENJAMIN and QUASTEL, 1972; VAN DEN BERG, 1973) of the particular amino acid. *In vitro* procedures involve stimulation of tissue slices or crude synaptosomal preparations with electric fields or high potassium ion concentrations (BRADFORD, 1972): these procedures may be useful for studying the ionic requirements of release and the effects of drugs, but considerable caution is necessary when relating the results to *in vivo* conditions.

2.3. Postsynaptic Action

The absolute necessity for establishing that a suspected transmitter has postsynaptic excitatory or inhibitory effects on central neurones identical to that of a synaptically released transmitter has been discussed previously in detail (WERMAN, 1966; CURTIS and JOHNSTON, 1970).

Although microelectrophoretic methods (CURTIS, 1964; KRNJEVIĆ, 1971) permit ready detection of excitation or depression of neuronal firing, the comparison of alterations of membrane properties induced synaptically (ECCLES, 1966) with those resulting from an artificially administered agent is difficult. It is essential that the synaptic action used in the comparison is purely excitatory or inhibitory in nature, preferably generated by volleys in one type of fibre converging on the neurones being studied. Intracellular recording is required for measurement of 'reversal' potentials (GINSBORG, 1967), and for determination of the effects of altered electrochemical gradients of ions possibly involved in the synaptic process. The correct interpretation of such analyses requires the activation by the transmitter and the administered amino acid of the same regions of cell membrane in relation to the recording microelectrode. The possibility must also be considered that the synaptic process being investigated may not directly involve changes in membrane ion permeability but alters the operation of ion pumps or intracellular metabolic processes (WEIGHT, 1971).

In addition to the many neurones from which prolonged intracellular recording is impossible, there remains a virtually insurmountable problem of how to investigate synaptic and amino acid induced changes in membrane properties at dendritic sites remote from the usually intrasomatic site of recording (see DIAMOND, 1968). It is fortuitous that the inhibitory synapses which are of considerable interest from the point of view of amino acid transmitters are located predominantly on the soma of neurones, but the study of excitatory synaptic mechanisms is complicated by the dendritic location of synapses (see SZENTAGOTHAI, 1971). The dendritic location of synapses is also a problem in relation to the investigation of remote or "presynaptic" inhibition (SCHMIDT, 1971). The effects of amino acids on primary afferent terminals have been studied *in vivo,* and by using isolated amphibian spinal cords: results and conclusions from these *in vitro* studies may not be completely applicable to the mammalian central nervous system.

2.4. Postsynaptic Antagonists

In the absence of methods which can be used with all types of neurone for comparing the actions of administered amino acids and synaptically released transmitters, considerable use has been made of amino acid antagonists, substances which selectively block the postsynaptic effects of particular amino acids. Such antagonism, together with the diminution of an appropriate synaptic process considered on other grounds to involve a certain amino acid, has been used as a basis for considering the participation of the amino acid at other synapses also sensitive to the antagonist. Thus antagonists, administered near single neurones or systemically, have proved convenient and useful in confirming the nature of the transmitter at certain synapses, or in a preliminary survey of synaptic events prior to detailed neurochemical and physiological studies.

Due consideration must be given to the fact that antagonism usually extends to close structural analogues of amino acid transmitters, that the one amino acid may have different antagonists at different receptors, and that an antagonist of an amino acid in one species may not necessarily be effective in another. Care is also required to exclude the formation of relatively stable amino acid/antagonist complexes as an explanation of antagonism. Although it may not be possible by microelectrophoretic methods alone to establish the mechanism of antagonism (CURTIS, DUGGAN, and JOHNSTON, 1971 c), the structure of antagonists may provide some evidence regarding the nature of amino acid receptors (SMYTHIES, 1971). Both this type of study, and proposals regarding the nature of the transmitter, are strengthened when there is a direct correlation between the potencies as amino acid antagonists of a series of substances of similar or different structure and their effectiveness in suppressing synaptic action.

One major problem concerned with the use of antagonists is the method of administration, both for establishing the mechanism and degree of selectivity of the antagonism, and for blocking synaptic transmission. A highly selective antagonist ejected from one orifice of a multibarrel micropipette may readily block the effect of one amino acid and not of others ejected from adjacent orifices, yet may apparently have little action on the effect of the same amino acid released synaptically because the antagonist gains access to too few synapses (CURTIS et al., 1971 c; CURTIS, 1971). On the other hand, if the antagonist is administered systemically, and local concentrations are sufficient to significantly modify synaptic transmission, these may not be adequate to influence the effects of localized and relatively high concentrations of electrophoretically administered amino acids. Potentially useful antagonists may not penetrate the blood brain barrier, and the dose of a systemically administered compound may be severely limited by general effects on the experimental animal and actions at synapses other than those being studied. Neither method of administration is ideal for determining the selectivity of an antagonist in such a complex tissue as the CNS. The estimations can best be considered no more than indications of the probability that the suppression of a synaptic event involves the participation of one amino acid, and not another, as the transmitter.

Accepting these restrictions, strychnine and some related compounds are useful antagonists at sites where glycine is the transmitter, and bicuculline and picrotoxinin, where GABA is involved. These and other agents are discussed in more detail in subsequent sections. At the present time no satisfactory antagonist has been found which distinguishes between aspartate and glutamate as neuronal excitants, and the specificity of those agents proposed as selective antagonists of excitant amino acids seems doubtful.

2.5. Inactivation and Removal

There is no evidence to indicate that amino acids are inactivated enzymically in the extracellular synaptic environment, and a number of observations suggest that carrier mediated cellular uptake (see WYSSBROD, SCOTT, BRODSKY, and SCHWARTZ, 1971), which may involve membrane bound enzymes as binding sites (MEISTER, 1973), might be responsible for limiting the immediate action of amino acids in the subsynaptic region and for preventing accumulation of amino acids in the extraneuronal space (IVERSEN and NEAL, 1968; CURTIS, DUGGAN, and JOHNSTON, 1970b; SNYDER et al., 1972). Uptake by synaptic terminals could provide a supply of transmitter for re-use, there are indications however that glial uptake can also take place (HENN and HAMBERGER, 1971). If local inactivation of amino acid transmitters occurs, selective inhibition of the uptake of an amino acid should lead to enhancement and prolongation of its effects after both synaptic release and artificial administration. This has yet to be achieved in a selective fashion for any of the amino acid transmitters, largely because close structural analogues which are themselves transported, so inhibiting the accumulation of the putative transmitter, have postsynaptic actions similar to that of the transmitter.

The uptake of radiolabelled amino acids has been studied using tissue slices or particulate matter in homogenates (COHEN and LAJTHA, 1972). Such *in vitro* methods, convenient for studying kinetics of the process and to assess the action of possible antagonists, may not, however, be directly applicable to processes occuring *in vivo*, particularly as CNS tissue has complex metabolic, ionic and organizational requirements. Thus analyses of the cellular distribution of radio labelled amino acids after direct *in vivo* injection (see Section 2.2) offer a better indication of the sites of uptake (HÖKFELT and LJUNGDAHL, 1972b).

Structurally specific uptake systems (COHEN and LAJTHA, 1972) fall into two groups on the basis of their kinetic parameters, subcellular and regional distributions. The first group is characterised by K_m's higher than $5 \times 10^{-4} M$, and appear to be similar to those found in other than CNS tissues. The second group has K_m's lower than $5 \times 10^{-5} M$: these "high affinity" systems, which are sodium dependent, have been found specific for the amino acids considered most likely to be transmitters, GABA, glycine, taurine and L-glutamate/L-aspartate, and appear to be associated with distinct populations of subcellular particles — nerve terminals and/or glial elements.

The cellular accumulation or uptake of amino acids is accompanied by a movement of ions across the membrane (CHRISTENSEN, 1970, 1972), but it has become apparent that this process differs in several respects from the ionic permeability increase responsible for excitation or inhibition by amino acids. In the first place, strychnine, picrotoxinin and bicuculline interfere with the inhibitory actions of glycine and GABA, but have not been found to modify the uptake of these amino acids by nervous tissue (Cat: BALCAR and JOHNSTON, 1973). Secondly, although groups of structurally similar amino acids are transported by common carrier mechanisms (COHEN and LAJTHA, 1972), and to some extent the structural requirements seem similar to those necessary for interaction with receptors for excitation or inhibition (CURTIS and WATKINS, 1965), a number of potent agonists of amino acid action are weak or relatively ineffective inhibitors of the uptake of the related naturally occuring amino acids. Thus D-homocysteate and N-methyl-D-aspartate are unlikely to be transported by the L-glutamate/aspartate system (BALCAR and JOHNSTON, 1972a), yet both are potent excitants (CURTIS and WATKINS, 1965). Additionally 3-aminopropane sulphonate has little effect on GABA uptake yet is a potent GABA-like depressant (BEART and JOHNSTON, 1973b). A direct relationship would be expected between uptake and postsynaptic action if excitation or inhibition by an amino acid directly involved its transport across the neuronal membrane. Finally, organic mercurials and particularly *p*-chloromercuriphenylsulphonate, enhance amino acid excitation and inhibition but block uptake (CURTIS et al., 1970b; BALCAR and JOHNSTON, 1973).

3. Neurochemistry and Neuropharmacology of Amino Acid Transmitters

This section outlines in general terms neurochemical and neuropharmacological aspects of the amino acid transmitters, and their related enzymes and transport systems.

Table 1 lists concentrations of the important amino acids in the plasma, cerebrospinal fluid and selected portions of the nervous system of the cat. Particularly noteworthy are the extremely low plasma and cerebrospinal fluid (CSF) levels of GABA, the low cerebrospinal levels of all amino acids considered as possible neurotransmitters (see also Human: GJESSING, GJESDAHL and SHAASTAD, 1972), the high concentrations of glutamate in the cerebellar and cerebral cortices, the high level of cerebellar taurine and the high spinal levels of glycine. These and other findings will be discussed in more detail in the subsequent text.

One important feature common to all of these amino acids is the concept of a blood-brain barrier: a barrier which hinders the free diffusion of a compound from the plasma to the extracellular fluid of the brain and cord, a fluid presumably

Table 1. Amino acid concentrations in cat plasma, cerebrospinal fluid, cerebral cortex, cerebellum and spinal cord (n.m.a. = no measurable amount)

	Aspartate	Glutamate	GABA	Glycine	Taurine	α-Alanine	Reference
Plasma (μmole/ml)		0.1		0.4		0.5	SNODGRASS et al. (1969)
	0.02	0.06	0.002–0.004	0.2		0.2	CROWSHAW et al. (1967)
Cerebrospinal fluid (μmole/ml)		0.04	0.01	0.02		0.05	SNODGRASS et al. (1969)
	0.02	0.06	0.002–0.004	0.2		0.2	CROWSHAW et al. (1967)
	0.005	0.03	n.m.a.	0.01	0.02	0.09	JASPER and KOYAMA (1969)
Cerebral cortex (μmole/g)	3.1	12.9	1.4	1.3	1.9	0.9	BATTISTIN et al. (1969)
Cerebellum (μmole/g)	2.9	12.6	1.5	1.5	3.1	0.9	BATTISTIN et al. (1969)
Spinal grey matter, mean of dorsal and ventral (μmole/g)	2.6	5.9	1.6	6.4			GRAHAM et al. (1967)
	4.4	5.2		6.0	0.1	1.3	JOHNSTON (1968)

of similar composition to cerebrospinal fluid (DAVSON, 1969). Single passage studies of the cerebral circulation indicate that glycine, GABA, aspartate and glutamate penetrate very little when blood levels are elevated in adult mammals, the barriers being less effective in immature animals (Rat: OLDENDORF, 1971. Dog: YUDILEVICH, DE ROSE, and SEPÚLVEDA, 1972. Mouse: SETA, SERSHEN, and LAJTHA, 1972). The movement of amino acids between blood and brain is, however, complicated by transport of both amino acids and precursors (RICHTER and WAINER, 1971), and by the anatomical and biochemical compartmentation of amino acids and metabolites within this tissue. From a practical point of view these processes are of significance only in relation to whether or not systemic administration of a particular amino acid or precursor can alter brain levels sufficiently to modify its transmitter function.

3.1. Glycine

Glycine is a zwitterion at neutral pH with pK values of 2.33 and 9.53 at 35° (KING, 1951). The rotational isomerism of glycine in solution has been studied by proton magnetic resonance spectroscopy (CAVANAUGH, 1967) and by potential energy calculations (PONNUSWAMY and SASISEKARAN, 1970).

Within the CNS, levels of glycine in excess of 3 μmole/g are found only in the medulla and spinal cord (Section 4.10 and Table 8). Intraperitoneal administration of glycine (13 μmole/g = 1000 mg/kg) raises the central levels of both glycine and serine (Rat: RICHTER and WAINER, 1971). The flux of glycine from the blood into adult rat CNS has been estimated to be of the order of 0.1 μmole/g/h (SHANK and APRISON, 1970), and to be 10 to 60 times higher in immature rats (BAÑOS, DANIEL, MOORHOUSE, and PRATT, 1971). Glycine has the lowest CSF to plasma ratio of the amino acids found in cat CSF (Table 1), perhaps due to a mediated transport of glycine from the CSF to the blood (Cat: MURRAY and CUTLER, 1970a, b; HENSCHEN and SÖDERBERG, 1968).

Glycine Metabolism. Glycine metabolism in the CNS is complex and poorly understood. Glycine synthesis from glucose appears to be via D-3-phosphoglycerate and/or D-glycerate (Fig. 1). Glycine inhibits D-glycerate dehydrogenase activity (EC. 1.1.1.29), but not D-3-phosphoglycerate dehydrogenase activity in extracts of rat brain, in a non-competitive manner suggestive of end-product inhibition (UHR and SNEDDON, 1971), and the regional distribution of this enzymic activity, but not that of D-3-phosphoglycerate dehydrogenase, in the cat CNS correlates with that of glycine (UHR and SNEDDON, 1972). These observations indicate that the rate controlling step in glycine metabolism may be the conversion of D-glycerate to hydroxypyruvate in the "non-phosphorylated" pathway. Two major possibilities exist for the production of glycine from hydroxypyruvate. The more obvious is that via serine by means of the enzymes hydroxypyruvate: α-alanine (glutamate/glutamine) transaminase and serine hydroxymethyltransferase (EC. 2.1.2.1). The required transaminase activity is, however, very low or absent in brain (Dog, ox: WALSH and SALLACH, 1966). The corresponding enzyme for the "phosphorylated" pathway, phosphohydroxypyruvate: glutamate transaminase is found in brain (Sheep: HIRSCH and GREENBERG, 1967), and there

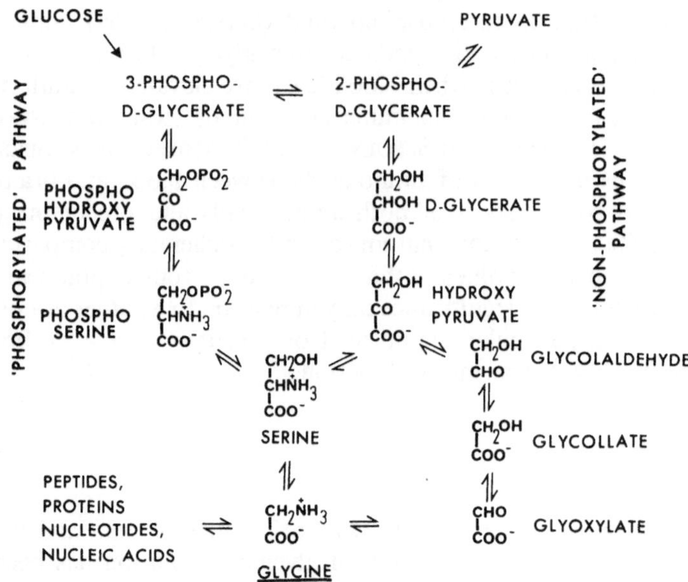

Fig. 1. Possible routes of glycine metabolism in the CNS

is evidence that serine is synthesised in the brain via this "phosphorylated" pathway with the feedback inhibition by serine of phosphoserine phosphatase being a major control point (Mouse: BRIDGERS, 1969).

Serine hydroxymethyltransferase has been purified from brain (Ox: BRODERICK, CANDLAND, NORTH, and MANGUM, 1972), and aspects of its development (Mouse: BRIDGERS, 1968), regional and subcellular distribution studied (Cat, rat: DAVIES and JOHNSTON, 1973). Glycine and serine are readily interconvertible *in vivo* in the CNS (Rat: SHANK and APRISON, 1970). An alternative pathway from hydroxypyruvate to glycine, not involving serine, is that via glycolaldehyde, glycollate and glyoxylate. The non-oxidative decarboxylation of hydroxypyruvate to glycolaldehyde by extracts of beef brain has been reported (HENDRICK and SALLACH, 1964), and the glycine-forming transamination of glyoxylate catalysed by extracts of cat and rat CNS tissue has been described (JOHNSTON and VITALI, 1969a, b; JOHNSTON, VITALI, and ALEXANDER, 1970; BENUCK, STERN, and LAJTHA, 1972). The conversion of glycolaldehyde to glyoxylate has however, not yet been reported. Experiments with $[1-^{14}C]$- and $[3,4-^{14}C]$ glucose indicate that *in vivo* labelling of the carbon atoms of glycine is some five times faster for a route via serine (presumably the "phosphorylated" pathway) than that via glyoxylate (Rat: SHANK, APRISON, and BAXTER, 1973), but these results provide little information about the net synthesis of glycine via either route.

Glycine is incorporated in the CNS, as in other tissues, into proteins, various peptides such as glutathione, and nucleotides. A number of peptide hydrolases capable of metabolising glycine peptides have been demonstrated in CNS tissues (MARKS, 1970). While glycine could be catabolised by oxidation (Cat, rat, sheep: DE MARCHI and JOHNSTON, 1969) or by oxidative decarboxylation (Cat, rat: UHR, 1973), it appears that glycine in nervous tissue is not readily converted into intermediates of either the tricarboxylic acid cycle or glycolysis.

Glycine Transport. At least two kinetically distinct transport systems mediate glycine uptake in the CNS: a "low affinity" system (K_m approx. 10^{-4} *M*), which is shared by other small neutral amino acids and is found in all areas of the CNS (Rat: SMITH, 1967. Mouse: BLASBERG, 1968), and a "high affinity" system (K_m approx. 10^{-5} *M*), which is specific for glycine, confined to the spinal cord, pons and medulla, and shows an absolute dependence on external sodium ions (Rat: NEAL, 1971; JOHNSTON and IVERSEN, 1971; LOGAN and SNYDER, 1972; BENNETT, LOGAN, and SNYDER, 1972. Cat: BALCAR and JOHNSTON, 1973). There is evidence from subcellular fractionation and autoradiographic studies to support a predominantly presynaptic location for the high affinity specific glycine uptake system, though glial components may also be important (Rat: IVERSEN and JOHNSTON, 1971; MATUS and DENNISON, 1972; ARREGUI, LOGAN, BENNETT, and SNYDER, 1972; HÖKFELT and LJUNGDAHL, 1972c). A variety of centrally-active drugs inhibit the high affinity specific glycine uptake system including chlorpromazine, imipramine and haloperidol (Rat: JOHNSTON and IVERSEN, 1971). *p*-Chloromercuriphenylsulphonate inhibits glycine uptake into slices of cat spinal cord (BALCAR and JOHNSTON, 1973) and, when administered microelectrophoretically, potentiates the depressant action of glycine on the firing of cat spinal neurones (CURTIS et al., 1970b).

Glycine Release. Evoked release of exogenous glycine has been demonstrated *in vivo* from cat spinal cord (JORDAN and WEBSTER, 1971), and *in vitro* from slices of rat spinal cord (HOPKIN and NEAL, 1971; HAMMERSTAD, MURRAY, and CUTLER, 1971) and isolated amphibian cords (APRISON, 1970; ROBERTS and MITCHELL, 1972). In the spinal cord tetanus toxin blocks glycine mediated inhibition probably by reducing the release of transmitter (Section 4.12.3.1).

Postsynaptic Action of Glycine. Intracellular studies of the mechanism of the depression by glycine of the firing of neurones in the spinal cord, medulla and cerebral cortex show that glycine hyperpolarises these neurones and increases the membrane conductance (Cat: WERMAN, DAVIDOFF, and APRISON, 1968; CURTIS, HÖSLI, JOHNSTON, and JOHNSTON, 1968b; BRUGGENCATE and ENGBERG, 1968, 1971; KELLY and KRNJEVIĆ, 1969; HÖSLI and HAAS, 1972). The ionic basis of the effects of glycine on spinal and Deiters' neurones cannot be distinguished from synaptically-induced inhibition by measurements of reversal potentials and the effects of intracellular injection of a series of anions and cations.

Glycine Antagonists. A variety of compounds (Fig. 2) selectively antagonise the depressant action of glycine on neurones (Section 4.12.3.1.). Strychnine is the most potent and the most widely investigated of these compounds; others include the alkaloids brucine, dendrobine, diaboline, gelsemine, laudanosine, morphine and thebaine, and the synthetic compounds N,N-dimethylmuscimol, levorphan, N-methylbicuculleine, and 4-phenyl-4-formyl-N-methylpiperidine (Cat: CURTIS, HÖSLI, and JOHNSTON, 1968a; CURTIS and DUGGAN, 1969; CURTIS, DUGGAN, FELIX, and JOHNSTON, 1971a; JOHNSTON, BEART, CURTIS, GAME, MCCULLOCH, and MACLACHLAN, 1972; CURTIS and JOHNSTON, unpublished observations).

Strychnine also antagonises the depressant actions on spinal neurones of α- and β-alanine, cystathionine, serine and taurine, which on this basis have been called "glycine-like" amino acids, but does not influence the actions of

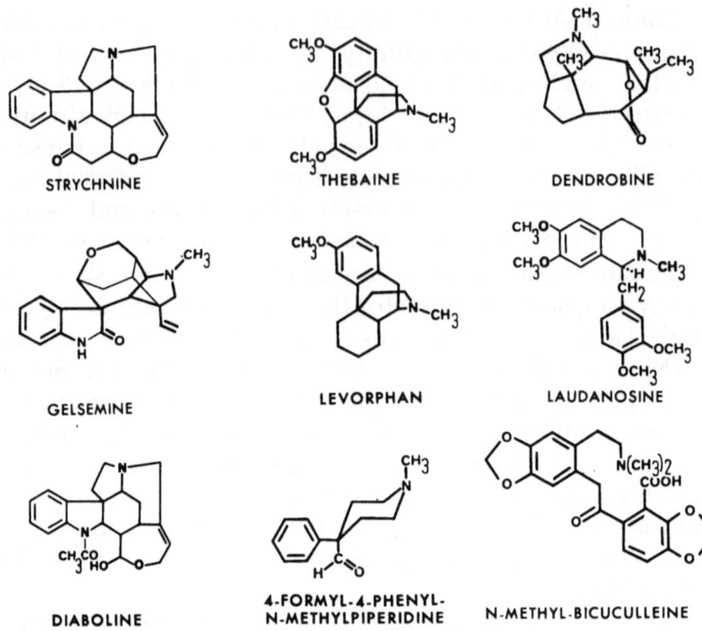

STRYCHNINE THEBAINE DENDROBINE

GELSEMINE LEVORPHAN LAUDANOSINE

DIABOLINE 4-FORMYL-4-PHENYL-N-METHYLPIPERIDINE N-METHYL-BICUCULLEINE

Fig 2. Antagonists of the postsynaptic action of glycine

the "GABA-like" amino acids, GABA, γ-amino-β-hydroxybutyric acid, 3-aminopropanesulphonic acid, δ-aminovaleric acid and ε-aminocaproic acid (Cat: CURTIS et al., 1968a). The antagonism of glycine by strychnine may be competitive in nature (Cat: JOHNSON, ROBERTS and STRAUGHAN, 1970; CURTIS et al., 1971c).

3.2. GABA

GABA is a flexible molecule: X-ray crystallography indicates that it exists in a partially folded conformation in the solid state (TOMITA, 1971; STEWARD, PLAYER, QUILLIAM, BROWN, and PRINGLE, 1971), while proton magnetic resonance studies (PARRY JONES, ROBERTS, and AHMED, 1971) favour extended conformations for GABA in solution. There is also evidence for extended conformations of GABA under *"in vivo"* conditions (BEART, CURTIS, and JOHNSTON, 1971; BEART, JOHNSTON, and UHR, 1972). The amino acid is a zwitterion at neutral pH with pK values of 4.03 and 10.27 at 35° (KING, 1954).

GABA occurs in mammals in appreciable amounts only in the CNS, with the highest levels in the substantia nigra and globus pallidus (BAXTER, 1970, 1972). While it is generally accepted that systemically administered GABA does not raise levels in the brain (ROBERTS and KURIYAMA, 1968), subcutaneous injection of large doses (60 μmole/g=6.2 g/kg) increases the brain levels in mice of all ages (LEVI, AMALDI, and MORISI, 1972). Brain levels of GABA are altered by a variety of drugs, many of which inhibit GABA-metabolising enzymes (BAXTER, 1970; IVERSEN, 1972; COLLINS, 1973). Although itself not incorporated into protein, GABA (and glycine) stimulate protein synthesis in cell free extracts of immature rat brain (BAXTER, TEWARI, and RAEBURN, 1972). An extensive review of GABA literature has been published (SYTINSKY, 1972).

GABA Metabolism. GABA is synthesised by the decarboxylation of L-glutamate. L-glutamate decarboxylase (GAD, EC. 4.1.1.15) exists in at least two forms: GAD I and GAD II. GAD I, the form more widely studied, requires pyridoxal phosphate as a cofactor and is strongly inhibited by anions such as chloride, and by carbonyl trapping reagents such as thiosemicarbazide (ROBERTS and KURIYAMA, 1968). GAD I is relatively insensitive to product inhibition by GABA, but GABA may repress the synthesis of GAD I (Mouse: SZE, 1970; SZE and LOVELL, 1970). GAD I appears to be localised predominantly within the axoplasm of particular nerve endings and is not found in appreciable amounts in non-neural tissues. GAD II is relatively insensitive to inhibition by anions, is stimulated by carbonyl trapping reagents and is found in various non-neural tissues (Human: HABER, KURIYAMA, and ROBERTS, 1970. Rabbit: KURIYAMA, HABER, and ROBERTS, 1970).

GABA is degraded by the action of a specific GABA transaminase (4-aminobutyrate:2-oxoglutarate aminotransferase, GABA-T, EC. 2.6.1.19) to succinic semialdehyde, which is in turn oxidised to succinate by a specific dehydrogenase. A number of multiple forms of GABA-T have been reported (Mouse, rat: WAKSMAN and BLOCH, 1968), and GABA-T activity has been found in non-neural tissues (Rat: CACIAPPO, PANDOLFO, and DI CHIARA, 1959). GABA-T transaminates some other substances related to GABA, including β-alanine and δ-amino-valeric acid, but not γ-amino-β-hydroxybutyric acid or taurine (Mouse: WAKSMAN and ROBERTS, 1965. Rat: SYTINSKY and VASILIJEV, 1970; BEART and JOHNSTON, 1973a). Many inhibitors of GABA-T activity, including amino-oxyacetic acid, hydroxylamine and 3-hydrazinopropionic acid, increase brain GABA levels (BAXTER, 1970; IVERSEN, 1972) but, as with inhibitors of GAD I, cause and effect are difficult to correlate.

The metabolic pathway from 2-oxoglutarate to succinate via L-glutamate, GABA and succinic semialdehyde constitutes the "GABA-shunt", i.e. an alternative pathway to that through succinyl-coenzyme A in the normal tricarboxylic acid cycle. It has been estimated that in rat brain slices the GABA-shunt constitutes about 8% of the total carbon flux of the tricarboxylic acid cycle (BÁLAZS, MACHIYAMA, HAMMOND, JULIAN, and RICHTER, 1970). *In vivo* experiments with rats indicate that the rate of GABA synthesis in the brain is about 5 µmole/g/h (VAN GELDER, 1966).

Histochemical methods have been described for the localization of GABA (WOLMAN, 1971), GABA-T (VAN GELDER, 1965a) and succinic semialdehyde dehydrogenase (SIMS, WEITSEN, and BLOOM, 1971). Use has also been made of the distribution of radiolabelled thiosemicarbazide as an indicator of GABA metabolism (CSILLIK, GEREBTZOFF, KISS, and KNYIHÁR, 1971).

GABA Transport. GABA is taken up into brain slices by a saturable "high affinity" process (K_m approx. $10^{-5} M$), which is structurally specific, dependent on temperature and requires sodium ions (ELLIOTT and VAN GELDER, 1958; ROBERTS and KURIYAMA, 1968; IVERSEN and NEAL, 1968; COHEN and LAJTHA, 1972). "Low affinity" uptake (K_m approx. $10^{-4} M$) of GABA has also been reported (LEVI and RAITERI, 1973). Various drugs inhibit GABA uptake in a relatively non-specific manner, including *p*-chloromercuriphenylsulphonate, chlorpromazine, imipramine, haloperidol, dibutyryl cyclic AMP and the protoveratrines (Rat: IVERSEN and JOHNSTON, 1971; GOTTESFELD and ELLIOTT, 1971; BEART and JOHNSTON, 1973b; HARRIS, HOPKIN, and NEAL, 1973). GABA transport appears to be dependent on pyridoxyl phosphate, and can be inhibited by relatively high concentrations of amino-oxyacetic acid (SNODGRASS and IVERSEN, 1973b). Autoradiographic studies indicate that a major site of uptake in rat cerebral cortex is a distinct population of nerve terminals (IVERSEN and BLOOM, 1972), but this is not true of all nervous tissue (Rat: NEAL and IVERSEN, 1972).

Synaptic vesicles isolated from brain bind GABA by a structurally specific, sodium dependent, process (Mouse: ROBERTS and KURIYAMA, 1968; KURIYAMA, ROBERTS, and KAKEFUDA, 1968) which is inhibited by imipramine (WEINSTEIN, VARON, and ROBERTS, 1971). This binding may be related to the storage of GABA in the nerve terminal.

GABA Release. The release of endogenous GABA has been demonstrated *in vivo* into the fourth ventricle and from the surface of the cerebral cortex following appropriate stimulation (Cat: OBATA and TAKEDA, 1969; JASPER and KOYAMA, 1969; IVERSEN, MITCHELL, and SRINIVASAN, 1971). In both the spinal cord and cerebellum tetanus toxin appears to block the synaptic release of this amino acid (CURTIS, FELIX, GAME, and MCCULLOCH, 1973a).

Postsynaptic Action of GABA. GABA depresses the firing of neurones throughout the mammalian nervous system and hyperpolarization with an increased membrane conductance has been demonstrated in Deiters' nucleus, the cerebral and cerebellar cortices and the spinal cord. (Cat: OBATA, ITO, OCHI, and SATO, 1967; KRNJEVIĆ and SCHWARTZ, 1967b; CURTIS et al., 1968b; BRUGGENCATE and ENGBERG, 1968. Rat: SIGGINS, OLIVER, HOFFER, and BLOOM, 1971). The reversal potentials for both synaptically- and GABA-induced hyperpolarizations are similar, and intracellular injection of ions has essentially the same effects upon both types of inhibitory hyperpolarizations. It is therefore generally accepted that GABA has transmitter-like effects at certain inhibitory synapses.

GABA Antagonists. Three series of compounds (Fig. 3) antagonise the postsynaptic effects of GABA: picrotoxin and related compounds, bicuculline and related compounds, and benzyl penicillin.

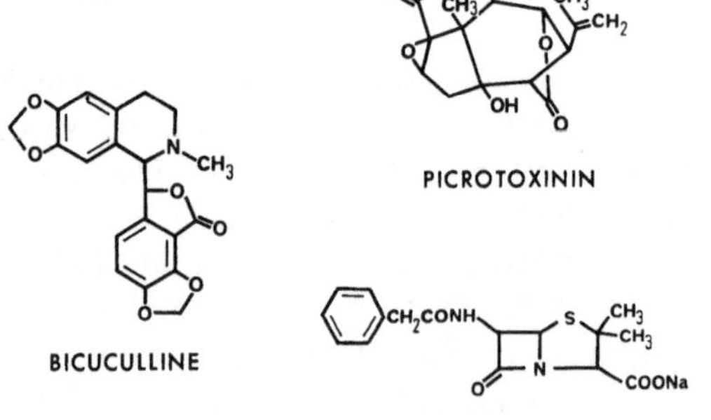

PICROTOXININ

BICUCULLINE

BENZYL PENICILLIN

Fig. 3. Antagonists of the postsynaptic action of GABA

Picrotoxin is an equimolar mixture of picrotoxinin and picrotin (PORTER, 1967), the former being about 50 times more active than the latter as a convulsant (Mouse: RAMWELL and SHAW, 1963; JARBOE, PORTER, and BUCKLER, 1968). Picrotoxin reversibly, but not consistently, antagonises the action of GABA in many areas of the CNS. The structurally related convulsant tutin (FASTIER, 1971) is also a GABA antagonist in the spinal cord of the cat.

Bicuculline, a convulsant phthalide isoquinoline alkaloid (Rabbit: WELCH and HENDERSON, 1934. Baboon, Rhesus monkey: MELDRUM and HORTON, 1971), is a relatively specific GABA antagonist (Cat: CURTIS et al., 1971 a). The structurally related compounds corlumine and bicucine methyl ester are similar in potency to bicuculline as convulsants and as GABA antagonists (Cat: JOHNSTON et al., 1972). Quaternisation of the heterocyclic nitrogen atom in bicuculline produces more active compounds. Bicuculline methiodide ("N-methyl-bicuculline", Rat: PONG and GRAHAM, 1972) is a more powerful convulsant than bicuculline hydrochloride when injected intracisternally. Bicuculline methochloride, which is about 100 times more soluble in water than bicuculline hydrochloride and is more active electrophoretically as a selective GABA antagonist (Cat: JOHNSTON et al., 1972), is less effective than bicuculline when administered systemically, as might be expected from its quaternary structure.

Benzyl penicillin is a specific but less effective GABA antagonist than bicuculline (Frog: DAVIDOFF, 1972b. Cat: DAVIDOFF, 1972c; CURTIS, GAME, JOHNSTON, MCCULLOCH, and MACLACHLAN, 1972b).

Other substances, which like bicuculline and picrotoxinin reduce inhibitions probably mediated by GABA and are therefore possibly GABA antagonists, include shikimin (KAJIMOTO, FUJIMORI, and HIROTA, 1955. Cat: prolonged spinal inhibition, CURTIS and JOHNSTON, unpublished observation), tetramethylenedisulphotetramine (HAGEN, 1950; D.I.B. KERR, personal communication. Cat: prolonged spinal inhibition, CURTIS and JOHNSTON, unpublished observations), diphenylamino-ethanol (Cat: cerebellar inhibition of Deiters' neurones, KEE, WELLS, and YIM, 1971). Additionally d-tubocurarine reduces the inhibitory effects of both GABA and glycine on cerebral cortical neurones (Cat: HILL, SIMMONDS, and STRAUGHAN, 1972).

3.3. Taurine

Taurine is a non-protein, sulphur containing amino acid, which is a zwitterion at neutral pH with pK values of approx. 1.5 and 8.82 at 35° (ALBERT, 1950; KING, 1953). In the solid state, it is in an extended conformation with a gauche relationship between the protonated amino and the sulphonate groups (OKAYA, 1966).

Taurine occurs in varying concentrations throughout the nervous system, the highest level being in the lateral geniculate nucleus (Cat: GUIDOTTI, BADIANI, and PEPEU, 1972), but, unlike glycine and GABA, higher levels of taurine exist in other than CNS tissue, (GAITONDE, 1970), particularly in the heart (Rat: AWAPARA, 1956). Levels are high during the period of rapid growth of the brain, and fall as the animal matures (AGRAWAL, DAVIS, and HIMWICH, 1966). Increasing the plasma levels of taurine 100 fold does not significantly increase brain levels (Rat: LEVI, 1968). The rate of entry of taurine into the brain from the blood is the same in immature and mature rats (BAÑOS et al., 1971). Repeated administration of taurine reduces both the seizure activity and the abnormalities of cortical amino acid levels induced by cobalt in the cortex (Cat, mouse: VAN GELDER, 1972).

Taurine Metabolism. Taurine is synthesised in the CNS via decarboxylation of cysteinesulphinic acid to hypotaurine and subsequent oxidation (GAITONDE, 1970). The decarboxylase activity appears to be associated with nerve endings in rat brain (AGRAWAL, DAVISON, and KACZMAREK, 1971). Taurine is degraded slowly in rat brain to isethionic acid (PECK and AWAPARA, 1967).

Taurine Transport. Taurine is taken up into brain slices by at least two kinetically distinct transport systems with K_m's of 5×10^{-5} *M* and 1×10^{-2} *M* (Rat: KACZMAREK and DAVISON, 1972; LÄHDESMÄKI and OJA, 1972). Uptake of taurine by the low K_m system is structurally specific for taurine and is inhibited by cyanide and ouabain (Cat: BATTISTIN, GRYNBAUM, and LAJTHA, 1969a. Rat: KACZMAREK and DAVISON, 1972).

Taurine Release. Endogenous taurine is released from the feline cerebral cortex *in vivo* as a result of stimulation of the reticular formation (JASPER and KOYAMA, 1969). Exogenous taurine is released from rat brain slices under a variety of conditions (KACZMAREK and DAVISON, 1972; LÄHDESMÄKI and OJA, 1972).

Postsynaptic Action of Taurine. Intracellular studies of the mechanism of taurine-induced depression of neuronal firing have not yet been reported, but an action similar to glycine seems very likely.

Taurine Antagonists. In the feline spinal cord and medulla, strychnine, but not bicuculline, is a taurine antagonist, but in the cerebral cortex and lateral geniculate nucleus both strychnine and bicuculline block the depression of neuronal firing by taurine. (CURTIS et al., 1968a, 1971a; CURTIS, DUGGAN, FELIX, JOHNSTON, and MCLENNAN, 1971b; CURTIS and TEBĒCIS, 1972).

3.4. Structurally Related Depressant Amino Acids

A number of depressant amino acids structurally related to glycine, GABA and taurine occur in central nervous tissue.

L-α-Alanine. This amino acid is present in moderate amounts throughout the CNS, showing little regional variation (TALLAN, 1962; Rat: SHAW and HEINE, 1965; SHANK and APRISON, 1970. Cat: BATTISTIN, GRYNBAUM, and LAJTHA, 1969b). After intraperitoneal administration L-α-alanine (11 μmole/g) raises the central levels of L-α-alanine and also of glycine (Rat: RICHTER and WAINER, 1971). This amino acid is taken up into brain slices by the "low affinity" small neutral amino acid transport system (COHEN and LAJTHA, 1972; Rat: SMITH, 1967; JOHNSTON and IVERSEN, 1971; LOGAN and SNYDER, 1972). Exogenous L-α-alanine is not released from slices of rat brain or spinal cord by electrical stimulation (MITCHELL, NEAL and SRINIVASAN, 1969; HOPKIN and NEAL, 1971). The depressant effect of L-α-alanine on the firing of feline spinal neurones, weaker than glycine but stronger than L-cystathionine and L-serine, is blocked by strychnine (CURTIS et al., 1968a).

L-Serine. L-Serine and glycine are interconvertible in the CNS. In the rat, serine concentrations are lower than glycine in all areas of the CNS, other than in the telencephalon and cerebellum

(Rat: SHAW and HEINE, 1965; SHANK and APRISON, 1970). L-Serine is also concerned with the metabolism of L-cystathionine (GAITONDE, 1970). Intraperitoneally administered glycine (13 μmole/g) raises the central levels of serine (Rat: RICHTER and WAINER, 1971).The rate of entry of serine from the blood into the brain of immature rats is 40–120 times that of mature rats (BAÑOS et al., 1971). When administered electrophoretically L-serine is a weak strychnine-sensitive depressant of the firing of feline spinal neurones (CURTIS et al., 1968a). L-Serine does not influence the high affinity uptake of glycine by slices of cat and rat spinal cord (JOHNSTON and IVERSEN, 1971; BALCAR and JOHNSTON, 1973), and is itself taken up by a "low affinity" system (K_m approx. $6 \times 10^{-4} M$) into slices and homogenates of rat cord (JOHNSTON and IVERSEN, 1971; LOGAN and SNYDER, 1972).

L-Cystathionine. The levels of L-cystathionine in the CNS vary over a wide range, with high levels in the cerebellum (Rat: SHAW and HEINE, 1965; KANDERA, LEVI, and LAJTHA, 1968) and the bovine pineal body (LABELLA, VIVIAN, and QUEEN, 1968). Spinal levels in cats (WERMAN, DAVIDOFF, and APRISON, 1966) were overestimated by a factor of ten (WERMAN, 1972). L-Cystathionine has a relatively weak, strychnine-sensitive, depressant action on the firing of feline spinal neurones (WERMAN, DAVIDOFF, and APRISON, 1966; CURTIS et al., 1968a). When intravenously administered, it is not transported into the brain but metabolic degradation products are incorporated into brain protein (Rat: BROWN and GORDON, 1971).

Hypotaurine. This sulphinic acid, a metabolic precursor of taurine, has been detected in extracts of rat brain (BERGERET and CHATAGNE, 1954). Hypotaurine depresses the firing of feline spinal neurones (CURTIS and WATKINS, 1965), and inhibits the uptake of taurine by rat brain slices (KACZMAREK and DAVISON, 1972).

β-Alanine. Very low levels of this amino acid occur in the CNS (TALLAN, 1962; Human: PERRY et al., 1971a. Cat: PERRY et al., 1972. Rat: SCHENCK, 1943; YOSHINO, DEFEUDIS, and ELLIOTT, 1970). The hyperpolarization by β-alanine of feline spinal motoneurones is similar to that of glycine (CURTIS et al., 1968b). In the feline spinal cord and lateral vestibular nucleus the depressant action of β-alanine is antagonised by strychnine and unaffected by bicuculline, but in the thalamus and in the cerebral, cerebellar and hippocampal cortices, this action is antagonised by both alkaloids (CURTIS et al., 1971a, b).

γ-Aminobutyrylcholine. This GABA derivative, present in the brain, has little or no depressant action in the cat on the firing of spinal interneurones or neurones of the cerebral cortex, but does influence the firing of Renshaw cells in the spinal cord. These and other observations suggest that the pharmacological actions of this compound in the CNS may be related to cholinergic systems rather than to those involving GABA (JOHNSTON and CURTIS, 1972).

γ-Amino-β-hydroxybutyric Acid. While the betaine carnitine undoubtedly occurs in the brain, controversy still surrounds the role (HAYASHI, 1966), and even the presence (BAXTER, 1970; Mouse, ox: SEILER and WIECHMANN, 1969. Rat: YOSHINO, DEFEUDIS, and ELLIOTT, 1970) of γ-amino-β-hydroxybutyric acid in nervous tissue. It is not a substrate for GABA-T (Rat: SYTINSKY and VASILIJEV, 1970; BEART and JOHNSTON, 1973a) and can be only a relatively weak substrate for the GABA uptake system in rat brain slices (IVERSEN and JOHNSTON, 1971). The amino acid is a weaker depressant of the firing of central neurones than is GABA, and this action is antagonised by bicuculline but not by strychnine (Cat: CURTIS et al., 1971a, b).

Imidazole-4-acetic Acid. This histamine metabolite (KAHLSON and ROSENGREN, 1971) structurally resembles both GABA and β-alanine. Following parental administration, imidazole-4-acetic acid has mixed excitant and depressant effects (Mice, rats: TUNNICLIFF, WEIN, and ROBERTS, 1972), although, on electrophoretic administration the firing of neurones in the feline spinal cord, cerebral cortex and medulla is depressed, an effect comparable in potency to GABA and suppressed by bicuculline (PHILLIS, TEBĒCIS, and YORK, 1968; CURTIS et al., 1971a, b; HAAS, ANDERSON, and HÖSLI, 1972). Imidazole-4-acetic acid is a weak inhibitor of GABA uptake into slices of spinal cord (Cat: BALCAR and JOHNSTON, 1973), and of rat brain GAD I (SMALL, HOLTON, and ANCILL, 1970).

2,4-Diaminobutyric acid. This amino acid has been detected in brain (Ox: NAKAJIMA, WOLFGRAM and CLARK, 1967), is a weak depressant of spinal interneurones (Cat: CURTIS and WATKINS, 1960) and a comparatively strong non-competitive inhibitor of the uptake of GABA by rat brain slices (IVERSEN and JOHNSTON, 1971).

Proline. This cyclic amino acid is present in brain (Human, cat: PERRY et al., 1971a, b, 1972) and is a weak strychnine-sensitive depressant of cat spinal interneurones. It is taken up into rat brain slices by a high affinity system.

3.5. L-Aspartate

L-Aspartate, like its higher homologue L-glutamate, is an optically active acidic amino acid, with three ionizable groups (pK's of 2.05, 3.87 and 10.00 at 25°: SIMMS, 1928), which can associate via intermolecular hydrogen bonds, both in the solid state (DERISSEN, ENDEMAN, and PEERDEMAN, 1968) and in aqueous solution (GLAGOLEVA, 1967).

The central levels of L-aspartate vary from region to region, but only within the feline spinal cord have the levels been correlated with a possible function as a transmitter of excitatory interneurones (DAVIDOFF, GRAHAM, SHANK, WERMAN, and APRISON, 1967).

When systemically administered to immature animals, L-aspartate produces degeneration of cells in the retina and hypothalamus (Mouse: OLNEY, HO and RHEE, 1971) and seizures (Rat: JOHNSTON, 1973). Intraventricular administration to mature mice also results in seizures (CRAWFORD, 1963).

L-Aspartate Metabolism. The metabolism of L-aspartate in the CNS resembles that of L-glutamate in being concerned with a variety of compartments. Exogenous L-aspartate rapidly enters the "small" L-glutamate pool where it is metabolised via the tricarboxylic acid cycle to L-glutamine, and does not equilibrate with a major portion of the endogenous L-aspartate. L-Aspartate transaminase, which catalyses the reversible transamination between L-aspartate and L-glutamate, and their corresponding oxo-acids oxalacetate and α-oxoglutarate, is present in two forms in brain differing in location and kinetic properties: neither form appears to be associated with nerve endings (Guinea pig, rat: FONNUM, 1968). High concentrations of N-acetyl-L-aspartate occur in the brain but the function of this amino acid derivative remains obscure (NADLER and COOPER, 1972).

L-Aspartate Transport. Slices and homogenates of rat brain and cat spinal cord transport L-aspartate by the same "high" and "low" affinity systems that transport L-glutamate (COHEN and LAJTHA, 1972; WOFSEY, KUHAR, and SNYDER, 1971; LOGAN and SNYDER, 1972; BALCAR and JOHNSTON, 1972a, b, 1973).

L-Aspartate Release. Endogenous L-aspartate is released from rat brain synaptosomes on electrical field stimulation (BRADFORD, 1970).

Postsynaptic Action of L-Aspartate. Feline spinal motoneurones are reversibly depolarized by L-aspartate (CURTIS, PHILLIS, and WATKINS, 1960) in a manner similar to that by L-glutamate (Section 3.6). Such a depolarizing action accounts for the excitant action of L-aspartate on other central neurones.

L-Aspartate Antagonists. All of the L-glutamate antagonists (Section 3.6; see Fig. 4), excepting LSD which was not tested against L-aspartate, have also been found to antagonise the excitant action of L-aspartate. Glutamate diethylester has been reported to be less effective as an aspartate than as a glutamate antagonist (HALDEMAN and MCLENNAN, 1972). This observation, together with the differential sensitivity of thalamic (Section 4.6.2) and spinal (Section 4.12.2) neurones to aspartate and glutamate, are the only indications to date that there may be different excitatory receptors for these two amino acids.

3.6. L-Glutamate

L-Glutamate is an optically active acidic amino acid with three ionizable groups with pK values of 2.16, 4.32 and 9.96 at 20° (NEUBERGER, 1936). It can associate in aqueous solution forming intermolecular hydrogen bonds (GLAGOLEVA, 1967).

L-Glutamate is the most abundant amino acid in the adult CNS, and regional variations in its distribution may reflect roles as a synaptic transmitter and as a major brain metabolite (VAN DEN BERG, 1970; JOHNSON and APRISON, 1971). Of the amino acids in feline spinal roots, only L-glutamate is more concentrated in the dorsal than in the ventral roots; this observation led to the proposal that it may be the excitatory transmitter released by the terminals of primary afferent fibres (GRAHAM, SHANK, WERMAN and APRISON, 1967). In other areas, however, it has proved difficult to correlate the levels of L-glutamate, and various L-glutamate-metabolizing enzymes, with suspected transmitter function (Cat: JOHNSON, 1972a, b).

Systemically administered L-glutamate produces degeneration of cells in the retina and hypothalamus (Mouse: OLNEY et al., 1971; PEREZ and OLNEY, 1972. Monkey: OLNEY, SHARPE and FEIGIN, 1972) and, in higher doses, seizures (Rat: HENNECKE and WIECHERT, 1970; STEWART, COURSIN, and BHAGAVAN, 1972; JOHNSTON, 1973). While immature animals are more susceptible than adult animals, factors other than age, for example the levels of pyridoxal 5′-phosphate and the activity of various metabolizing enzymes, may influence the efficiency of the blood-brain barrier(s) to L-glutamate (Rat: BHAGAVAN, COURSIN, and STEWART, 1971). L-Glutamate and related excitant amino acids inhibit protein synthesis in rat brain slices (ORREGO and LIPMANN, 1967), although glutamate and aspartate stimulated protein synthesis in cell free preparations of immature rat brain (BAXTER and TEWARI, 1970). The toxic effects of L-glutamate, despite rapid conversion to L-glutamine, appear not to be due to a disturbance of ammonia metabolism, since other amino acids which have a direct excitant action on central neurones but are not concerned with ammonia metabolism (for example D-glutamate and N-methyl-D-aspartate) exhibit similar toxicity (Mouse: CRAWFORD, 1963; OLNEY et al., 1971. Rat: MUSHAHWAR and KOEPPE, 1971; JOHNSTON, 1973). It seems very likely that the depolarizing action of these excitant amino acids on central neurones is the basis of their toxic effects.

L-Glutamate Metabolism. The metabolism of L-glutamate in the brain is exceedingly complex (VAN DEN BERG, 1973; BALÁZS, PATEL and RICHTER, 1973; JOHNSON, 1972a; WATKINS, 1972). There are at least two major metabolic "pools" of L-glutamate. *In vivo,* and under appropriate conditions *in vitro* (Rat: BERL, NICKLAS and CLARKE, 1968), administration of radioactive L-glutamate results in rapid labelling of L-glutamine, such that the specific activity of L-glutamine exceeds that of L-glutamate. The exogenous L-glutamate apparently does not readily equilibrate with a major portion of the endogenous L-glutamate (the "large" L-glutamate pool), but enters the "small" pool in which it is converted to L-glutamine by the action of L-glutamine synthetase. The L-glutamate in the "large" pool appears to be synthesised predominantly from D-glucose and intermediates of glycolysis, and may represent "transmitter" L-glutamate in addition to L-glutamate involved in energy metabolism related to electrical and general metabolic activity. The "small" pool may be associated with glial cells and may represent "inactivated" L-glutamate taken up from the synaptic cleft after synaptic release.

L-Glutamate Transport. At least two kinetically distinct transport systems have been described for L-glutamate in isolated brain tissue. The system of higher K_m, the "low affinity" system, seems to be a general property of many subcellular particles, whereas the system of lower K_m, the "high affinity" system, is associated with a unique population of subcellular particles (Rat: WOFSEY et al., 1971; LOGAN and SNYDER, 1972). Neither system is specific for L-glutamate, but may transport a variety of acidic amino acids including L-aspartate and L-cysteate (COHEN and LAJTHA, 1972; Rat: BALCAR and JOHNSTON, 1972a, b). The "high affinity" system shows an absolute dependence on sodium ions (Rat: BENNETT et al., 1972). The possible relationship between the high and low affinity uptake systems and the two pools of L-glutamate metabolism is at present unknown.

L-Glutamate uptake is inhibited by various drugs including juglone, chlorpromazine and *p*-chloromercuriphenylsulphonate (Cat, rat: BALCAR and JOHNSTON, 1972a, b, 1973): the latter when administered microelectrophoretically enhances the excitant action of L-glutamate on feline spinal interneurones (CURTIS et al., 1970b).

L-Glutamate Release. The release of endogenous L-glutamate *in vivo* from the surface of the feline cerebral cortex is increased by brain stem stimulation (JASPER and KOYAMA, 1969), and *in vitro* that from synaptosomes by electrical field stimulation (Rat: BRADFORD, 1970). The release of exogenous L-glutamate *in vitro* from slices of rat brain and spinal cord following electrical field stimulation or stimulation by increased extracellular potassium ions has also been reported (MITCHELL et al., 1969; ARNFRED and HERZ, 1971; HOPKIN and NEAL, 1971).

Postsynaptic Action of L-Glutamate. L-Glutamate and structurally related acidic amino acids excite feline central neurones by a reversible depolarization accompanied by an increase in membrane conductance (CURTIS et al., 1960; KRNJEVIĆ and SCHWARTZ, 1967a; ZIEGLGÄNSBERGER and PUIL, 1973). D-Homocysteate, N-methyl-D-aspartate and β-N-oxalyl-L-α,β-diaminopropionate are more powerful excitants than L-aspartate and L-glutamate, which in turn are slightly stronger than their corresponding D-isomers (CURTIS and WATKINS, 1963; WATKINS, CURTIS, and BISCOE, 1966). Analysis of the complete ionic mechanism of amino acid-induced depolarization, or indeed of synaptic excitation, raises considerable experimental difficulties since the majority of excitatory synapses upon neurones are located on the dendrites. However comparisons of the "reversal" potentials for amino acid depolarization and synaptic excitation (CURTIS, 1965; ZIEGLGÄNSBERGER and PUIL, 1973) indicate that the two types of depolarization could involve the same change of membrane ion conductance. Furthermore, the extracellular administration of tetrodotoxin, or the intracellular injection of sufficient potassium or chloride ions to convert the hyperpolarizing action of glycine into depolarization, failed to modify the depolarization of feline spinal motoneurones induced synaptically or by microelectrophoretic DL-homocysteate or L-glutamate (CURTIS, DUGGAN, FELIX, JOHNSTON, TEBĒCIS, and WATKINS, 1972 a; ZIEGLGÄNSBERGER and PUIL, 1972). These experiments indicate that depolarization may involve a simultaneous increase in membrane permeability to sodium and potassium ions which is unaffected by tetrodotoxin. L-Glutamate and related excitant amino acids promote an influx of sodium ions into brain slices *in vitro* which is inhibited by tetrodotoxin — such ion movements are thus distinct from the initial depolarizing action of the amino acids, and probably are those linked with action potentials arising from the depolarization (HARVEY and MCILWAIN, 1969; CURTIS, 1970; BALCAR and JOHNSTON, 1972a).

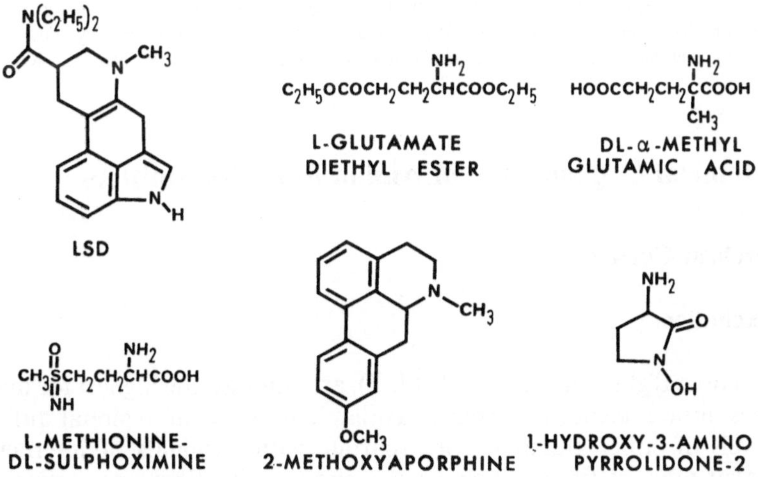

LSD

$C_2H_5OCOCH_2CH_2CH(NH_2)COOC_2H_5$
L-GLUTAMATE DIETHYL ESTER

$HOOCCH_2CH_2C(NH_2)(CH_3)COOH$
DL-α-METHYL GLUTAMIC ACID

$CH_3S(O)(NH)CH_2CH_2CH(NH_2)COOH$
L-METHIONINE-DL-SULPHOXIMINE

2-METHOXYAPORPHINE

1-HYDROXY-3-AMINO PYRROLIDONE-2

Fig. 4. Substances reported to antagonize the postsynaptic action of L-glutamate

L-Glutamate Antagonists. The following compounds have been reported as L-glutamate antagonists under certain conditions with varying degrees of selectivity in the feline CNS: LSD (BOAKES, BRADLEY, BRIGGS, and DRAY, 1970), L-glutamate diethyl ester, DL-α-methyl glutamic acid (HALDEMAN, HUFFMAN, MARSHALL, and MCLENNAN, 1972), 2-methoxyaporphine, L-methionine-DL-sulphoximine (CURTIS et al., 1972a) and 1-hydroxy-3-amino-pyrrolidone-2 (DAVIES and WATKINS, 1972). The structures of these compounds (Fig. 4) may provide clues to more useful antagonists.

3.7. Structurally Related Excitant Amino Acids

At least two excitant amino acids, in addition to L-glutamate and L-aspartate, occur in normal CNS tissue, albeit in relative low concentrations: L-cysteate and L-cysteine sulphinate. It is possible that others also are present as there are unidentified peaks in amino acid analyses of rat and mouse brain in the regions where the known acidic amino acids are eluted (SHAW and HEINE, 1965; SHIMADA et al., 1971). Furthermore the natural occurrence of D-glutamate and D-aspartate in CNS tissue has not been excluded (CORRIGAN, 1969).

L-Cysteate. The sulphonic acid analogue of L-aspartate is present in low concentration (approx. 0.1 μmole/g) in extracts of rat brain (MUSSINI and MARCUCCI, 1962; MANDEL and MARK, 1965). This amino acid depolarizes feline spinal motoneurones and is comparable in potency to L-aspartate (CURTIS et al., 1960). In general, L-cysteate may be metabolised by two pathways: decarboxylation to taurine or transamination with α-oxo acids (GAITONDE, 1970). L-Cysteate is a strong competitive inhibitor of the high affinity uptake of L-glutamate by rat brain slices and thus could be a substrate for this transport system (BALCAR and JOHNSTON, 1972b). L-Cysteate produces acute neuronal degeneration in the hypothalamus of infant mice following subcutaneous administration (OLNEY et al., 1971).

L-Cysteine Sulphinate. This acidic amino acid, which has been detected in rat brain (BERGERET and CHATAGNE, 1954) is a powerful excitant of spinal neurones (Cat: CURTIS and WATKINS, 1963) and a powerful competitive inhibitor of the high affinity uptake of L-glutamate by rat brain slices (BALCAR and JOHNSTON, 1972b). The enzyme L-cysteine sulphinate decarboxylase, which produces hypotaurine from L-cysteine sulphinate, occurs in rat brain associated with synaptosomes (AGRAWAL et al., 1971). The complex metabolic interrelations of L-cysteine sulphinate, L-cysteate, taurine and hypotaurine are discussed by GAITONDE (1970). Subcutaneous administration of L-cysteine sulphinate is followed by degeneration in the hypothalamus of infant mice (OLNEY, HO and RHEE, 1971).

4. Anatomical Organization of Amino Acid Transmitters

4.1. Cerebral Cortex

4.1.1. Excitation

Although cortical glutamate levels (Table 2), are amongst the highest for mammalian tissues, little evidence is currently available to associate regional differences in the levels of either glutamate or aspartate with terminals of interneurones or particular afferent fibres. Such an association with fibres projecting to the cortex is suggested by the reduced levels of both amino acids (and GABA) which follows undercutting of the suprasylvian gyrus in the cat (BERL and MCMURTRY, 1967), although the reduced glutamate levels of the motor cortex of the cat as a result of tetanic stimulation of the thalamic ventrolateral nucleus have been interpreted largely in terms of energy metabolism (CONSTANTINESCU, FLORESCU, and HATEGANU, 1970). The moderate levels of glutamate in the corpus callosum and the posterior portion of the internal capsule (Table 2), which exceed those of the dorsal root, may indicate the presence of glutamate releasing fibres; the low level in the peduncles suggest that the fibres of the pyramidal tract may not be of this type.

Subcellular distribution studies have hitherto failed to demonstrate any selective storage of either aspartate or glutamate within nerve endings (synaptosomes) of the cerebral cortex (Rat: RYALL, 1964. Guinea pig: RYALL, 1964; MANGAN and WHITTAKER, 1966; see also KRNJEVIĆ and WHITTAKER, 1965; WHITTAKER, 1968). Both amino acids, however, are accumulated from low extracellular concentrations by a subcellular fraction of homogenates of this tissue which may be largely synaptosomal in nature, and can be distinquished from fractions accumulating other amino acids (Rat: WOFSEY et al., 1971). This "high affinity"

Tabel 2. Excitant amino acids in regions of cerebral cortex (μmole/g, *Biopsy)

Region	Animal	Aspar-tate	Gluta-mate	Reference
Frontal	Human	1.9	8.9	PERRY et al. (1971a)
Occipital	Human	2.6	8.6	PERRY et al. (1971a)
Temporal	Human	1.9	10.8	PERRY et al. (1971a)
Temporal	Cat	3.1	12.9	BATTISTIN et al. (1969)
Suprasylvian	Cat	2.9	10.5	BERL and MCMURTRY (1967)
Post Cruciate	Cat	2.7	10.3	VAN GELDER (1972)
Association	Cat	1.5*	6.0*	PERRY et al. (1972)
Motor	Cat	1.6*	5.5*	PERRY et al. (1972)
Visual	Cat	1.8*	5.9*	PERRY et al. (1972)
Primary auditory	Cat		10.7	JOHNSON and APRISON (1971)
Primary somatosensory	Cat		10.2	JOHNSON and APRISON (1971)
Sylvian ectosylvian	Cat		13.0	JOHNSON and APRISON (1971)
Various areas	Cat	2.3*	9.7*	KOYAMA (1972)
Corpus callosum	Cat		5.9	JOHNSON and APRISON (1971)
Corpus callosum	Cat	1.4	10.6	BATTISTIN et al. (1969)
Peduncle	Cat		3.7	JOHNSON and APRISON (1971)
Internal capsule (posterior)	Cat		5.6	JOHNSON and APRISON (1971)

uptake, which has also been studied in slices of cerebral cortex (Rat: BALCAR and JOHNSTON, 1972b) is absolutely sodium dependent (BENNETT et al., 1972).

Both glutamate and aspartate are released from the exposed and superfused cerebral cortex of neuroaxially intact cats, the rates being higher in aroused than in sleeping animals (JASPER, KHAN, and ELLIOTT, 1965; JASPER and KOYAMA, 1969). Rates of release in resting encephale isolé preparations were comparable with those of intact sleeping animals, and were considerably enhanced by stimulation of the mesencephalic reticular formation, which simulated arousal, but not by stimulation of the medial thalamic recruiting system. Since both kinds of stimulation increase the cortical release of acetylcholine, separate cholinergic and excitant amino acid-releasing "activating" pathways have been proposed (JASPER and KOYAMA, 1969). However, reciprocal alterations in the rate of GABA release were also recorded, and changes in amino acid release from the cortical surface may not necessarily be a direct measure of that released synaptically. Glutamate is also released from the cortical surface during spreading depression (VAN HARREVELD and KOOIMAN, 1965), and there is an extensive literature concerning a possible association between excitant amino acids and this abnormal state in both the cortex and retina (VAN HARREVELD, 1970; VAN HARREVELD and FIFKOVA, 1972; PHILLIS and OCHS, 1971; DO CARMO and LEÃO, 1972).

The excitatory effect of glutamate and aspartate on the cortex was first demonstrated after intracortical or intracarotid injection (OKOMOTO, 1951; ITO, TANI, SATO, and ODA, 1952; HAYASHI, 1952). Such an action also accounts for the effects of topically (Rabbit: VAN HARREVELD, 1959. Cat: PURPURA, GIRADO, SMITH, CALLAN, and GRUNDFEST, 1959) or intraventricularly (see CURTIS and WATKINS, 1965) administered acidic amino acids, and direct excitation was subsequently confirmed at the cellular level in microelectrophoretic studies (Cat: KRNJEVIĆ and PHILLIS, 1963; KRNJEVIĆ, 1964; CRAWFORD and CURTIS, 1964). The depolarization of cortical neurones by L-glutamate (KRNJEVIĆ, 1964) is associated with an increased membrane conductance (KRNJEVIĆ and SCHWARTZ, 1967a).

As elsewhere in the feline CNS, the effects of electrophoretic glutamate and aspartate are characteristically rapid in both onset and recovery, the potencies of the L-isomers are similar and only slightly greater than those of the D-isomers. No systematic variations have been found in the relative sensitivity of different cortical neurones to glutamate, aspartate and DL-homocysteate (CRAWFORD, 1970), although such a study should be repeated with anatomical/physiological identification of cell types. Surface stimulation of the cruciate cortex inhibits the firing of pyramidal cells excited by L-glutamate for much longer periods than firing at the same frequency produced by DL-homocysteate (MCLENNAN, 1970). Although this may indicate differences in the nature of L-glutamate and DL-homocysteate receptors (MCLENNAN, 1970), an alternative explanation has been provided in terms of the distribution of amino acid receptors activated by these excitants (CURTIS et al., 1972a).

A number of possible antagonists of amino acid excitation have been tested in the feline pericruciate cortex, particularly on cells also fired by acetylcholine. Selective antagonism of excitation by aspartate and glutamate (and DL-homocysteate) has been demonstrated using L-methionine-DL-sulphoximine, 2-methoxy-aporphine (CURTIS et al., 1972a) and 1-hydroxy-3-aminopyrrolidone-2 (DAVIES and WATKINS, 1972, 1973). None of these agents appeared to modify the spontaneous firing of cortical neurones, and the effects on synaptic firing have yet to be determined.

4.1.2. Inhibition

The extensive literature concerning the possible role of neutral amino acids, particularly GABA, as inhibitory transmitters in the mammalian cerebral cortex has been the subject of numerous reviews (ROBERTS, 1960, 1968; CURTIS and WATKINS, 1965; KRNJEVIĆ, 1970; SYTINSKY, 1972). The essential evidence can be summarized as follows.

Cortical glycine levels are relatively low (Table 3) and the content in the cerebral hemispheres of the cat (1.3 μmole/g) and rat (0.8–1 μmole/g) are lower than those of spinal grey matter (SHAW and HEINE, 1965; APRISON, SHANK, and DAVIDOFF, 1969; see also SHANK and APRISON, 1970). A high affinity uptake system for glycine has not been found in cortical tissue (Rat: JOHNSTON and IVERSEN, 1971; NEAL, 1971; BENNETT et al., 1972).

There are considerable species differences in cortical levels of taurine (Table 3; see also GAITONDE, 1970). High levels are found in the cerebral hemispheres of the rat (7–8 μmole/g, KACZMAREK, AGRAWAL, and DAVISON, 1971; SHANK and APRISON, 1970) and the mouse (VAN GELDER, 1972) although levels in the cerebral cortex of the cat are lower. No particular association with synaptosomes has been demonstrated. There seems to be little correlation between regional distributions of taurine and the enzyme responsible for its synthesis although cysteine sulphinic acid decarboxylase is present in synaptosomes (KACZMAREK, AGRAWAL, and DAVISON, 1970). Two uptake systems for taurine have been demon-

Table 3. Depressant amino acids in regions of cerebral cortex (μmole/g, *Biopsy)

Region	Animal	GABA	Glycine	Taurine	α-Alanine	β-Alanine	Cyst-athionine	Reference
Frontal	Human	1.9					0.5	GJESSING and TORVICK (1966)
Frontal	Human	2.1	1.5	1.3	1.4	0.02	0.4	PERRY et al. (1971a)
Frontal	Human		1.4					BOEHME et al. (1973)
Occipital	Human	2.3	1.8	1.2	1.3	0.02	0.9	PERRY et al. (1971a)
Temporal	Human	2.1	2.0	1.6	1.4	0.02	0.3	PERRY et al. (1971a)
Temporal	Cat	1.4	1.3	1.9	0.9		0.3	BATTISTIN et al. (1969)
Temporal	Cat			2.1				GUIDOTTI et al. (1972)
Frontal	Cat			2.1				GUIDOTTI et al. (1972)
Parietal	Cat			2.3				GUIDOTTI et al. (1972)
Occipital	Cat			2.1				GUIDOTTI et al. (1972)
Post cruciate	Cat	1.2	0.7		0.2			VAN GELDER (1972)
Various areas	Cat	1.8*	0.7*		0.6*			KOYAMA (1972)
Association	Cat	0.9*	0.6*	1.3*	0.5*	tr.*	0.2*	PERRY et al. (1972)
Motor	Cat	0.8*	0.5*	1.3*	0.5*	tr.*	0.2*	PERRY et al. (1972)
Visual	Cat	1.0*	0.6*	1.7*	0.6*	tr.*	0.3*	PERRY et al. (1972)
Post cruciate	Mouse	1.5	0.9		0.4			VAN GELDER (1972)
Neocortex	Mouse	2.1	1.4		1.8			SHIMADA et al. (1972)
Motor	Monkey	2.1						FAHN and CÔTÉ (1968)
Occipital	Monkey	2.7						FAHN and CÔTÉ (1968)
Frontal	Monkey	2.1						FAHN and CÔTE (1968)
Frontal	Baboon	2.4						OKADA et al. (1971)
Frontal	Guinea pig	2.6						OKADA et al. (1971)
Frontal	Rabbit	2.1						OKADA et al. (1971)
Frontal	Rat	2.9						OKADA et al. (1971)
Corpus callosum	Cat	1.0	0.6	3.0	0.7		0.7	BATTISTIN et al. (1969)

strated in rat cortical slices (KACZMAREK and DAVISON, 1972; LÄHDESMÄKI and OJA, 1972), both differing from that associated with GABA transport. The efflux of taurine from rat cortical slices has also been studied (KACZMAREK and DAVISON, 1972; LÄHDESMÄKI and OJA, 1972), and under *in vivo* conditions the efflux of taurine and of glycine from the surface of the cortex of intact or encephale isolé cats was enhanced during electrocortical arousal (JASPER and KOYAMA, 1969).

Although GABA levels in the cortical grey matter are not high (Table 3) the values exceed that of white matter (Human: 0.7 μmole/g, SYTINSKY, 1969. Rhesus monkey: 0.3 μmole/g, FAHN and CÔTÉ, 1968; 0.5 μmole/g, DEFEUDIS, DELGADO, and ROTH, 1970). Furthermore the GABA content of single Betz cells (Cat: 2,5 mM, OTSUKA, OBATA, MIYATA, and TANAKA, 1971) is higher than that of spinal motoneurones (0.9) but not as high as that of Deiters' neurones (6.3). In the motor cortex of adult monkey (*Macaca mulatta*) highest levels of GABA occur in the outer two layers, whereas in the visual cortex the highest level is found in layer 4 (HIRSCH and ROBINS, 1962). GABA has not been shown to be specifically associated with synaptosomes prepared from cortical tissue (RYALL, 1964; MANGAN and WHITTAKER, 1966; WHITTAKER, 1968; NEAL and IVERSEN, 1969; TACHIKI et al., 1972), although a high proportion of cortical GAD is present in these particles (DE ROBERTIS, 1968). The level of this enzyme in the cerebral cortex is moderately high (LOWE, ROBINS, and EYERMAN, 1958; ALBERS and BRADY, 1959; MÜLLER and LANGEMANN, 1962), in the occipital cortex levels in layers 3 and 4a exceed those of other cortical layers (ALBERS and BRADY, 1959). Similarly, cortical levels of GABA-T are moderate, in both the occipital and motor areas the highest levels occur in layers 2 and 3 (SALVADOR and ALBERS, 1959).

A number of investigations have been concerned with the specificity, kinetics, mechanism and effects of drugs on the carrier mediated uptake of GABA by homogenates or slices of cortical tissue (see Section 3.2). The subcellular distribution of labelled GABA appears to be similar to that of the endogenous amino acid (NEAL and IVERSEN, 1969, but see TACHIKI, et al., 1972), the principal sites of uptake by homogenates and slices being nerve terminals (IVERSEN and BLOOM, 1972; HÖKFELT and LJUNGDAHL, 1971b). Following direct injection of labelled GABA into the parietal cortex of rats, the amino acid is accumulated by stellate cells of layers II and III (HÖKFELT and LJUNGDAHL, 1972a).

The efflux of GABA from the surface of the cerebral cortex of the cat has been correlated with the presence of slow wave and spindle EEG activity, being considerably reduced by reticulo-cortical activation (JASPER, KHAN, and ELLIOTT, 1965; JASPER and KOYAMA, 1969). GABA efflux, however, is enhanced by electrical stimulation of medial brain stem structures (Cat: KOYAMA and JASPER, 1972, personal communication) and a calcium-dependent release of both endogenous and labelled GABA (presumably from inhibitory synaptic terminals) from the cat visual cortex has been demonstrated to accompany cortical inhibition induced by electrical stimulation of the cortical surface or of the lateral geniculate nucleus (IVERSEN et al., 1971; IVERSEN, MITCHELL, NEAL, and SRINIVASAN, 1970). Both the spontaneous and evoked release of GABA from the visual cortex tended to be greater in animals pretreated with amino-oxyacetic acid (IVERSEN et al., 1971).

Neurochemical data for the participation of other neutral amino acids in cortical inhibition, for example imidazole-4-acetate and β-alanine, are not as detailed as that available for GABA (see Section 3.4).

Important evidence for the transmitter function of GABA has been obtained in microelectrophoretic experiments of the effects of amino acids on the firing of single cortical neurones. Most experiments have been performed on physiologically identified pyramidal tract cells of the feline pericruciate cortex, and the findings account for earlier observations of the effects of topically or systemically administered amino acids on cortical function (see CURTIS and WATKINS, 1965). The depression of the firing of cortical neurones by GABA (Cat: KRNJEVIĆ and PHILLIS, 1963; CRAWFORD and CURTIS, 1964; KRNJEVIĆ, 1964. Rat: BISCOE, DUGGAN, and LODGE, 1972. Rabbit: CLARKE and HILL, 1972) is associated with membrane hyperpolarization and an increased conductance (KRNJEVIĆ and SCHWARTZ, 1967b) which, like the membrane changes accompanying inhibition at synapses activated by electrical stimulation of the cortical surface (KELLY, KRNJEVIĆ, MORRIS, and YIM, 1969), appears to involve predominantly an increased chloride ion permeability (DREIFUSS, KELLY, and KRNJEVIĆ, 1969). Since most comparisons between synaptic and GABA inhibition have hitherto been made on cortical neurones with relatively low resting potentials, further examination of the ionic mechanism of inhibition is warranted, including that of inhibitions evoked by more physiological stimuli. Nevertheless the postsynaptic action of GABA is generally accepted to closely resemble that of the inhibitory transmitter released upon the bodies of cortical pyramidal cells.

Of the naturally occuring neutral amino acids, GABA is the most effective depressant of the firing of cortical neurones, glycine, α-alanine and taurine being much less active (KRNJEVIĆ and PHILLIS, 1963; CRAWFORD and CURTIS, 1964; JOHNSON et al., 1970; CURTIS et al., 1971b; BISCOE et al., 1972). Furthermore the postsynaptic inhibitory action of glycine may differ from that of GABA (KELLY and KRNJEVIĆ, 1969). Although electrophoretically administered amino-oxyacetic acid did not modify the time course of either the action of similarly administered GABA, or the synaptic inhibition of cortical neurones, the effect of GABA was enhanced, and inhibition resulting from stimulation of the cortical surface or the pyramidal tract prolonged, following intravenous administration of amino-oxyacetic acid (GOTTESFELD, KELLY, and RENAUD, 1972). These effects were interpreted in terms of an interference with the removal of GABA after synaptic or electrophoretic release (see also SNODGRASS and IVERSEN, 1973b), but the selectivity of amino-oxyacetic acid in relation to other depressant amino acids was not determined (GOTTESFELD et al., 1972).

Further evidence for an inhibitory transmitter role of GABA (or a GABA-like substance) has been obtained by the use of strychnine and bicuculline. The effects of these convulsants on the firing and responses of cortical neurones are complex, particularly after systemic administration. However much valuable information has been obtained from microelectrophoretic experiments, and the controversies which have arisen are perhaps mainly associated with methodological difficulties and differences of technique rather than reflecting serious conflicts in the understanding of amino acid-receptor interactions. In numerous electrophoretic experiments strychnine has been demonstrated to be a selective antagonist of the depression of feline cortical neurones by glycine (Curtis et al., 1968a, 1971b; Johnson et al., 1970), β-alanine and taurine (Curtis et al., 1968a, 1971b), the action of GABA being relatively insensitive. The same high degree of selectivity appears not to occur when rat cortical neurones were studied (Biscoe et al., 1972). Furthermore the inhibition of the firing of pyramidal tract cells in the cat by local cortical stimulation (Krnjević, Randić and Straughan, 1966; Crawford, Curtis, Voorhoeve, and Wilson, 1963), by recurrent volleys in the pyramidal tract (Crawford et al., 1963; Biscoe and Curtis, 1967), by stimulation of thalamic nuclei (Krnjević et al., 1966), and by direct chemical stimulation of intracortical interneurones (Biscoe and Curtis, 1967) was not blocked by strychnine.

In contrast, electrophoretically administered bicuculline selectively reduced the inhibitory effect of GABA and some related amino acids (β-alanine, taurine and imidazole acetic acid) on cortical neurones of the cat (Curtis et al., 1971b; see also Straughan, Neal, Simmonds, Collins, and Hill, 1971). Similar results have been observed using bicuculline methochloride (Curtis, Game, Johnston, and McCulloch, unpublished observations). The same degree of selectivity has not been demonstrated in the rat cortex (Biscoe et al., 1972) where in a high proportion of tests bicuculline reduced the effects of both glycine and GABA. These observations are of considerable relevance to the reduction by bicuculline of the synaptic inhibition of pyramidal tract neurones by recurrent volleys, direct cortical electrical stimulation or chemical excitation of intracortical neurones (Cat: Curtis and Felix, 1971). Additionally, antagonism of the inhibitory effects of β-alanine and taurine on cortical neurones by both strychnine and bicuculline probably excludes either amino acid as a transmitter of inhibitory synapses selectively blocked by bicuculline.

Picrotoxin has also been reported to reduce the inhibition of cortical neurones in the cat by GABA (Hill, Simmonds, and Straughan, 1972b), and to reduce the effects of both GABA and glycine in the rat cortex (Biscoe et al., 1972). Further investigation is warranted since the recurrent inhibition of pyramidal cells in the cat has been reported to be diminished by relatively low doses (0.1–0.3 mg/kg, Brooks and Asanuma, 1965), but inhibition evoked by direct cortical stimulation appeared insensitive (0.6–2 mg/kg, Krnjević et al., 1966). This latter inhibition was also apparently not modified by tetanus toxin (Krnjević et al., 1966), although abolition of the recurrent and transcallosal inhibition of pericruciate neurones (Brooks and Asanuma, 1965) suggests that the effects of tetanus toxin may be similar to those found in the cerebellar cortex (Curtis et al., 1973a). Penicillin has also been demonstrated to antagonize the inhibitory effect of GABA on cortical neurones, an action which may account for the epileptogenic activity of this compound (Curtis et al., 1972b).

As a consequence of these studies, a proportion of stellate cells in the cerebral cortex may be considered inhibitory in nature, releasing GABA at axosomatic synapses on pyramidal cells (see Curtis and Felix, 1971). Further studies are required to ascertain the location of the interneurones activated by volleys in particular inhibitory pathways, and to determine the significance of the inhibitory control by this amino acid in the functioning of pyramidal and stellate cells in various cortical areas. The importance of glycine as a cortical transmitter may be minimal, the effects of strychnine on cortical activity (Ajmone-Marsan,

1969) probably not being associated with a specific action at glycinergic synapses (see CURTIS et al., 1971c).

4.2. Cerebellum

4.2.1. Excitation

Although aspartate and glutamate levels are relatively high in the cerebellum (Table 4), the intracerebellar distributions of these amino acids have not been determined, apart from the estimation of glutamate in the cerebellar cortex (Cat: 10.8 μmole/g, JOHNSON and APRISON, 1971).

Table 4. Cerebellar excitant amino acids (μmole/g, *Biopsy)

Animal	Aspartate	Glutamate	Reference
Human	2.3	9.6	PERRY et al. (1971a)
	0.8*	7.2*	PERRY et al. (1971b)
Cat	2.9	12.6	BATTISTIN et al. (1969)
		10.8	JOHNSON and APRISON, 1971
	1.8*	8.5*	PERRY et al. (1972)
Rat	2.2	8.4	SHAW and HEINE (1965)
	2.0	10.3	KANDERA et al. (1968)
	2.0	9.7	SHANK and APRISON (1970)

Purkinje, basket and granule cells, and cells of the intracerebellar nuclei are excited by glutamate and DL-homocysteate (CRAWFORD, CURTIS, VOORHOEVE, and WILSON, 1966; CHAPMAN and MCCANCE, 1967; MCCANCE and PHILLIS, 1968; KAWAMURA and PROVINI, 1970), and the sensitivity of developing Purkinje cells to glutamate precedes the presence of dendritic synapses (WOODWARD, HOFFER, SIGGINS, and BLOOM, 1971).
Doubt has recently been cast on the identity of ergothioneine as the "cerebellar excitatory factor"(BRIGGS, 1972); when administered electrophoretically ergothioneine did not excite cerebellar neurones in the cat (CRAWFORD et al., 1966).

4.2.2. Inhibition

Although the overall levels of GABA in the cerebellum are not high (Table 5), the highest values of both the amino acid and GAD occur in the Purkinje cell layer (Rabbit: KURIYAMA, HABER, SISKEN, and ROBERTS, 1966. Monkey and rat: HIRSCH and ROBINS, 1962; see also LOWE et al., 1958; ALBERS and BRADY, 1959).

The GABA-T content of the molecular layer is double that of the granular layer (Monkey: SALVADOR and ALBERS, 1959). In the cat particularly high levels of GABA are associated with both Purkinje cells (5.5 ± 1.1 μmole/g, cerebellar

Table 5. Cerebellar depressant amino acids (μmole/g, *Biopsy)

Region	Animal	GABA	Glycine	Taurine	α-Alanine	Cyst-athionine	Reference
Cere-bellum	Human	3.0				0.9	GJESSING and TORVICK (1966)
	Human	2.3	2.1	3.3	1.7	0.09	PERRY et al. (1971a)
	Human	0.8*	0.5*	2.5*	0.7*	3.2*	PERRY et al. (1971b)
	Human		1.7				BOEHME et al. (1973)
	Cat	1.5	1.5	3.1	0.9	0.2	BATTISTIN et al. (1969)
	Cat	0.8*	0.5*	2.9*	0.3*	0.2*	PERRY et al. (1972)
	Cat		0.8				APRISON, SHANK, and DAVIDOFF (1969)
	Rat	1.6	1.0	4.9	0.9	0.2	SHAW and HEINE (1965)
	Rat	1.5	0.6	5.6	0.6	0.3	KANDERA et al. (1968)
	Rat	1.6	0.6	6.2	0.6		SHANK and APRISON (1970)
	Mouse	2.2	2.2		0.8		SHIMADA et al. (1972)
Cortex (not pure grey)	Rhesus monkey	2.0					FAHN and CÔTÉ (1968)
Vermal cortex	Baboon	2.1					OKADA et al. (1971)
	Rat	3.0					OKADA et al. (1971)
	Rabbit	1.8					OKADA et al. (1971)
	Guinea pig	2.5					OKADA et al. (1971)
	Mouse	2.5	2.6		1.3		SHIMADA et al. (1972)

cortex 1.0 ± 0.1) and cells of intracerebellar nuclei (8.0 ± 1.1, nuclei 1.8 ± 0.2; OTSUKA et al., 1971). Since the relatively high levels of GAD in the rat cerebellar cortex (46.8 ± 3.6 μmole/h/g dry weight) and nucleus interpositus (151.8 ± 20.5) are not significantly reduced by transection of the peduncles, GAD activity is largely associated with neurones *intrinsic* to the cerebellum (FONNUM, 1972). Furthermore the enzyme seems roughly evenly distributed between synaptic endings and other cell components (FONNUM, 1972). Recent calculations, based on amino acid and enzyme measurements, subcellular fractionation and quantitative morphological analyses, suggest that the concentration of GABA in Purkinje cell terminals in the nucleus interpositus is 86–138mM, the activity of GAD permitting the synthesis of GABA at a maximum rate of 613–980 mM/hr (Cat: FONNUM and WALBERG, 1973).

There is very strong evidence that the Purkinje cells, which are inhibitory in function, release GABA as a transmitter at terminals in the lateral vestibular nucleus (see Section 4.9). Impulses in Purkinje cell axons also inhibit monosynaptically neurones of intracerebellar nuclei (ITO, YOSHIDA, OBATA, KAWAI, and UDO, 1970). These nuclei lie close to the roof of the fourth ventricle and, in cats pretreated with amino-oxyacetic acid, electrical stimulation of the cerebellar cortex enhances the spontaneous efflux of GABA into this ventricle by a factor of approximately three (OBATA and TAKEDA, 1969). Some success has also been achieved in detecting the release of GABA into intracerebellar nuclei by means of a push-pull cannula (OBATA, 1972b).

The distribution of GAD also suggests that GABA could be a transmitter within the cerebellar cortex and nuclei. Such a proposal gains support from pharmacological evidence cited below, and observations that [^{3}H]-GABA, after direct injection into the cerebellum (Rat: HÖKFELT and LJUNGDAHL, 1972a) is accumulated in stellate cells, basket cell axons and terminals, and possibly also by Golgi cells, all of inhibitory function (ECCLES, ITO, and SZENTAGOTHAI, 1967). Similarly, superficially located stellate cells accumulate [^{3}H]-GABA after injection into the lateral ventricle (Rat: SCHON and IVERSEN, 1972), no such specific localization could be shown for [^{3}H]-glycine. The uptake of GABA by rat cerebellar cortical tissue is approximately 50% of that by cerebral cortical tissue (HÖKFELT, JONSSON, and LJUNGDAHL, 1970; IVERSEN and JOHNSTON, 1971). Labelled GABA is also accumulated by neurones of cultured rat cerebellar tissue, uptake by glia, granule cells and macrophages being minimal (SOTELO, PRIVAT, and DRIAN, 1972).

Of the other amino acids present in the cerebellum (Table 5), glycine and α-alanine levels are low, whereas those of taurine are high: the intracerebellar distribution of this amino acid is unknown. Only a low affinity uptake process for glycine has been found in cerebellar tissue (Rat: JOHNSTON and IVERSEN, 1971).

The firing of Purkinje cells is readily depressed by electrophoretically administered glycine, GABA and related amino acids including β-alanine and taurine (Cat: KAWAMURA and PROVINI, 1970; CURTIS et al., 1971b). Of these substances GABA was the most effective, and hyperpolarizes and increases the membrane conductance of Purkinje cells (Rat: SIGGINS et al., 1971. Frog: WOODWARD, HOFFER, SIGGINS, and OLIVER, 1971). It is thus highly significant to the transmitter role of GABA that the inhibition of Purkinje cells which follows excitation of cerebellar basket (and stellate cells) is not affected by strychnine (Cat: *Intravenous*, ANDERSEN, ECCLES, LØYNING, and VOORHOEVE, 1963; CRAWFORD et al., 1963; BISTI, IOSIF, MARCHESI, and STRATA, 1971. *Electrophoretic,* CURTIS and FELIX, 1971. Also frog: WOODWARD et al., 1971), an alkaloid which does not influence the inhibitory effect of GABA (CURTIS, et al., 1971b) but is blocked by bicuculline (CURTIS and FELIX, 1971; BISTI et al., 1971), bicuculline methochloride (CURTIS and JOHNSTON, unpublished observations) or relatively large doses of picrotoxinin (BISTI et al., 1971). Furthermore the Golgi cell inhibition of cerebellar granule cells is not influenced by strychnine (intravenous) but is reduced by bicuculline or picrotoxinin (BISTI et al., 1971). Basket cell inhibition of Purkinje cells is also suppressed by tetanus toxin injected into the cerebellar folium, but as the cells retain sensitivity to electrophoretic GABA the toxin probably blocks synaptic release of the amino acid rather than its postsynaptic action (CURTIS, FELIX, GAME, and MCCULLOCH, 1973a).

In the absence of evidence for strychnine-sensitive inhibition in the cerebellar cortex, the role of glycine as an inhibitory transmitter is doubtful. Furthermore, although direct tests with taurine have not been carried out, the suppression of the depressant effect of β-alanine on Purkinje cells by both bicuculline and strychnine (CURTIS et al., 1971b) probably indicates that taurine is unlikely to function as an inhibitory transmitter of bicuculline-sensitive cerebellar inhibition. Thus, although these studies have been mainly carried out on the cerebellar vermis it seems probable that all inhibitory Purkinje, basket, stellate and Golgi cells within the mammalian cerebellum operate by releasing GABA as an inhibitory transmitter.

4.3. Rhinencephalon

4.3.1. Excitation

Little is known regarding excitant amino acid transmitters, apart from many observations of excitation of neurones by L-glutamate (STEFANIS, 1964; HERZ and GOGOLAK, 1965; HERZ and NACIMIENTO, 1965; STRAUGHAN and LEGGE, 1965; LEGGE, RANDIĆ and STRAUGHAN, 1966; BISCOE and STRAUGHAN, 1966; STEINER and RUF, 1967). One noteworthy feature is the ease with which seizure-like activity can be triggered (BISCOE and STRAUGHAN, 1966; STEFANIS, 1969) by exciting relatively few neurones within the hippocampal cortex. A study of the sensitivity of neurones in hippocampal slices (Guinea pig, cat: DUDAR, 1972) suggests that there are glutamate receptors on both cell bodies and dendrites. The latter presumably are associated with axodendritic synapses, but this type of experiment, both *in vitro* and *in vivo*, requires correlation with the distribution of enzymes involved in glutamate metabolism such as aspartate aminotransferase (BORRE and GENESER-JENSEN, 1972).

4.3.2. Inhibition

GABA is present in the hippocampus (Rabbit: 2.4 μmole/g. Rat: 3.6. Guinea pig: 2.7. Baboon: 2.8, OKADA, NITSCH-HASSLER, KIM, BAK, and HASSLER, 1971), and although detailed studies have not been reported of the distribution of depressant amino acids within various regions of the rhinencephalon, several investigations have been concerned with the distributions of GAD and GABA-T. In contrast to low activity in the fimbria, high levels of GAD occur in the molecular and pyramidal layers of area CA1, and in the molecular and granular layers of the area dentata (Rat: STORM-MATHISEN and FONNUM, 1971, 1972; see also ALBERS and BRADY, 1959). Futhermore surgical interruption of afferent pathways to these areas did not reduce GAD levels in either CA1 or area dentata (STORM-MATHISEN, 1972), suggesting that the enzyme is associated with cells *intrinsic* to the hippocampal cortex. A high proportion of the GAD from homogenized hippocampal tissue can be recovered in the synaptosomal fraction, the particles being denser than those containing choline acetyltransferase (FONNUM, 1972). All of these results are consistent with the presence of GAD in the bodies of inhibitory basket cells and their axosomatic terminals upon hippocampal pyramidal cells (STORM-MATHISEN and FONNUM, 1971; STORM-MATHISEN, 1972; FONNUM, 1972). Supporting neurochemical evidence is provided by the presence of GABA-T in the hippocampal cortex (SALVADOR and ALBERS, 1959) which in the mouse surrounds pyramidal cell bodies (VAN GELDER, 1965b), and the similar distribution of radio-label when slices of hippocampal tissue are incubated with [^{3}H]-GABA in the presence of amino-oxyacetic acid (Rat: HÖKFELT and LJUNGDAHL, 1971 b).

Electrophoretically administered GABA depresses the firing of neurones in the many regions of the cat rhinencephalon which have been examined—CA1, CA2, CA3 (STEFANIS, 1964; BISCOE and STRAUGHAN, 1966; CURTIS et al., 1971b): dentate gyrus (STEFANIS, 1964): septum (HERZ and GOGOLAK, 1965): amygdala (STRAUGHAN and LEGGE, 1965): pyriform cortex (LEGGE, RANDIĆ, and STRAUGHAN, 1966); and similar results have been observed in the rat (STEINER and RUF, 1967). Although GABA was a more effective depressant of the firing of pyramidal cells than either glycine or β-alanine (CURTIS et al., 1971b), hyperpolarization by GABA has yet to be demonstrated. However, bicuculline blocks the inhibition of pyramidal cells by GABA (and by β-alanine, CURTIS et al., 1971b) and reduces the prolonged synaptic inhibition (ANDERSEN, ECCLES, and LØYNING, 1964) of these neurones which follows excitation of basket cells (CURTIS, FELIX, and MCLENNAN, 1970c). This synaptic inhibition is not affected by strychnine (ANDERSEN et al., 1963), an alkaloid which appeared to have little selectivity as an amino acid antagonist in this region (CURTIS et al., 1971b). The effects of systemic or topical convulsants on hippocampal responses are complex (BAKER, KRATKY and BENEDICT, 1965; GESSI, RABINI, and VOLTA, 1967).

4.4. Retina

Although a number of amino acids are present in the vertebrate retina, together with relevant enzymes and transport processes (LOLLEY, 1969), the precise nature of the neurones synthesising and releasing particular amino acids has yet to be determined. The organization and structure of the vertebrate retina has been reviewed recently by DOWLING (1970).

The high levels of both GABA and GAD in the ganglion cell layer of the rabbit were considered to reflect technical difficulties in the physical separation of retinal layers rather than an association of the amino acid with excitatory neurones (KURIYAMA, SISKEN, HABEN and ROBERTS, 1968). The distribution of the enzyme and amino acid in the frog retina led to the suggestion that amacrine, and probably also horizontal, cells may synthesise GABA (GRAHAM, 1972). Glutamate is evenly distributed between the receptor, bipolar and ganglion cell layers of the frog retina, and the levels are not significantly different in retinae adapted to light or to darkness. In contrast GABA levels in both bipolar and ganglion cell layers are *decreased* in dark adapted frog retina (GRAHAM, BAXTER, and LOLLEY, 1970). Similarly, in the goldfish, light stimulation enhances the synthesis of GABA from glutamate (LAM, 1972).

Retinal tissue takes up amino acids, both *in vivo* and *in vitro*, and use has been made of radiolabelled compounds to determine both the kinetics of the processes (GABA-Rat: GOODCHILD and NEAL, 1973: high affinity, STARR and VOADEN, 1972a; NEAL and STARR, 1973. Glycine-Rabbit: high affinity, BRUUN and EHINGER, 1972. Taurine-Rat: low affinity, STARR and VOADEN, 1972b; Chicken: PASANTES-MORALES, KLETHI, URBAN, and MANDEL, 1972. Glutamate-Rat: high affinity, NEAL and WHITE, 1971) and sites of amino acid accumulation.

Although GABA accumulated by rat retinae, *in vitro* in the presence of aminooxyacetic acid, is located predominantly in Müller type glial cells (NEAL and IVERSEN, 1972), the major location in the rabbit, either *in vitro* or after intravitreal injection *in vivo* appears to be amacrine cells with some located in retinal ganglion cells (EHINGER and FALCK, 1971). On the other hand the major site of uptake in the goldfish retina is horizontal cells (LAM and STEINMAN, 1971). Glycine is also accumulated by some amacrine cells (Rabbit: EHINGER and FALCK, 1971; BRUUN and EHINGER, 1972; EHINGER, 1972a. Human, rat, guinea pig: EHINGER, 1972b) whereas aspartate and glutamate are found in Müller and retinal pigment cells (EHINGER and FALCK, 1971; EHINGER, 1972a).

In the goldfish, both *in vitro* and after intravitreal injection, little GABA is located in amacrine cells, the major site of accumulation being external and internal horizontal cells, particularly when the retinae are stimulated with light (LAM and STEINMAN, 1971). Following the uptake of labelled GABA by rat retinae *in vitro,* although exposure to light does not enhance GABA efflux, more GABA is released by dark adapted retinae than by those in the light (VOADEN and STARR, 1972).

When administered electrophoretically L-glutamate enhances and GABA depresses the firing of retinal ganglion cells (Rabbit: NOELL, 1959. Cat: STRASCHILL and PERWEIN, 1969), effects consistent with observations made after intra-arterial (Cat: STRASCHILL, 1968), intravenous (Chicken: SCHOLES and ROBERTS, 1964), or topical (Limulus: ADOLPH, 1966. Bullfrog: KISHIDA and NAKA, 1967. Rabbit: AMES and POLLEN, 1969. Carp: horizontal cells, MURAKAMI, OHTSU, and OHTSUKA, 1972) administration of these amino acids. Intravitreal GABA decreases the sensitivity of the rat retina to light stimuli, whereas picrotoxin and bicuculline increase light sensitivity (GRAHAM and PONG, 1972a) and produce rhythmic potentials originating in the retina (GRAHAM and PONG, 1972b).

Conflicting results have been obtained of the influence of strychnine and picrotoxin (BORNSCHEIN and HEISS, 1966; CHU, 1968; VORKEL and HANITZSCH, 1971; BURKHARDT, 1972) on retinal responses to light, and although GABA may be the inhibitory transmitter of amacrine cells, and probably also of horizontal cells, further investigation is required of the role of this and related amino acids in the retina. A function for glutamate as an excitatory transmitter is suggested by the association between glutamate and retinal spreading depression (VAN HARREVELD and FIFKOVA, 1970, 1971, 1972).

4.5. Olfactory Bulb

The main physiological and pharmacological interest in amino acid transmitters in the olfactory bulb involves the participation of GABA as an inhibitory transmitter released at granule cell synapses on mitral cells. The rat olfactory bulb (POPOV, POHLE, RÖSLER, and MATTHIES, 1967) contains high levels of glutamate (13 μmole/g), moderate levels of GABA (3 μmole/g) and relatively low levels of aspartate (2 μmole/g).

Glycine, β-alanine and GABA depress the firing rate of mitral cells, and GABA is usually more potent than either glycine or β-alanine (Cat: FELIX and

McLennan, 1971; Rabbit: Nicoll, 1971). Electrophoretic strychnine reduces the inhibitory action of glycine but not that of GABA (Felix and McLennan, 1971; Nicoll, 1971), whereas bicuculline (Felix and McLennan, 1971; Nicoll, 1971) and picrotoxin (Nicoll, 1971) are antagonists of GABA but not of glycine. Although some bulb neurones are excited by aspartate, glutamate and DLH, these amino acids often inhibit mitral cell firing (Rabbit: von Baumgarten, Bloom, Oliver and Salmoiraghi, 1963; Nicoll, 1971. Cat: Felix and McLennan, 1971; McLennan, 1971). Such an effect is considered to be mediated via GABA-releasing inhibitory interneurones, presumably granule cells, since the inhibition is occasionally preceded by brief excitation of mitral cells and is often associated with excitation of neighbouring neurones having small action potentials. Moreover, the inhibition can be blocked or even reversed into excitation by raising the extracellular magnesium concentration (Nicoll, 1971), or by bicuculline (Felix and McLennan, 1971; McLennan, 1971; Nicoll, 1971) or picrotoxin (Nicoll, 1971).

The antidromic excitation of mitral cells by electrical stimulation of the lateral olfactory tract (LOT) is followed by prolonged inhibition which is considered to arise mainly from the inhibitory action of granule cells at reciprocal dendrodendritic synapses (Shepherd, 1972). Strychnine does not affect this inhibition (Rabbit: Green, Mancia, and von Baumgarten, 1962; Nicoll, 1971. Cat: Felix and McLennan, 1971), nor that evoked by stimulating the anterior commissure (Rabbit: von Baumgarten, Green, and Mancia, 1962). The effects of picrotoxin and bicuculline on LOT inhibition of mitral cells are complex. Neither agent has been shown to reduce the duration of inhibition of single mitral cells after electrophoretic administration (Felix and McLennan, 1971; Nicoll, 1971) and in the cat the duration was frequently enhanced (Felix and McLennan, 1971; McLennan, 1971). A similar enhancement occurred when inhibition was evoked by stimulating the olfactory mucosa (McLennan, 1971) or when bicuculline was administered intravenously (LOT inhibition, McLennan, 1971). In contrast, when LOT inhibition of mitral cells is detected as a reduction in the amplitude of LOT-evoked field potentials (N_2 wave) by a previous LOT stimulus, both bicuculline and picrotoxin, systemically administered, effectively reduce the inhibition (McLennan, 1971; Nicoll, 1971).

A number of explanations, in terms of anatomical complexities within the bulb, have been offered to account for the discrepancy between the effects of GABA antagonists upon synaptically induced inhibition of mitral cells and that which follows amino acid excitation of granule cells (Nicoll, 1971; McLennan, 1971; Felix and McLennan, 1971), but all results seem to be consistent with mediation of granule cell inhibition of mitral cells by GABA. The inhibition of mitral cells by noradrenaline has been explained as secondary to the excitation of granule cells (McLennan, 1971).

The inhibition of field potentials recorded in the external plexiform layer by a preceding LOT volley is enhanced and prolonged by pentobarbitone and other anaesthetics (Westecker, 1971; Nicoll, 1972). Such an effect, considered to be related to the enhancement of spinal prolonged inhibition by these agents, may arise from facilitation of the release of GABA from inhibitory terminals, modification of processes inactivating released GABA, or an enhancement of the postsynaptic action of the transmitter (NICOLL, 1972). Interference with GABA metabolism appears unlikely since the effect could not be produced by administering amino-oxyacetic acid (NICOLL, 1972).

4.6. Thalamus

4.6.1. Neurochemistry

Few detailed studies have been made of the amino acid content of various thalamic nuclei. In general glycine, taurine and GABA levels are relatively low (Table 6), although the taurine content of the feline lateral geniculate nucleus is higher than that of other nuclei (GUIDOTTI et al., 1972).

Table 6. Thalamic amino acid levels (μmole/g)

Region	Animal	Aspartate	Glutamate	GABA	Glycine	Taurine	α-Alanine	Cystathionine	Reference
Thalamus	Human	2.5	8.4	2.5	2.3	1.1	1.8	1.0	PERRY et al. (1971 a)
	Human				2.4				BOEHME et al. (1973)
Thalamus medial	Rhesus monkey			3.0					FAHN and CÔTÉ (1968)
lateral				2.7					FAHN and CÔTÉ (1968)
anterior				2.5					FAHN and CÔTÉ (1968)
Thalamus	Baboon			3.5					OKADA et al. (1971)
	Cat	2.7	12.4	3.7	1.7	1.1	0.6	0.4	BATTISTIN et al. (1969)
	Cat			4.5					KRŽALIĆ et al. (1962)
	Cat					2.3			GUIDOTTI et al. (1972)
Lateral geniculate	Cat					4.3			GUIDOTTI et al. (1972)
Medial geniculate	Cat					1.7			GUIDOTTI et al. (1972)
	Cat			≈4.8					UTLEY (1963)
Diencephalon	Cat				1.6				APRISON et al. (1969)
	Rat	2.4	8.6	3.5	0.9	4.5	0.5		SHANK and APRISON (1970)
Thalamus	Ox			2.1					LOVELL and ELLIOTT (1963)
	Mouse			2.3	2.2		1.2		SHIMADA et al. (1972)

Thalamic levels of GAD are not high when compared with that of the substantia nigra (LOWE et al., 1958; MÜLLER and LANGEMANN, 1962; ALBERS and BRADY, 1959), whereas the activity of GABA-T in the monkey is relatively high, the levels being higher in the medial than in the lateral thalamus (SALVADOR and ALBERS, 1959). As in other areas, glutamate is the most abundant amino acid present in the rat diencephalon (8.6 μmole/g, SHANK and APRISON, 1970). The aspartate and glutamate contents of optic nerves (Rat: 2.8, CASOLA and DI MATTEO, 1972. Cat: 3.5, JOHNSON and APRISON, 1971) are similar to those of dorsal roots (Table 8). The relatively high optic nerve content of taurine (CASOLA and DI MATTEO, 1972), and also that of the lateral geniculate nucleus (Cat: GUIDOTTI et al., 1972), may be related to the high retinal levels of this amino acid (KUBIČEK and DOLENÉK, 1958).

Section of the optic nerve alone produces no significant alterations in lateral geniculate levels of either GABA, aspartic acid, glutamine or glutamic acid in the rabbit (MARGOLIS, HELLER, and MOORE, 1968). Observations of the effect of ablation of the visual cortex suggested that aspartate and glutamate are largely contained in lateral geniculate cell bodies rather than in terminals of optic fibres (MARGOLIS et al., 1968). An association of GABA with intrathalamic interneurones rather than with neurones projecting from or to thalamic nuclei is suggested by the minimal changes in GABA levels which follow optic nerve section and visual cortex ablation (Rabbit: lateral geniculate, MARGOLIS et al., 1968) and removal of cortical auditory projection areas (Cat: medial geniculate, UTLEY, 1963).

4.6.2. Excitation

Thalamic neurones are excited by aspartate, glutamate and related amino acids (Cat: CURTIS and DAVIS, 1962; CURTIS and WATKINS, 1963; ANDERSEN and CURTIS, 1964a; MCCANCE, PHILLIS, TEBĒCIS, and WESTERMAN, 1968; MCLENNAN, HUFFMAN, and MARSHALL, 1968; TEBĒCIS, 1970). Differences in relative amino acid sensitivity have been reported in the ventrolateral nucleus, where neurones excited synaptically by impulses in the brachium conjunctivum were more sensitive to glutamate, relative to DL-homocysteate and N-methyl-DL-aspartate, than those outside the nucleus (MCLENNAN et al., 1968; HALDEMAN et al., 1972). A gradient in sensitivity to aspartate and glutamate has also been demonstrated in the lateral geniculate nucleus: cells receiving an input from near the centre of the visual field tending to be more sensitive to L-glutamate than to L-aspartate, those with more peripheral receptive fields were more sensitive to L-aspartate than L-glutamate (Cat: MORGAN, VRBOVA, and WOLSTENCROFT, 1972). Such differences suggest the possible involvement of both aspartate and glutamate as excitatory transmitters within thalamic nuclei, which, in the lateral geniculate nucleus, may be associated with different types of optic nerve fibre (HOFFMANN, STONE, and SHERMAN, 1972).

The participation of L-glutamate as a transmitter has also been suggested from a study of the effects of excitant amino acid antagonists. Both DL-α-methylglutamic acid and L-glutamate diethylester reversibly reduced the excitant effects of L-glutamate on neurones of the ventralis lateralis nucleus to a greater extent than the effects of either L-aspartate, DL-homocysteate or acetylcholine (HALDEMAN et al., 1972; HALDEMAN and MCLENNAN, 1972). Furthermore, L-glutamate diethylester reversibly depressed the synaptic excitation of nucleus ventralis posterolateralis neurones by impulses from cutaneous nerves (HALDEMAN et al., 1972), and the synaptic excitation of ventralis lateralis neurones by brachium conjunctivum volleys (HALDEMAN and MCLENNAN, 1972). In contrast, neither of these glutamate analogues were found to be selective glutamate antagonists in the ventrobasal and lateral geniculate nuclei, and, although some selectivity towards glutamate and aspartate was demonstrated for L-methionine-DL-sulphoximine and 2-methoxyaporphine, this was considered inadequate to be of assistance in elucidating possible transmitter functions of the amino acids (CURTIS et al., 1972a).

The synaptic excitation of feline lateral geniculate neurones by impulses in the optic nerve is blocked by low concentrations of 5-hydroxytryptamine and a number of structurally related substances, in the absence of a change in neurone sensitivity to L-glutamate (CURTIS, and DAVIS, 1962; TEBĒCIS and DI MARIA, 1972a) and other excitatory amino acids, including D-glutamate, D-, L-aspartate, D-, L-cysteate, DL-homocysteate and N-methyl-D-aspartate (TEBĒCIS, 1973a). Since 5-hydroxytryptamine may suppress the release of optic nerve transmitters rather than block their postsynaptic action these observations do not necessarily exclude the transmitters being excitant amino acids. Such a selective action of 5-hydroxytryptamine has not been demonstrated on the excitation of ventrobasal neurones by impulses in the medial lemniscus (ANDERSON and CURTIS, 1964b) or of medial geniculate neurones by impulses in the brachium of the inferior colliculus (TEBĒCIS, 1973a). The transmitters of these afferent pathways may thus differ from those of the optic nerve.

4.6.3. Inhibition

The firing of thalamic neurones is readily depressed by neutral amino acids, and in general GABA is more effective than either β-alanine or glycine (Cat: Ventrobasal thalamus, ANDERSEN and CURTIS, 1964b; CURTIS et al., 1971b; DUGGAN and MCLENNAN, 1971. Lateral geniculate, TEBĒCIS and DI MARIA, 1972a. Medial geniculate, TEBĒCIS, 1970). Furthermore, antagonism has been demonstrated between bicuculline and GABA, and between strychnine and glycine (TEBĒCIS, 1970; Curtis et al., 1971b; DUGGAN and MCLENNAN, 1971; CURTIS and TEBĒCIS, 1972). Since the prolonged postsynaptic inhibition of thalamic neurones which follows either an orthodromic or antidromic volley is not blocked by strychnine (Ventrobasal complex, ANDERSEN et al., 1963; DUGGAN and MCLENNAN, 1971. Lateral geniculate, CURTIS and TEBĒCIS, 1972) but can be depressed by bicuculline (DUGGAN and MCLENNAN, 1971; CURTIS and TEBĒCIS, 1972), GABA seems most likely to be the inhibitory transmitter released at the terminals of thalamic interneurones excited by impulses in axon collaterals of thalamo cortical relay neurones in the feline ventrobasal complex and lateral geniculate nucleus. Since the inhibitory effects of β-alanine on ventrobasal neurones

(CURTIS et al., 1971 b), and β-alanine and taurine on lateral geniculate neurones (CURTIS and TEBĒCIS, 1972), appear to be reduced by both bicuculline and strychnine, these amino acids seem unlikely to be transmitters at inhibitory synapses selectively sensitive to bicuculline.

4.7. Hypothalamus

Although GABA levels in the mammalian hypothalamus are relatively high (Dog: 5.7 μmole/g, HAULICĂ, CĂPÂLNĂ, NESTIANU, BORDEIANU, and BĂDESCU, 1964. Rhesus monkey: 6.2, FAHN and CÔTÉ, 1968. Baboon: 4.48; rabbit, 5.33, OKADA et al., 1971) the distribution of this and other amino acids within various nuclei has yet to be determined. Aspartate, glutamate and GABA were released from electrically stimulated synaptosomes isolated from the hypothalamus (Sheep: EDWARDSON, BENNETT, and BRADFORD, 1972).

Neurones of the paraventricular and supraoptic nuclei are excited by L-glutamate and related amino acids, and firing is depressed by GABA and glycine (Cat: BRODY, DEFEUDIS, and DEFEUDIS, 1969; BARKER, CRAYTON, and NICOLL, 1971; NICOLL and BARKER, 1971. Rat: MOSS, DYBALL, and CROSS, 1971; DREIFUSS and KELLY, 1972. Rabbit: MOSS, URBAN, and CROSS, 1972). Excitation by L-glutamate and depression by GABA has also been observed in the ventromedial and lateral hypothalamic nuclei (Rat: OOMURA, OOYAMA, YAMAMOTO, ONO, and KOBAYASHI, 1969). The effect of glycine on supraoptic neurones is blocked by strychnine, that of GABA by picrotoxin and bicuculline, and since none of these antagonists (microelectrophoretic, intravenous) reduce the recurrent inhibition of supraoptic neurones evoked by stimulating their axons within the neurohypophysis, neither glycine nor GABA appear to be involved as transmitters in the inhibitory pathway (NICOLL and BARKER, 1971). Similar selective antagonism by strychnine and bicuculline has been demonstrated in the ventromedial hypothalamus (Rat: DREIFUSS and MATTHEWS, 1972). The effects of intraventricular GABA on the activity of reticulo-hypothalamic-hypophysical pathways had earlier been interpreted in terms of a metabolic rather than a transmitter function (HAULICĂ et al., 1964).

4.8. Basal Ganglia and Red Nucleus

4.8.1. Substantia Nigra

The very high levels of both GABA (Table 7) and GAD (Human: MÜLLER and LANGEMANN, 1962. Monkey: ALBERS and BRADY, 1959. Cat and human: MCGEER, MCGEER, WADA, and JUNG, 1971) in this region contrasts with only moderate amounts of GABA-T (Monkey: SALVADOR and ALBERS, 1959). In the human substantia nigra the highest GABA levels (5.1–5.5 μmole/g) occur in the middle portion of pars reticulata, bordering on the pars compacta (KANAZAWA, MIYATA, TOYOKURA, and OTSUKA, 1973).

Most (80%) of the GAD in the cat substantia nigra seems to be localized in nerve endings, and the levels are only slightly higher in pars reticulata than pars compacta (FONNUM, 1972). Unilateral subthalamic hemisection involving strio-nigral pathways in the rat reduces nigral GABA levels by approximately 50%, unilateral striatal destruction lowering nigral levels by only 30% (KIM, BAK, HASSLER, and OKADA, 1971). Additionally, after striatal coagulation in the rat, degenerating boutons have been observed in the proximal axodendritic and axosomatic regions of substantia nigral neurones, some of which in aldehyde fixed tissue contain flattened vesicles (KIM et al., 1971). There is thus neurochemical and morphological support for a GABA-releasing strio-nigral inhibitory pathway in the rat.

Table 7. Basal ganglia GABA levels (μmole/g, * Biopsy)

Region	Animal	Level	Reference
Substantia nigra	Human	5.8	GJESSING and TORVICK (1966)
	Human	5.3	PERRY et al. (1971a)
	Human	4.2	KANAZAWA et al. (1973)
	Rhesus monkey	9.7	FAHN and CÔTÉ (1968)
	Baboon	9.6	OKADA et al. (1971)
	Rabbit	8.5	OKADA et al. (1971)
	Rat	10.1	OKADA et al. (1971)
	Guinea pig	9.7	OKADA et al. (1971)
	Ox	2.7	LOVELL and ELLIOTT (1963)
Caudate	Human	1.9	GJESSING and TORVICK (1966)
	Human	3.0	PERRY et al. (1971a)
	Rhesus monkey	3.2	FAHN and CÔTÉ (1968)
	Cat	1.3*	PERRY et al. (1972)
	Ox	3.8	LOVELL and ELLIOTT (1963)
Caudate/Putamen	Mouse	2.7	SHIMADA et al. (1972)
Striatum	Rabbit	3.5	OKADA et al. (1971)
	Rat	3.6	OKADA et al. (1971)
	Guinea pig	2.9	OKADA et al. (1971)
Putamen	Human	2.9	GJESSING and TORVICK (1966)
Putamen/Globus pallidus	Human	5.7	PERRY et al. (1971a)
Putamen	Rhesus monkey	3.6	FAHN and CÔTÉ (1968)
Globus pallidus	Human	7.3	GJESSING and TORVICK (1966)
	Rhesus monkey	9.5	FAHN and CÔTÉ (1968)
	Baboon	8.9	OKADA et al. (1971)
	Rabbit	6.4	OKADA et al. (1971)
	Rat	7.7	OKADA et al. (1971)
	Guinea pig	8.2	OKADA et al. (1971)
Red nucleus	Human	1.9	PERRY et al. (1971a)

In cats, however, destruction of the striatum, excluding the globus pallidus, has been reported to result in no substantial reduction in nigral GAD activity (FONNUM, 1972). Unilateral lesions in the globus pallidus region lowers nigral GAD levels in cats by almost 60%, without significantly affecting those of the caudate nucleus, an observation which has been correlated with low levels of GAD in both the globus pallidus and substantia nigra of patients with Parkinson's disease, and may indicate the involvement of GABA in a pallidal-nigral inhibitory pathway (McGeer et al., 1971). In the presence of amino-oxyacetic acid, slices of rat substantia nigra accumulate GABA to a greater extent than either glycine or leucine (Okada and Hassler, 1973).

Cells of the feline substantia nigra are excited by electrophoretic glutamate and aspartate, and firing is inhibited by GABA, the cells being relatively insensitive to glycine (Feltz, 1971). Inhibitory hyperpolarizing potentials have been recorded from nigral neurones after stimulation within the caudate nucleus, the long latency IPSP's (14.6–20 msec) being considered monosynaptic with a conduction velocity of the afferent fibres approximating 0.9 ms^{-1} (Cat: Yoshida and Precht, 1971). The finding that the accompanying positive extracellular field potentials and the inhibition of spontaneously active neurones were reduced by intravenous picrotoxin (2.5 mg/kg), but not by strychnine (0.6 mg/kg), (Precht and Yoshida, 1971) supports the participation of GABA as the inhibitory transmitter of this caudato-nigral pathway.

4.8.2. Caudate and Other Nuclei

Table 7 indicates that in a number of species GABA levels are moderate in the caudate and putamen and higher in the globus pallidus. These figures are consistent with the levels of GAD in the caudate (16.2 µmole GABA/g dry weight/hour, putamen 16.4, and globus pallidus 33.4, Monkey: Lowe et al., 1958). Similarly high enzyme levels have been reported in the human (Müller and Langemann, 1962; McGeer et al., 1971). In contrast, however, the levels of GABA-T are relatively low in the monkey globus pallidus, but higher in the putamen and caudate (Salvador and Albers, 1959). High levels of GABA-T have been found in the human basal ganglia (Sheridan, Sims, and Pitts, 1967).

The caudate content of other amino acids is listed by DeFeudis et al., (1970), Perry et al., (1971 a, b, 1972) and Shimada et al., (1972). The uptake of GABA, aspartate and glutamate by subcellular particles of the rat striatum has been studied by Gfeller, Kuhar, and Snyder (1971).

Striatal GABA levels in the rat are not altered by transection of nigro-striatal pathways (Kim et al., 1971), and the GAD content of the cat caudate is not altered after mid-brain lesions in the vicinity of the substantia nigra (Hockman, Lloyd, Farley, and Hornykiewicz, 1971). Apart from these observations no information is available regarding the possible origin of GABA releasing inhibitory pathway projecting to these basal ganglia nuclei, although if caudato-pallidal fibres are axon collaterals of the inhibitory caudato-nigral pathway, GABA may be involved as an inhibitory transmitter within the pallidum as well as in the substantia nigra (see Yoshida, Rabin, and Anderson, 1972).

Caudate neurones are readily excited by L-glutamate or DL-homocysteate (Cat: Bloom, Costa, and Salmoiraghi, 1965; McLennan and York, 1967; Connor, 1970. Rabbit: Herz and von Freytag-Loringhoven, 1968) and depressed by GABA (Bloom et al., 1965; Herz and von Freytag-Loringhoven, 1968). The excitant amino acids are also effective in the globus pallidus-putamen (Cat: York, 1968. Monkey: York, 1972).

4.8.3. Red Nucleus

Dissected red nucleus neurones contain moderate levels of GABA (Cat: 3.5 mM, MIYATA, OBATA, TANAKA, and OTSUKA, 1970), and overall levels of other amino acids in the human nucleus are listed by PERRY et al. (1971 a). L-glutamate and L-aspartate excite neurones of the nucleus (Cat: DAVIS and VAUGHAN, 1969; ALTMANN, STEINBERG, BRUGGENCATE, and SONNHOF, 1972. Baboon: parvicellular and magnocellular neurones, DAVIS and HUFFMAN, 1969), and further investigation seems warranted of inhibitory pathways projecting to this region as glycine was a more potent depressant than GABA in the baboon (DAVIS and HUFFMAN, 1969), but the inhibitory actions of GABA were more pronounced than those of glycine in the cat (ALTMANN et al., 1972).

4.9. Vestibular Nuclei

There is very acceptable evidence that GABA is the inhibitory transmitter released at the terminals of Purkinje cell axons in the dorsal portion of the lateral vestibular nucleus (Deiters' nucleus), although the origin of pathways which could account for the inhibitory effect of glycine upon Deiters' cells remains unknown.

In the cat, axons of anterior vermal Purkinje cells terminate in Deiters' nucleus, the ventral portion of the lateral vestibular nucleus (LVN) being devoid of such afferents (WALBERG and JANSEN, 1961). The terminal portion of Purkinje axons form basket-like structures around Deiters' neurones (BRODAL, POMPEIANO, and WALBERG, 1962), the synaptic terminals are somatic and dendritic upon all types of neurone in the nucleus and contain flattened vesicles (MUGNAINI and WALBERG, 1967).

4.9.1. Neurochemistry

When dissected free from freeze dried tissue, the mean GABA content of single neurones and adhering terminals from the dorsal LVN is 6.3 μmole/g, that of ventral neurones being 2.7 μmole/g, the overall level in the vestibular nuclei being 1.8 μmole/g (Cat: OTSUKA et al., 1971). Similar analyses performed 9 and 40 days after total removal of the vermis yielded mean values of 1.7 μmole/g for dorsal neurones, the levels for ventral neurones being unaltered (OTSUKA et al., 1971). The suggestion that GABA is concentrated in Purkinje terminals on Deiters' neurones gained support from estimations of the levels of GAD in the vestibular nucleus before and after cerebellar ablation (FONNUM, STORM-MATHISEN, and WALBERG, 1970; see also FONNUM, 1972). Using tissue samples dissected from freeze-dried material, the activity of GAD in the dorsal portion of the nucleus (16.1 ± 3.4 μmoles CO_2/h/g dry weight) exceeded that in the ventral portion by a ratio of 2.54, and was approximately three times the activity of the enzyme in white matter dorsal to the LVN, a site expected to be rich in Purkinje axons. Subsequent to destruction of vermal Purkinje cells, the dorsal/ventral ratio of GAD activities fell to approximately 1, activity in the ventral portion remaining unaltered and there being no major disturbance of the distributions of choline-acetyltransferase, acetylcholinesterase, lactate dehydrogenase or succinate dehydrogenase. The major proportion (approx. 60%) of GAD was found in the particulate fraction of LVN homogenates (FONNUM et al., 1970), and calculations based on these studies, and those of OTSUKA and his colleagues, suggest that the concentration of GABA within Purkinje cell terminals on Deiters' neurones is approximately 58 mM, and the GAD activity is capable of synthesizing 350 mM GABA/h (Cat: FONNUM and WALBERG, 1973). Similar high levels of the amino acid and enzyme have been proposed for Purkinje terminals in the nucleus interpositus, and these values are comparable with GABA concentrations of 100–150 mM proposed for invertebrate inhibitory axon terminals (KRAVITZ and POTTER, 1965).

Detailed analyses are not available for the distribution of other amino acids or their associated enzymes in the LVN.

4.9.2. Inhibition

Monosynaptic inhibitory hyperpolarizations have been recorded from Deiters' neurones in response to impulses in Purkinje cell axons evoked by electrical stimulation of the anterior lobe of the cerebellum; the reversal potential of these IPSP's is more negative than the resting potential and the IPSP's are reversed to depolarizations by elevation of the intracellular chloride ion concentration (Cat: ITO and YOSHIDA, 1966; see also ECCLES et al., 1967). Electrophoretic hydroxylamine, but not amino-oxyacetic acid, occasionally enhanced the amplitude of IPSP's without any modification of time course (OBATA et al., 1967). Electrophoretic GABA, which depresses the firing of Deiters' neurones, also hyperpolarizes the membrane, an effect associated with an increased conductance and a reversal potential similar to that of IPSP's evoked by cerebellar stimulation (Cat: OBATA et al., 1967; OBATA, TAKEDA, and SHINOZAKI, 1970; BRUGGENCATE and ENGBERG, 1971). A hyperpolarizing action has also been demonstrated for glycine, β-alanine, δ-aminovaleric and ε-aminocaproic acids, the reversal potential for hyperpolarization by glycine being indistinguishable from that by GABA (OBATA et al., 1970; BRUGGENCATE and ENGBERG, 1971). In all studies, however, GABA has been the most potent depressant of Deiters' neurones (see also CURTIS et al., 1971 b).

A clear association between inhibition by Purkinje cell impulses and GABA has been established by the effects of strychnine, picrotoxin, and bicuculline. The synaptic inhibition (Purkinje cell) of Deiters' neurones is insensitive to strychnine (Cat: *Intravenous*, to 0.7 mg/kg, OBATA et al., 1967. *Electrophoretic*, OBATA et al., 1970) but is reduced or blocked completely by picrotoxin (Cat: *Intravenous*, 5–10 mg/kg, OBATA et al., 1970) or bicuculline (Cat: *Intravenous*, 0.05–0.1 mg/kg, CURTIS, DUGGAN, and FELIX, 1970a). Stychnine reduces the inhibitory effect of glycine without modification of that of GABA (*Electrophoretic, intravenous*, OBATA et al., 1970. *Electrophoretic*, BRUGGENCATE and ENGBERG, 1971; CURTIS et al., 1970a). In contrast, selective antagonism of GABA but not of glycine has been demonstrated with bicuculline (CURTIS et al., 1971 b) and with a proportion of cells tested with picrotoxin (OBATA et al., 1970; BRUGGENCATE and ENGBERG, 1971). Diphenylaminoethanol has also been proposed as an antagonist of both synaptic and GABA inhibition of Deiters' cells (HUFFMAN and YIM, 1969; KEE et al., 1971).

These pharmacological observations thus provide strong support for GABA as a major inhibitory transmitter regulating the firing of Deiters' neurones. It is probable that a proportion of GABA released into the fourth ventricle by stimulation of the cerebellar cortex originates in Deiters' nucleus (OBATA and TAKEDA, 1969), release of the amino acid in the nucleus has also been detected with push-pull cannulae (OBATA, 1972b). GABA also appears to be the transmitter released on neurones of the superior vestibular nucleus by axons of Purkinje cells from the flocculus since this inhibition is blocked by picrotoxin (10–15 mg/kg) but not by strychnine (0.3 mg/kg. Rabbit: FUKUDA, HIGHSTEIN, and ITO, 1972). It is of interest that whereas Deiters' neurones are considered excitatory in nature (WILSON, 1971, 1972), axons of the superior vestibular nucleus probably inhibit both oculomotor and trochlear motoneurones by releasing GABA as a transmitter (Rabbit: HIGHSTEIN, ITO, and TSUCHIYA, 1971. Cat: PRECHT, BAKER, and OKADA, 1973; see Section 4.10.1), and the reduction by strychnine (0.15–0.6 mg/kg) and the insensitivity to picrotoxin (6 mg/kg) and bicuculline (0.4 mg/kg) of the inhibition of cervical motoneurones following activation of cells in the medial vestibular nucleus suggest that cells of this nucleus release glycine as an inhibitory transmitter (Cat: FELPEL, 1972).

4.10. Brain Stem

Because of the complex nature of the brain stem and medulla it is not unexpected that little data exists regarding the detailed distribution of particular amino acids and related enzymes. Levels of free amino acids for the rat are given by SHANK and APRISON (1970) and for the mouse by SHIMADA et al., (1972).

In the rat GABA levels are highest in the midbrain (3.6 μmole/g), the levels falling in the pons (1.7 μmole/g) and medulla (1.6 μmole/g, SHANK and APRISON, 1970). In contrast, glycine levels progressively increase in a caudal direction towards the relatively high values of spinal tissue both in the cat (APRISON et al., 1969, midbrain 2 μmole/g, pons 3 and medulla 3.4) and rat (SHANK and APRISON, 1970, 1.6, 3 and 4 respectively). The presence in the pons and medulla, but not in the midbrain and more cranial tissue, of a high affinity uptake system for glycine, similar to that of the cord, has suggested the importance of glycine as a brain stem transmitter (JOHNSTON and IVERSEN, 1971; ARREGUI et al., 1972). Taurine levels fall in a caudal direction whereas aspartate and glutamate are fairly evenly distributed throughout the brain stem.

4.10.1. Cranial Motor Nuclei

The motor nuclei of the *extraocular muscles* contain relatively high levels of GABA (Rabbit: oculomotor, 4.5 μmole/g, OKADA et al., 1971. Cat: trochlear,

3.6 μmole/g, PRECHT et al., 1973) and isolated single oculomotor cells contain higher concentrations (5 mM, assuming uniform distribution in dissected neurones with attached synaptic terminals. Cat: MIYATA et al., 1970) than either trigeminal motoneurones (4.2 mM), nucleus ambiguus cells (3.0 mM), facial motoneurones (2.4 mM) or spinal motoneurones (1 mM. Cat: MIYATA et al., 1970; OTSUKA and MIYATA, 1972). GABA levels in the region of the oculomotor and trochlear nuclei are reduced by section of the ipsilateral medial longitudinal fasciculus, and there is physiological evidence for an inhibitory tract within this fasciculus which originates in the superior vestibular nucleus (Cat: PRECHT, et al.,1973). Seventy five percent of the GAD of the oculomotor nucleus can be isolated in particulate fractions, and levels of the enzyme in both this and the trochlear nucleus fall after lesions in vestibular nuclei (FONNUM, 1972), observations which suggest that a high proportion of GABA and GAD in these ocular motor nuclei could be in terminals of fibres from the vestibular complex.

Electrical stimulation of the vestibular nerve inhibits neurones of the ipsilateral oculomotor (Rabbit: HIGHSTEIN, ITO and TSUCHIYA, 1971) and trochlear (Cat: PRECHT and BAKER, 1972) nuclei, an effect mediated by inhibitory neurones within the superior vestibular nucleus. The IPSP's of these motoneurones were reversed by elevation of intracellular chloride concentrations, and in the case of trochlear neurones axosomatic inhibitory synapses were proposed, activation of which selectively increased membrane chloride conductance (LLINAS and BAKER, 1972). The firing of oculomotor neurones was depressed by electrophoretic glycine and GABA, the effect of glycine being selectively blocked by electrophoretic strychnine, that of GABA by picrotoxin (Rabbit: OBATA and HIGHSTEIN, 1970). Since strychnine (Rabbit: 0.2 mg/kg intravenous, HIGHSTEIN et al.,1971) does not affect vestibular inhibition of oculomotor neurones, whereas this inhibition was reduced by picrotoxin (Rabbit: *Intravenous,* 3–10 mg/kg, abolished by 10–20 mg/kg, HIGHSTEIN et al., 1971. *Electrophoretic,* OBATA and HIGHSTEIN, 1970), and both picrotoxin (4 mg/kg) and bicuculline (1 mg/kg) block vestibular inhibition of trochlear neurones (Cat: PRECHT, BAKER, and OKADA, 1973), the synaptically released transmitter seems most likely to be GABA.

Both glycine and GABA may be involved in the inhibition of *masseter motoneurones:* short latency inhibitions by impulses in the inferior dental nerve (Cat: KIDOKORA, KUBOTA, SHUTO, and SUMINO, 1968) or following electrical stimulation of the contralateral orbital gyrus (Cat: SAUERLAND and MIZUNO, 1969) were reduced by strychnine, whereas longer latency inhibitions from these sources were sensitive to picrotoxin (see also GOLDBERG, 1972). A picrotoxin-sensitive inhibition of lingual motoneurones by impulses in high threshold masseter afferents has also been described (NAKAMURA and WU, 1970).

There are relatively high levels of GABA in the *inferior colliculi* of various mammals (Rhesus monkey: 4.7 μmole/kg, FAHN and CÔTÉ, 1968. Baboon: 5.5, Rabbit: 3, Rat: 5, and Guinea pig: 4.7, OKADA et al., 1971), and alterations in the response of inferior collicular neurones to auditory stimuli by locally injected picrotoxin (WATANABE and SIMADA, 1971) indicate that GABA may be an inhibitory transmitter in this region.

Both glycine and GABA inhibit the firing of *hypoglossal motoneurones* (Cat, rabbit: BRUGGENCATE and SONNHOF, 1971. Rat: DUGGAN, LODGE, and BISCOE,

1973), glycine usually being more effective than GABA when the amino acids are administered near cell bodies, with a reverse order of relative potency when the amino acids were administered more dorsally near dendrites (Rabbit: ALTMANN, BRUGGENCATE, and SONNHOF, 1972). Both of these amino acids, and β-alanine and δ-aminovaleric acid, increase the membrane conductance of hypoglossal motoneurones and hyperpolarize these cells (BRUGGENCATE and SONNHOF, 1972). Furthermore the inhibitory effect of glycine is selectively blocked by strychnine (Cat: BRUGGENCATE and SONNHOF, 1972. Rat: DUGGAN et al.,1973) and that of GABA by picrotoxin (BRUGGENCATE and SONNHOF, 1972) and bicuculline (DUGGAN et al., 1973). Since IPSP's recorded from hypoglossal motoneurones in response to impulses in the lingual nerve are reduced by strychnine (Cat: *Intravenous* 0.08 mg/kg, MORIMOTO, TAKATA, and KAWAMURA, 1968), and inhibition evoked by impulses in the glossopharyngeal nerve, and associated with chloride sensitive IPSP's, can be reduced by intravenous or electrophoretic strychnine but not by electrophoretic bicuculline (Rat: DUGGAN et al., 1973), glycine can be assumed to be an inhibitory transmitter influencing hypoglossal motoneurones most probably at axosomatic synapses. The relatively higher sensitivity of the dendritic regions of these cells to GABA may be associated with an inhibitory transmitter function of this amino acid at axodendritic synapses (see ALTMANN et al., 1972). It is thus significant that impulses in high threshold hypoglossal afferents have been reported to produce long latency and duration chloride-sensitive IPSP's in hypoglossal motoneurones, an inhibitory effect blocked by picrotoxin (2 mg/kg) but not by strychnine (0.2 mg/kg, Cat: MORIMOTO and KAWAMURA, 1972).

4.10.2. Reticular Formation

In the brain stem and medullary reticular formation most investigations have been concerned with anatomically located "reticular" neurones, although the sensitivity of antidromically identified reticulospinal neurones to depressant amino acids has been correlated with the operation of certain inhibitory pathways.

Medullary reticulospinal neurones are approximately equally sensitive to L-glutamate and L-aspartate (Cat: TEBĒCIS, 1973b) as are reticular neurones (Cat: HÖSLI and TEBĒCIS, 1970). In both of these studies DL-homocysteate was a more potent excitant than either naturally occurring acidic amino acid, although another report suggests that unlike the situation elsewhere in the nervous system, medullary reticular neurones excited by DL-homocysteate were not always excited by L-glutamate, and *vice versa* (Cat: BOAKES et al., 1970).

Lysergic acid diethylamide (LSD) has been reported to block the excitation of medullary reticular neurones by 5-hydroxytryptamine and L-glutamate, the excitatory action of acetylcholine, DL-homocysteate, noradrenaline and of L-glutamate on cells *depressed* by 5-hydroxytryptamine, being unaffected (Cat: BOAKES et al., 1970). In contrast, in the rat, LSD reduced the effect of L-glutamate on neurones within the dorsal and median raphé nuclei which were depressed by 5-hydroxytryptamine (Rat: AGHAJANIAN, HAIGLER, and BLOOM, 1972).

Of the depressant amino acids, glycine was usually more effective than GABA in reducing the firing rate of reticular (HÖSLI and TEBĒCIS, 1970) and reticulospinal (TEBĒCIS and DI MARIA, 1972b) neurones in the cat. When administered extracellulary glycine increased the membrane conductance of reticular neurones, producing a chloride sensitive hyperpolarization: reticular neurones (HÖSLI, TEBĒCIS, and HAAS, 1971; HÖSLI and HAAS, 1972); reticulospinal neurones (TEBĒCIS and ISHIKAWA, 1973). The latter cells are also hyperpolarized by GABA. The inhibitory effect of glycine (and of β-alanine) is readily reduced by strychnine which shows little antagonism towards GABA (HÖSLI and TEBĒCIS, 1970; TEBĒCIS and DI MARIA, 1972). On the other hand selective antagonism of either amino acid has not been demonstrated using picrotoxin (*Electrophoretic, intravenous,* Cat: HÖSLI and TEBĒCIS, 1970) and both bicuculline (reticular neurones, TEBĒCIS, HÖSLI, and HAAS, 1971) and bicuculline methochloride (reticulospinal neurones, TEBĒCIS, 1973b) reduced the depressant action of glycine and GABA, although the effect of glycine was usually less sensitive to these alkaloids than that of GABA. The selectivity of bicuculline towards GABA receptors on reticular neurones is thus less than that demonstrated in the spinal cord. The relatively weak inhibitory effect of imidazole-4-acetate on reticular neurones was not affected by strychnine, but was reduced by bicuculline (HAAS et al., 1972). Furthermore strychnine, but not bicuculline, selectively blocked the inhibitory effect of taurine on medullary reticular neurones (HAAS and HÖSLI, 1973a).

On the basis of the effects of strychnine on the inhibition of medulary reticulospinal neurones by stimulation of reticulospinal axons (and presumably mediated via axon collaterals: ITO, UDO, and MANO, 1970) glycine has been proposed as an inhibitory transmitter released by one type of neurone in the ventrocaudal portion of the medial medullary reticular formation (TEBĒCIS and DI MARIA, 1972; TEBĒCIS and ISHIKAWA, 1973). A strychnine sensitive inhibition of reticulospinal neurones could also be evoked by stimulation in the medial pontine reticular formation (TEBĒCIS and DI MARIA, 1972). These types of relatively short latency and duration inhibition were usually insensitive to bicuculline (but see TEBĒCIS, 1973b) and some evidence has also been provided for a picrotoxin-sensitive (2 mg/kg intravenous) long duration inhibition of reticulospinal neurones, evoked by ventral spinal cord stimulation, which may involve GABA (Cat: TEBĒCIS and ISHIKAWA, 1973). Strychnine also suppressed the inhibition of medullary reticular neurones by impulses in limb nerves (HAAS and HÖSLI, 1973b).

4.11. Dorsal Column Nuclei

4.11.1. Excitation

In the cat, glutamate levels within the dorsal medullary region containing primary sensory nuclei are significantly higher (6.93 ± 0.53 μmole/g) than those of the ventral medulla (5.94 ± 0.30, JOHNSON and APRISON, 1970), findings consistent with the high levels of this amino acid in dorsal roots and columns.

Neurones of the gracile and cuneate nuclei are readily excited by L-glutamate (Cat: MEYER, 1965; STEINER and MEYER, 1966; GALINDO, KRNJEVIĆ, and

SCHWARTZ, 1967, 1968), the effect of D-glutamate being somewhat weaker (GALINDO et al., 1967). Although both the excitatory actions of L-glutamate, L-aspartate and the synaptic firing of cuneate neurones were reduced by 1-hydroxy-3-aminopyrrolidone-2, supporting an excitatory transmitter function for the amino acid (Cat: DAVIES and WATKINS, 1973), the specificity of the antagonist towards other excitants of these neurones has not been determined. When administered topically to these nuclei L-aspartate (10^{-2} M) was apparently without effect whereas, L-glutamate (10^{-4}–10^{-2} M) depolarized terminals of afferent fibres (Rat: DAVIDSON and SOUTHWICK, 1971). Since this effect was not observed after the administration of sufficient picrotoxin to suppress the depolarization of terminals by conditioning volleys in afferent fibres, the depolarization of by L-glutamate has been considered to be unrelated to "presynaptic" inhibition in the cuneate nucleus (DAVIDSON and SOUTHWICK, 1971).

4.11.2. Inhibition

Although the levels of individual amino acids in the gracile and cuneate nuclei have not been determined, both glycine and GABA have been considered as possible inhibitory transmitters. When administered electrophoretically these amino acids depress the activity of cuneate neurones; in general glycine is as effective as GABA although proprioceptive cells appear more sensitive to the latter amino acid (Cat: GALINDO et al., 1967). The effect of glycine can be blocked by strychnine, that of GABA by picrotoxin (Cat: GALINDO, 1969) or bicuculline (Cat: KELLY and RENAUD, 1971).

A number of studies have been concerned with the effects of amino acids on the excitability of cuneate presynaptic terminals, particularly in relation to the possible role of GABA in "presynaptic" inhibition. Although both glycine and GABA (electrophoretic) have been shown to diminish the excitability of the terminals of primary afferent fibres within the cuneate nucleus (GALINDO, 1969), the effect of GABA being blocked by electrophoretic picrotoxin, most experiments have involved topical administration of amino acid solutions. Under these conditions glycine (10^{-2} M) reduced (Rat: DAVIDSON and SOUTHWICK, 1971), whereas GABA (10^{-4}–10^{-1} M) enhanced presynaptic excitability (Cat: GALINDO, 1969. Rat: DAVIDSON and SOUTHWICK, 1971). This effect of GABA was blocked by picrotoxin (GALINDO, 1969); alternatively topical GABA blocked the suppression by picrotoxin of primary afferent depolarization (PAD) induced by afferent volleys (DAVIDSON and SOUTHWICK, 1971).

Both short (20 msec) and long (about 200 msec) duration postsynaptic inhibition, and prolonged "presynaptic" inhibition have been proposed to account for the depression of transmission through the cuneate nucleus which follows cutaneous afferent volleys (ANDERSEN, ECCLES, OSHIMA, and SCHMIDT, 1964; ANDERSEN, ETHOLM, and GORDON, 1970). "Presynaptic" inhibition has frequently been assessed by the presence of afferent terminal depolarization, and the involvement of GABA in these inhibitory processes is suggested by the following findings: semicarbazide lowered central GABA levels and reduced afferent terminal depolarization (Cat: BANNA and JABBUR, 1971); afferent terminal depolarization was reduced by picrotoxin (Cat: BANNA and JABBUR, 1969. Rat: DAVIDSON and REISINE, 1971) and bicuculline (DAVIDSON and REISINE, 1971. Cat: BANNA, NACCACHE, and JABBUR, 1972); the inhibition of the firing of cuneate neurones by a prior cutaneous volley was blocked by picrotoxin (Cat: BOYD, MERITT and GARDNER, 1966; BANNA and JABBUR, 1969) and bicuculline (Cat: KELLY and RENAUD, 1971); bicuculline also blocked the inhibition of cuneate neurones by volleys in the pyramidal tract (KELLY and RENAUD, 1971). Although strychnine has been reported to diminish short latency inhibition in the cuneate nucleus (BOYD et al., 1966; BANNA and JABBUR, 1969) no clear evidence has been provided for a glycine-releasing inhibitory pathway terminating in this nucleus.

The consensus of opinion seems to be that intra-nuclear inhibitory interneurones release GABA as a hyperpolarizing inhibitory transmitter upon cuneate relay neurones, and that GABA possibly also depolarizes the terminals of afferent fibres within the nucleus, an effect which will be discussed in more detail in the next section.

4.12. Spinal Cord

The considerable body of information concerned with amino acid transmitters in the cord is largely the consequence of the relative ease with which this region can be studied chemically, physiologically and pharmacologically. Most investigations have been concerned with lumbar segments of the cat.

4.12.1. Neurochemistry

Overall amino acid levels are of value only for comparison with other regions of the nervous system, and more useful information relative to spinal mechanisms has been derived from analyses of intraspinal amino acid distributions. The major observations are summarized for lumbar segments of the cat in Table 8, limited but similar studies have been made on other mammals (APRISON et al., 1969; SHANK and APRISON, 1970; DUGGAN and JOHNSTON, 1970b). Some of the discrepancies between figures reported by different groups may be associated with different methods of amino acid extraction and estimation. The distribution of enzymes associated with amino acid metabolism is given in Table 9.

Table 8 also includes figures for peripheral nerves (see also MARKS, DATTA, and LAJTHA, 1970), these levels may represent amino acids required for metabolic purposes, the relatively high levels within dorsal root ganglia possibly being associated with protein synthesis by cell bodies. It should also be recalled that DALE (1935) anticipated an association between the transmitter of axon-reflex vasodilation and that at central synapses of the same afferent fibres.

One useful technique applicable to the cord is the analysis of alterations of amino acids after selective lesions of neurones or pathways. Hindlimb rigidity can be produced in the cat by temporary occlusion of the thoracic aorta which results in a loss of small neurones in the central portion of the lumbar grey matter, involving dorsal and ventral grey, with little destruction of motoneurones

Table 8. Distribution of amino acids in the spinal cord of the cat (μmole/g)

Amino acid	PNM	PNC	DSR	DRG	DR	DW	DG	VG	VW	VR	Reference
Aspartate					1.0	1.1	2.1	3.1	1.3	1.3	Graham et al. (1967)
						(1.2)	(1.5)	(2.4)	(1.3)		Davidoff et al. (1967)
					1.5	1.9	3.9	4.8	2.5	1.0	Johnston (1968)
	0.9	0.7	1.1	4.4	1.5/1.2						Duggan and Johnston (1970)
			1.0	2.3	0.7					0.7	Johnson and Aprison (1970)
Glutamate					4.6	4.8	6.5	5.3	3.9	3.1	Graham et al. (1967)
						(4.2)	(4.7)	(4.5)	(3.4)		Davidoff et al. (1967)
					4.0	3.7	5.9	4.4	3.3	1.8	Johnston (1968)
	2.0	1.8	2.0	4.5	4.5/3.4						Duggan and Johnston (1970)
			2.1	3.2	2.8		5.9			1.7	Johnson and Aprison (1970)
Glutamine					1.9	3.6	5.3	5.4	3.8	1.9	Graham et al. (1967)
						(3.6)	(5.1)	(5.0)	(3.7)		Davidoff et al. (1967)
					0.9	4.0	5.5	4.8	3.7	0.7	Johnston (1968)
	2.2	1.8	2.1	5.1	2.5/2.4						Duggan and Johnston (1970)
Glycine					0.6	3.0	5.7	7.1	4.4	0.6	Graham et al. (1967)
						(2.3)	(4.1)	(5.0)	(3.3)		Davidoff et al. (1967)
					0.6	2.9	5.4	6.5	4.6	0.6	Johnston (1968)
	1.4	0.8	1.1	1.9	0.9/0.8						Duggan and Johnston (1970)
			0.9	2.1	0.5						Johnson and Aprison (1970)
GABA					0.1	0.4	2.2	1.0	0.4	0.1	Graham et al. (1967)
						(0.4)	(2.1)	(1.3)	(0.3)		Davidoff et al. (1967)
			0.1	0.1	0.1					0.1	Johnson and Aprison (1970)
				<0.1				1.0			Otsuka et al. (1971)
α-Alanine					0.7	0.8	1.3	1.2	0.9	0.7	Johnston (1968)
	0.7	0.6	0.8	3.6	0.9/0.8						Duggan and Johnston (1970)
Serine					0.5	0.6	0.9	0.9	0.6	0.7	Johnston (1968)
Cystathionine					0.03	1.4	1.6	1.3	1.7	0.04	Johnston (1968)

Levels are given for peripheral nerve, muscular (PNM) and cutaneous (PNC), sensory root distal to ganglion (DSR), dorsal root ganglion (DRG), dorsal root (DR – when two figures are given the first is distal, the second proximal to the cord), dorsal white (DW, dorsal columns plus dorsolateral white), dorsal grey (DG), ventral grey (VG), ventral white (VW – ventromedial and ventrolateral) and ventral roots (VR).
The figures in brackets are those obtained after aortic occlusion (Davidoff et al., 1967)

Table 9. Enzyme levels in extracts of cat spinal cord and roots (μmole/hr/g wet tissue)

Enzyme	Dorsal roots	Dorsal white	Dorsal grey	Ventral grey	Ventral white	Ventral roots	Reference
Aspartate transaminase	545	2454	5412	5418	2480	537	GRAHAM and APRISON (1969)
Glutamate dehydrogenase	17	149	446	448	139	16	GRAHAM and APRISON (1969)
Glutamine synthetase	7	20	67	67	19	4	GRAHAM and APRISON (1969)
Glutaminase	12	67	397	331	66	13	GRAHAM and APRISON (1969)
Glutamate decarboxylase	4	0	13[a]	7	0	3	GRAHAM and APRISON (1969)
GABA transaminase	1	3	14	13	3	1	GRAHAM and APRISON (1969)
D-Amino acid oxidase		0.8	2.2	2.1	0.4		DE MARCHI and JOHNSTON (1969)
Glycine transaminase		6.6	9.9	9.6	7		JOHNSTON, VITALI and ALEXANDER (1970)
Serine hydroxymethyltransferase		2.5	4.3[a]	5	2.5		DAVIES and JOHNSTON (1973)

[a] Levels in dorsal grey significantly different from those in ventral grey.

(DAVIDOFF et al., 1967). In these animals, which showed no response to nociceptive cutaneous stimulation in the affected segments, spinal monosynaptic reflexes were enhanced, polysynaptic reflexes were reduced, and although some evidence has been obtained for antidromic inhibition mediated by Renshaw cells, details have not been published regarding other types of spinal inhibition (MURAYAMA and SMITH, 1965). In the rat, however, hindlimb rigidity produced in the same fashion is accompanied by reduced responses to painful cutaneous stimulation, enhanced monosynaptic and reduced polysynaptic reflexes, and reduction of two types of inhibition of ankle extensor motoneurones: short latency ("postsynaptic") inhibition by impulses in the deep peroneal nerve and prolonged ("presynaptic") inhibition by repetitive impulses in the posterior biceps nerve (MATSUSHITA and SMITH, 1970). There is thus evidence in these studies of a loss of both excitatory and inhibitory spinal mechanisms associated with interneurones.

Useful information might also be provided by transection of dorsal roots, although subsequent alterations in amino acid levels and associated enzymes may be complicated by disturbances of the blood supply (Rat: SUGAR and GERARD, 1940. Cat: CHAMBERS, ELDRED, and EGGETT, 1972) in addition to loss of primary afferent fibres. With both aortic occlusion and de-afferentation, histological control and unchanged levels of some metabolites are required to exclude "non-specific" loses of amino acids produced by infarction, oedema, gliosis or cavity formation (see DAVIDOFF et al., 1967).

4.12.1.1. Aspartate

Within the cord ventral grey levels are higher than dorsal grey. An association with small neurones, presumably interneurones, has been suggested by the significant reduction in aspartate produced by temporary aortic occlusion, the reduced levels of this amino acid in dorsal and ventral grey matter correlating with the reduced count of small neurones in the central grey region (DAVIDOFF et al., 1967).

As shown in Table 9 the activity of aspartate aminotransferase does not seem to correspond closely with that of the amino acid (Table 8).

4.12.1.2. Glutamate

L-Glutamate is the only putative amino acid transmitter for which dorsal root levels exceed those of the ventral root (Table 8), as might be expected of a transmitter released by primary afferent fibres. Similar differences occur in the dog and rat (DUGGAN and JOHNSTON, 1970b). The high level of the amino acid in dorsal root ganglia, and the gradient along the root (Cat: ganglion 4.5 μmole/g, dorsal root distal to cord 4.5, proximal to cord 3.4, DUGGAN and JOHNSTON, 1970a), which contrasts to the approximately equal levels of other amino acids in peripheral nerves and dorsal roots (Table 8), suggests that glutamate synthesised in ganglion cells is transported towards central terminals. No such dorsal root gradient was present in the dog (DUGGAN and JOHNSTON, 1970b). Further investigation is warranted of the relationship to the possible transmitter function of glutamate of the influx into, and particularly the efflux from, stimulated peripheral nerves (WHEELER, BOYARSKY, and BROOKS, 1966; DEFEUDIS, 1971; WHEELER and BOYARSKY, 1971).

After aortic occlusion, and loss of spinal interneurones, glutamate levels fall significantly in all segments of the cord, especially in the dorsal grey, but no correlation could be established between neurone loss and change in amino acid concentration (DAVIDOFF et al., 1967). Furthermore spinal glutamine levels were not modified (Table 8). The pattern of activity of glutamine synthetase and glutaminase (Table 9), enzymes associated with glutamate-glutamine interconversion, and of glutamate dehydrogenase, correspond with the distribution of glutamine, but not with that of glutamate (Table 8).

Both high and low affinity uptake mechanisms are present in spinal tissue for acidic amino acids, the same high affinity process transporting L-glutamate and L-aspartate (Rat: LOGAN and SNYDER, 1972. Cat: BALCAR and JOHNSTON, 1973). Osmotically sensitive particles of density similar to that of nerve endings are in part responsible for this uptake, which in the cat is less sensitive to inhibition by *p*-chloromercuriphenylsulphonate than the high affinity uptake of either glycine or GABA. In the rat the particles accumulating aspartate and glutamate are denser than those accumulating glycine (ARREGUI et al., 1972).

4.12.1.3. Glycine

In contrast to relatively low levels in dorsal and ventral roots, higher levels of glycine are found in the spinal grey matter, particularly ventrally (Table 8).

Furthermore, glycine levels are higher in the cord, and medulla, than elsewhere in the feline (SHANK and APRISON, 1970) and other vertebrate nervous systems (APRISON et al., 1969), and the highest intraspinal levels of glycine in the cat and other vertebrates correspond to regions of high grey matter to white matter ratio, regions containing neurones associated with limb innervation (also Human: BOEHME, FORDICE, MARKS and VOGEL, 1973).

Spinal white matter contains long descending and ascending tracts as well as the axons of propriospinal fibres. The dorsal column content of the latter is minimal (NATHAN and SMITH, 1959; SZENTAGOTHAI, 1964), and hence the low levels of glycine in this region (Cat: lumbar, 1.2 µmoles/g) compared with dorsolateral (4.6), ventrolateral (5.6) and ventromedian white regions (3.5) (APRISON et al., 1969) suggests that glycine-containing propriospinal axons account for the relatively high levels of this amino acid in the lateral and ventromedial tracts.

A strong association between glycine and spinal interneurones was established by the very significant reduction observed in dorsal grey, ventral grey and ventral white segments (Table 8) following temporary aortic occlusion in the cat and destruction of neurones in the central region of the cord (DAVIDOFF et al., 1967). The ventrolateral white area is rich in propriospinal fibres, and there was a statistically significant correlation between the concentration of glycine in the grey matter and the number of small neurones remaining after anoxic destruction.

Of enzyme activities associated with glycine metabolism, spinal tissue contains glycine transaminase (glycine: 2-oxoglutarate transaminase: JOHNSTON and VITALI, 1969 a, JOHNSTON et al., 1970; BENUCK et al., 1972), serine hydroxymethyltransferase (SHANK and APRISON, 1970; DAVIES and JOHNSTON, 1973), D-glycerate dehydrogenase and D-3-phosphoglycerate dehydrogenase (UHR and SNEDDON, 1971, 1972) and glycine decarboxylase (UHR, 1973). Glycine is also a poor substrate of D-aminoacid oxidase (Cat: DE MARCHI and JOHNSTON, 1969). The regional distributions of neither glycine transaminase nor serine hydroxymethyltransferase correlate with that of glycine in the cord (Table 9). *In vivo* studies of glycine metabolism in the rat cord suggest that glycine is synthesised only slowly from glucose (SHANK and APRISON, 1970; SHANK et al., 1973).

A number of investigations have been concerned with the high affinity structurally specific uptake of glycine by spinal tissue, a process mainly confined to the cord, medulla and pons (Rat: NEAL, 1971; IVERSEN and JOHNSTON, 1971; JOHNSTON and IVERSEN, 1971; LOGAN and SNYDER, 1972; ARREGUI, et al., 1972; APRISON and MCBRIDE, 1973. Cat: BALCAR and JOHNSTON, 1973). The subcellular particles associated with this uptake are denser than those accumulating GABA but less dense than those taking up aspartate and glutamate (Rat: IVERSEN and JOHNSTON, 1971; ARREGUI et al., 1972). Radioautographic analyses of the uptake of labelled glycine by spinal cord slices *in vitro* shows that the uptake is highest in the ventral horn particularly around motoneurone bodies, with intense uptake into nerve terminals (Rat: MATUS and DENNISON, 1972; HÖKFELT and LJUNGDAHL, 1971 a) which differ from those accumulating GABA (Rat: IVERSEN and BLOOM, 1972). In one study all labelled terminals contained flat vesicles, some 60% of flat vesicle synapses accumulating the labelled amino acid (MATUS and DENNISON, 1972). Under *in vivo* conditions, following direct injection of labelled glycine into the cat spinal cord, the amino acid is present in nerve terminals containing flat vesicles, in glial cells, interneurones which appear not to accumulate the amino acid in slices, whereas motoneurones show little activity (LJUNGDAHL and HÖKFELT, 1973). Uptake into glia and neurones has also been found in cultures of rat spinal cord (HÖSLI, LJUNGDAHL, HÖKFELT, and HÖSLI, 1972), and all of these observations are consistent with the uptake of glycine by inhibitory nerve terminals and interneurones.

An enhanced efflux of glycine has been shown to follow direct electrical stimulation of spinal cord slices (Rat: HOPKIN and NEAL, 1971; CUTLER, HAMMERSTAD, CORNICK, and MURRAY, 1971; HAMMERSTAD et al., 1971). A calcium dependent efflux of glycine (and GABA, glutamate and asparate) was produced by electrical stimulation of the rostral portion of isolated, hemisected frog or toad spinal cords (ROBERTS and MITCHELL, 1972), or of dorsal roots (APRISON, 1970). It is difficult to relate these studies directly to mammalian cord *in vivo*, and of more significance is the release of exogenous glycine into the central canal of cats, in the presence of *p*-hydroxymercuribenzoate, and in association with acetylcholine, by stimulation of the femoral and sciatic nerves (JORDAN and WEBSTER, 1971). The organic mercurial was essential to demonstrate this enhanced release of glycine, suggesting that a highly efficient uptake mechanism was present for the removal of this amino acid after synaptic release.

4.12.1.4. GABA

The levels of GABA in the dorsal grey exceed those ventrally (Table 8), and detailed analysis of small blocks of tissue show that the highest levels (2.2–2.6 μmole/g) occur in laminae III and IV (Cat: MIYATA and OTSUKA, 1972; OTSUKA and MIYATA, 1972). Although low relative to the concentration within Deiters' (6.3 mM) and Purkinje (6.0 mM) cells, the GABA concentration of dissected motoneurones (0.9) exceeds that of spinal ganglion cells (0.2, Cat: OTSUKA et al., 1971), and the technique almost certainly underestimates any contribution made by axodendritic synapses.

The high dorsal concentrations of GABA correspond to those of GAD (Table 9, also Monkey: ALBERS and BRADY, 1959) whereas GABA-T is more uniformly distributed in the grey matter (Table 9). In the rhesus monkey, however, transaminase levels are somewhat higher dorsally than ventrally (SALVADOR and ALBERS, 1959).

Following aortic occlusion, no significant alterations occurred in spinal GABA levels (Table 8), which is consistent with preservation of neurones in the superficial layers of the dorsal horn. On the other hand cauterization of the vessels supplying the dorsal horn, which suppressed generation of dorsal root potentials by afferent volleys, significantly reduced the dorsal horn content of GABA, the levels of other amino acids not being reported (MIYATA and OTSUKA, 1972). Extensive deafferentation of the rat's cervical cord resulted in reduced GABA levels, a marked reduction of GAD activity, reduced GABA transaminase activity and a reduced uptake of GABA into synaptosomes of tissue located in the dorso-lateral portion of the cord. There was also considerable reduction of GAD activity in ventral segments, and although the findings are suggestive of a loss of GABA-containing nerve terminals of either fibres in the dorsal root or of inhibitory interneurones degenerating as a consequence of deafferentation (GOTTESFELD, KELLY, and RAYNER, 1973a), changes resulting from damage to the vascular supply need to be excluded. Only a high affinity uptake mechanism specific for GABA has been demonstrated in cat spinal tissue (BALCAR and JOHNSTON, 1973), into particles having a density characteristic of nerve terminals (Rat: IVERSEN and JOHNSTON, 1971; ARREGUI et al., 1972) which can be distinguished from those accumulating other amino acids. In autographic studies of spinal cord slices the principal site of GABA uptake appears to be nerve terminals (Rat: IVERSEN and BLOOM, 1972) which could not be classified morphologically but differ from those taking up glycine. In spinal cord cultures GABA is accumulated mainly by small neurones (HÖSLI et al., 1972).

4.12.1.5. Other Amino Acids

The distribution of the other amino acids present in spinal tissue does not suggest major roles as synaptic transmitters: α-alanine, cystathionine and serine (Table 8), and taurine (Rat: spinal grey 1.6 μmole/g, white 1.9, SHANK and APRISON, 1970). There are additionally other unidentified amino acids which may require further investigation (see BRIEL and NEUHOFF, 1972). A polypeptide has recently been extracted from bovine dorsal roots, but not from ventral roots, which is similar chemically and pharmacologically to physalaemin (OTSUKA, KONISHI, and TAKAHASHI, 1972; see also LEMBECK and ZETLER, 1962) and which, like physalaemin (KONISHI and OTSUKA, 1971) has a direct depolarizing action on motoneurones of the isolated frog spinal cord and dilates blood vessels.

4.12.2. Excitation

Since the first demonstration of the direct depolarizing action on spinal motoneurones of electrophoretic L-glutamate, L-aspartate and L-cysteate (Cat: CURTIS et al, 1960), a number of investigations have been concerned with the excitation of spinal motoneurones, interneurones and Renshaw cells by these and related amino acids. Most emphasis has naturally been placed on L-glutamate and L-aspartate because of the neurochemical evidence discussed earlier. Differences in relative potencies have been related to the effectiveness of the amino acid-receptor interaction (CURTIS and WATKINS, 1960, 1963) and to the inactivation of amino acids in the tissue (CURTIS et al., 1970b). However, whilst in general L-glutamate and L-aspartate are approximately equipotent as excitants of spinal dorsal horn interneurones, L-aspartate is significantly more active than L-glutamate as an excitant of Renshaw cells (Cat: DUGGAN, 1974). This important observation is consistent with the operation of L-glutamate as the transmitter of primary afferent fibres, which do not synapse directly with Renshaw cells (see CURTIS and RYALL, 1966a), whereas L-aspartate may be the transmitter of spinal excitatory interneurones (DAVIDOFF et al., 1967).

Sympathetic (Cat: HONGO and RYALL, 1966) and parasympathetic (DE GROAT and RYALL, 1968) preganglionic spinal neurones are excited by DL-homocysteate, and both L-glutamate and DL-homocysteate increase the excitability of the terminal portions of primary afferent fibres (Cat: CURTIS and RYALL, 1966b). Depolarization of primary afferent fibres by L-glutamate has also been demonstrated in isolated preparations of amphibian spinal cord (SCHMIDT, 1963; TEBĒCIS and PHILLIS, 1969; DAVIDOFF, 1972d; BARKER and NICOLL, 1973).

The depolarization of spinal neurones by excitant amino acids is accompanied by an increased membrane conductance (CURTIS et al., 1960; ZIEGLGÄNSBERGER and PUIL, 1973) which seems to involve an increased permeability to at least sodium ions. A number of factors complicate studies of the ionic mechanism of neuronal depolarization, particularly the technical difficulty of uniformly altering either intracellular ion concentrations or the membrane potential. These factors are also important when attempting to compare amino acid and synaptic depolarizations, particularly of neurones of complicated shape such as motoneurones: the site of amino acid action during electrophoretic administration may be very restricted, synaptic excitation by impulses in group Ia afferents occurs predominantly at dendrites (see CURTIS et al., 1972a). Nevertheless the available evidence suggests that the ionic mechanisms for amino acid depolarization and synaptic excitation may be similar, presumably involving both sodium and potassium ions (ECCLES, 1966). Thus elevation of intracellular chloride concentrations does not modify amino acid depolarization (CURTIS et al., 1972a), and both the reversal potential for the depolarization, measured by passing current through an intracellular microelectrode, and the maximally depolarized membrane potential, achieved by high concentrations of L-glutamate or N-methyl-DL-aspartate, are of the order of –20 to –30 mV (CURTIS, 1965; CURTIS et al., 1972a; ZIEGLGÄNSBERGER and PUIL, 1973). The movement of sodium ions, however, is not that associated with the generation of action potentials since concentrations of tetrodotoxin sufficient to block the latter process have little or no influence on either amino acid or synaptic depolarizations (CURTIS et al., 1972a; ZIEGLGÄNSBERGER and PUIL, 1972).

No evidence has been obtained that enzymic modification is of major importance in the inactivation of electrophoretically administered amino acids, indeed the similarity of the time course of action of optical isomers, the failure of thiosemicarbazide and L-methionine-DL-sulphoximine to enhance the action of glutamate, and the observed enhancement of the actions of glutamate and aspartate by organic mercurials (CURTIS et al., 1970b; BALCAR and JOHNSTON, 1973) suggest the importance of transport processes in lowering extracellular amino acid levels. The potentiation of amino acid excitation by *p*-chloromercuriphenylsulphonate, known to suppress amino acid uptake *in vitro*, excludes ion movements associated with this uptake as a direct explanation of the depolarization. Under *in vivo* conditions organic mercurials are difficult to use electrophoretically and as yet no specific, non-excitatory, antagonist of the uptake of glutamate and aspartate has been found which could be used to compare the inactivation of amino acids with that of synaptically released transmitters.

Further evidence of the nature of a spinal excitatory transmitter would be provided by a selective antagonist capable of blocking both synaptic and amino acid excitation without influencing the effects of other transmitters such as acetylcholine. The spinal Renshaw cell, being excited by acetylcholine, L-glutamate and L-aspartate has proved a useful test situation for assessing the selectivity of such antagonists prior to testing on dorsal horn interneurones excited synaptically and by the amino acids. Little success, however, has been achieved in tests on a large number of anaesthetics, amino acid analogues and other substances selected on the basis of their structure or depressant effect when administered systemically (CURTIS et al., 1972a; and unpublished observations). A major difficulty, necessitating a relatively high degree of selectivity of an antagonist, is the problem of achieving adequate concentrations at axodendritic synapses whilst preserving specificity with higher concentrations of the antagonist closer to the site of electrophoretic administration (see CURTIS, DUGGAN, and JOHNSTON, 1971c). The most promising compounds include L-glutamate diethylester (HALDEMAN and MCLENNAN, 1972), 2-methoxyaporphine, L-methionine-DL-sulphoximine (CURTIS et al., 1972a) and 1-hydroxy-3-aminopyrrolidone-2 (CURTIS, JOHNSTON, GAME, and MCCULLOCH, 1973b). All four compounds were found to antagonise excitation by L-glutamate and L-aspartate (and DL-homocysteate) to the same extent (CURTIS et al., 1972a, 1973), although L-glutamate diethylester has been reported to be more effective as an antagonist of the excitation of spinal interneurones by L-glutamate than of the excitation by the other amino acids (HALDEMAN and MCLENNAN, 1972). This diethylester reduced the synaptic excitation of dorsal horn interneurones (HALDEMAN and MCLENNAN, 1972; CURTIS et al., 1972a), as also did L-methionine-DL-sulphoximine (CURTIS et al., 1972a). additionally 1-hydroxy-3-aminopyrrolidone-2 reduced non-cholinergic synaptic excitation of Renshaw cells to a slightly greater extent than the excitation produced by ventral root volleys which release acetylcholine at axon collateral terminals (CURTIS et al., 1973b).

These observations may be taken to support proposals that aspartate and glutamate are excitatory transmitters in the cord, but relatively low or inconsistent degrees of selectivity (see CURTIS et al., 1972a), technical problems associated with electrophoretic ejection, and the need for relatively high currents, limit the usefulness of all of these antagonists, and further investigation is clearly necessary in order to find more selective excitant amino acid antagonists.

4.12.3. Inhibition

Of the neutral amino acids present in spinal tissue, α-alanine, serine, glycine, taurine, cystathionine and GABA depress the firing of spinal neurones. However the available neurochemical findings favour the involvement of only glycine and GABA as inhibitory transmitters. The firing of spinal neurones is regulated by two types of inhibition- "postsynaptic" and "presynaptic", descriptions which presuppose the mechanism of the inhibitory process. There is, however, considerable debate regarding the mechanism of "presynaptic" inhibition, although the participation of GABA as a transmitter seems generally accepted. On the evidence to be discussed below the terms "postsynaptic", and strychnine-sensitive are used synonymously to describe inhibition mediated by glycine, and "presynaptic", prolonged, remote and picrotoxin (or bicuculline)-sensitive to describe inhibition in which GABA is involved.

4.12.3.1. Glycine

When administered electrophoretically glycine and structurally related α- and β-amino acids depress the firing of spinal interneurones, parasympathetic preganglionic neurones, motoneurones and Renshaw cells (Cat: CURTIS, PHILLIS, and WATKINS, 1959; CURTIS and WATKINS, 1960; WERMAN et al., 1966, 1968; CURTIS et al., 1968b; BRUGGENCATE and ENGBERG, 1968; RYALL and DE GROAT, 1972. Rat: BISCOE et al., 1972. Toad: ENDO and ARAKI, 1969). In part these observations account for the effects of intrathecal (DHAWAN, SHARMA, and SRIMAL, 1972) or systemic glycine WERMAN, DAVIDOFF, and APRISON, 1967; STERN and HADŽOVIČ, 1970; TAKANO and NEUMANN, 1972) and β-alanine (MUNEOKA, 1961) on spinal reflexes.

In general electrophoretic glycine was a more effective depressant of the firing of feline interneurones, and particularly of motoneurones, than GABA (CURTIS et al., 1968a, b; WERMAN et al., 1968; BRUGGENCATE and ENGBERG, 1968) whereas the amino acids were approximately equipotent when tested on Renshaw cells (CURTIS et al., 1968a).

The depression of neurone excitability by glycine is associated with a membrane hyperpolarization and conductance increase (WERMAN et al., 1968; CURTIS et al., 1968b; BRUGGENCATE and ENGBERG, 1968). These changes are rapid in both onset and offset, with prolonged ejection of glycine the hyperpolarization becomes reduced in magnitude in the presence of a sustained increase in conductance, probably because protracted activation of glycine receptors results in altered ion concentrations in both the intra- and extracellular phases, so changing equilibrium potentials for individual ions. Intracellularly administered glycine appears not to have a hyperpolarizing effect on the motoneurone membrane (GLOBUS et al., 1968). The reversal potential for the glycine hyperpolarization of motoneurones was at a slightly more hyperpolarized level than the maximum hyperpolarization which could be achieved, and was either approximately identical to or at a slightly less hyperpolarized level than the reversal potential for short latency, direct IPSP's (WERMAN et al., 1968; CURTIS et al., 1968b). A close similarity was also demonstrated for the reversal potentials of the glycine hyperpolarization of interneurones and that of IPSP's evoked by impulses in hind limb afferents and descending fibres in the cord (BRUGGECATE and ENGBERG, 1968). Such observations would suggest that the equilibrium potentials for the ionic permeability changes induced by the amino acid and the synaptically released transmitter were identical, and hence that the changes in ion permeability were the same; the slight differences in reversal potentials would be expected from differences in the geometric distribution of membrane affected by glycine and by the transmitter in relation to the intracellular polarizing electrode. Close similarities in these two mechanisms were also indicated by a study of the effects of intracellularly injected anions and potassium on the glycine hyperpolarization and the IPSP. In particular, synaptic membrane activated by glycine and the transmitter appear to discriminate between chloride, bromide and iodide in the same ratio (WERMAN et al., 1968). Furthermore those anions known to convert hyperpolarizing IPSP's into depolarizations—bromide, iodide, chloride, nitrite, thiocyanate, chlorate and formate (ECCLES, 1966) had a similar effect on glycine hyperpolarization, whereas intracellular injection of bromate, sulphate and citrate were without effect on either type of potential (CURTIS et al., 1968b).

Thus, within the limits of the available techniques, glycine and the transmitter at certain spinal inhibitory synapses initiate the same change in the permeability of the postsynaptic membrane to chloride, and possibly also to potassium ions

(ECCLES, 1966). The types of inhibition used in these comparisons were of relatively short latency and duration, such as "direct" IPSP's evoked by impulses in group Ia muscle afferents which are readily reversed in direction by polarizing currents or ions passed from an intracellular microelectrode, and hence are generated at synapses located predominantly on the soma and proximal portion of motoneurone dendrites (ECCLES, 1966; SMITH, WUERKER and FRANK, 1967; JANKOWSKA and ROBERTS, 1972). There is also evidence for a similar location of inhibitory terminals mediating the inhibition of motoneurones by volleys in groups Ib and III muscle afferents, cutaneous afferents and the recurrent inhibition of motoneurones (COOMBS, ECCLES, and FATT, 1955a), although synapses mediating recurrent inhibition appear to be more distal from the soma than those of direct inhibition (BURKE, FEDINA, and LUNDBERG, 1971).

On the foregoing evidence it might readily be concluded that the transmitter at these synapses was glycine, particularly if neurochemical evidence of the presence and distribution of this amino acid is considered together with the association established with small spinal neurones in aortic occlusion experiments. However, similar changes in the membrane properties of motoneurones have been observed to result from electrophoretic administration of β-alanine, GABA (CURTIS et al., 1968b) and δ-aminovaleric acid (BRUGGENCATE and ENGBERG, 1968). Of these substances only glycine and GABA warrant serious consideration as transmitters on neurochemical grounds, although such a function for L-α-alanine and taurine cannot be fully excluded as these depressant amino acids presumably have postsynaptic effects similar to those of glycine.

One very important characteristic of these types of short latency spinal inhibition is sensitivity to strychnine. In particular "direct" and recurrent inhibition of motoneurones are reversibly blocked by relatively low concentrations of strychnine administered intravenously (Cat: BRADLEY, EASTON, and ECCLES, 1953; ECCLES, FATT, and KOKETSU, 1954) or electrophoretically (CURTIS, 1962), in the absence of significant modification of the activity of the appropriate inhibitory interneurones. A close analysis of the effects of strychnine on recurrent IPSP's supports previous proposals that the alkaloid probably hinders the access of transmitter to receptor sites rather than interferes with the flow of ions through the activated membrane (LARSON, 1969). Other types of inhibition of these cells are also affected, including those of spinal (COOMBS, ECCLES, and FATT, 1955b, CURTIS, 1963) and supraspinal (see CURTIS, 1969) origin. The recurrent inhibition of gamma motoneurones is also blocked by strychnine (ELLAWAY, 1971), as is also the inhibition of Renshaw cells by impulses originating in hind paw afferents and from the medullary reticular formation (BISCOE and CURTIS, 1966). It is thus very significant in establishing glycine as a spinal transmitter that strychnine selectively and reversibly blocks the inhibitory effect of glycine (and of "glycine-like" α- and β-amino acids) on spinal motoneurones, interneurones, Renshaw cells and autonomic neurones, but shows little or no antagonism towards the action of GABA and "GABA-like" γ- and higher ω-amino acids (Cat: CURTIS et al., 1968b, 1971c; LARSON, 1969; DE GROAT, 1970b. Rat: BISCOE et al., 1972). Even the one apparently dissenting report indicates that the effect of glycine on spinal neurones is suppressed more completely by strychnine than is that of GABA (DAVIDOFF, APRISON, and WERMAN, 1969).

Within the limits of *in vivo* experiments, using both electrophoretic and intravenous administration, low concentrations of strychnine probably interfere competitively with the interaction between glycine and postsynaptic receptors (CURTIS et al., 1971a). Additionally a number of alkaloids and a series of synthetic compounds which block this type of short latency spinal inhibition, and hence are convulsants when administered intravenously, are also glycine antagonists, the relative potencies for these two actions being similar. The substances (Fig. 2) include brucine, thebaine (FATT, 1954; PINTO CORRADO, and LONGO, 1961; CURTIS, 1962); morphine (CURTIS and DUGGAN, 1969); diaboline (see WEST, 1937); Weiland-Gumlich aldehyde; laudanosine, dendrobine, gelsemine (CURTIS et al., 1971 a); 4-phenyl-4-formyl-N-methylpiperidine (LONGO and PINTO CORRADO, 1961; CURTIS, 1962); 5,7-diphenyl-1,3-diazadamantan-6-ol (LONGO, SILVESTRINI, and BOVET, 1959; CURTIS, 1962); and hexahydro-2′-methylspiro [cyclohexane-1,8′-(6H)-oxazino (3,4-A)pyrazine] (CHEN and HAUCK, 1961). It is also relevant that other convulsants which are selective GABA antagonists in the cord, bicuculline (Cat: CURTIS et al., 1971a. Rat: BISCOE et al., 1972) and penicillin (Cat: DAVIDOFF, 1972c; CURTIS et al., 1972b), do not block spinal inhibitions of the strychnine sensitive type. Picrotoxinin has no effect on "direct" inhibition (ECCLES, SCHMIDT, and WILLIS, 1963) and although picrotoxin-sensitive recurrent IPSP's of motoneurones have been reported (KELLERTH, 1968), and there is some controversy regarding the effectiveness and specificity of electrophoretic picrotoxinin as an amino acid antagonist in the cord (Cat: DAVIDOFF and APRISON, 1969; CURTIS, DUGGAN, and JOHNSTON, 1969; DE GROAT, 1970; ENGBERG and THALLER, 1970. Rat: BISCOE et al., 1972), the general consensus of opinion seems to be that this substance is a GABA antagonist (see Section 2.4).

There is thus very good evidence for the presence in the cord of glycinergic inhibitory interneurones of several types: "Ia" interneurones mediating "direct" or reciprocal inhibition of motoneurones (JANKOWSKA and ROBERTS, 1972; JANKOWSKA and LINDSTRÖM, 1972); Renshaw cells mediating recurrent inhibition (ECCLES et al., 1954; JANKOWSKA and LINDSTRÖM, 1971), and other interneurones involved in the relatively short latency inhibition of motoneurones, other interneurones and Renshaw cells by impulses of muscular and cutaneous origin. It is relevant to the disturbances of spinal glycine levels, reflexes and inhibitions produced by temporary aortic occlusion that Ia interneurones and Renshaw cells lie dorsomedial and ventromedial respectively to motoneurones, and are possibly not affected to the same extent by such a procedure. The location of other inhibitory interneurones, and the connections between them and the convergence upon them of pathways, of both local spinal and descending origin, remain to be determined.

Evidence has recently been provided that in addition to motoneurones and "Ia" interneurones (HULTBORN, JANKOWSKA, and LINDSTRÖM, 1971) Renshaw cells inhibit other Renshaw cells (RYALL, 1970). This "mutual" inhibition is probably related to the "pause" in the firing of Renshaw cells and the reduced chemical excitability which follows the initial cholinergic excitation evoked by a ventral root volley (CURTIS and RYALL, 1966a). In contrast to the sensitivity to strychnine of the inhibition of Renshaw cells by volleys in hind limb afferents (BISCOE and CURTIS, 1966; WILSON and TALBOT, 1963), this "mutual" inhibition is resistant to strychnine and to bicuculline (RYALL, PIERCEY and POLOSA, 1972; CURTIS, GAME and McCULLOCH, unpublished observations). Although an explanation has been offered in terms of the presence on Renshaw cells of strychnine-insensitive glycine receptors which would be compatible with all Renshaw cells being glycinergic (RYALL et al., 1972), the finding that tetanus toxin abolishes the inhibition of Renshaw cells by hind limb impulses, but does not modify "mutual" inhibition, (CURTIS, GAME, and McCULLOCH, unpublished observations) tends to exclude glycine (and GABA) as an inhibitory transmitter for this process. Further investigation of both "mutual" inhibition and the pause seems warranted, particularly as hyperpolarizing IPSP's have not been demonstrated for either phenomenon.

The effects of tetanus toxin on spinal inhibition are relevant to transmitter functions of both glycine and GABA. With respect to glycine, spinal inhibitions sensitive to strychnine are also reduced by tetanus toxin (Cat: BROOKS, CURTIS and ECCLES, 1957; CURTIS and DE GROAT, 1968). However, unlike the situation after strychnine, spinal neurones retain sensitivity to glycine (CURTIS and DE GROAT, 1968; GUSHCHIN, KOZHECHKIN and SVERDLOV, 1969). Since the levels of glycine within the cord are little influenced (Cat: JOHNSTON, DE GROAT, and CURTIS, 1969; FEDINEC and SHANK, 1971; see also SEMBA and KANO, 1969), tetanus toxin apparently interferes with the release of this amino acid from inhibitory terminals.

4.12.3.2 GABA

The depression of the firing of spinal interneurones, parasympathetic preganglionic neurones, motoneurones and Renshaw cells by electrophoretic GABA (Cat: CURTIS et al., 1959; DE GROAT, 1970. Rat: BISCOE et al., 1972) is associated with an increased membrane potential and a conductance increase similar to that induced by glycine (CURTIS et al., 1968b). This, together with the effects of intrathecal and systemic GABA on spinal reflexes (KUNO and MUNEOKA, 1962; BHARGAVA and SRIVASTAVA, 1964; DHAWAN et al., 1972), and the intraspinal distribution of the amino acid, suggests that GABA is an inhibitory transmitter in the cord. Furthermore, the relative insensitivity of depression of spinal neurone firing by GABA to strychnine (CURTIS et al., 1968a, b; DE GROAT, 1970; LARSON, 1969; CURTIS et al., 1971c; BISCOE et al., 1972), and the antagonism of the postsynaptic effects of GABA by bicuculline (CURTIS et al., 1971a; BISCOE et al., 1972), related alkaloids (JOHNSTON et al., 1972), and penicillin (DAVIDOFF, 1972c; CURTIS et al., 1972b), indicate that GABA is unlikely to be the transmitter of inhibitions suppressed by strychnine and strychnine-like glycine antagonists.

Consideration of the transmitter role of GABA in the spinal cord is closely related to the controversial subject of "presynaptic" inhibition, a process reviewed recently in depth (SCHMIDT, 1971). Although there is general agreement that this type of inhibition is of longer latency and duration than that discussed in the previous section, and is insensitive to strychnine but reduced by picrotoxinin and bicuculline, there is still debate regarding the mechanism of the process. The inhibition has been measured by the reduction in amplitude of either intracellularly recorded EPSP's or monosynaptic reflexes, and is diminished by picrotoxin

(ECCLES et al., 1963; KELLERTH and SZUMSKI, 1966; SCHMIDT, 1971), bicuculline (CURTIS et al., 1971a; HUFFMAN and MCFADIN, 1972) and penicillin (DAVIDOFF, 1972c).

Prolonged spinal inhibition has been held to be essentially entirely *presynaptic* in nature, depolarization at axo-axonic synapses on the terminals of primary afferent fibres being responsible for a reduced release of excitatory transmitter and hence a diminished effectiveness of excitatory volleys (SCHMIDT, 1971). Primary afferent depolarization (PAD), detected by direct intrafibre recording, measurement of fibre and terminal excitability or the recording of dorsal root potentials and reflexes, is thus considered a primary event, generated at axo-axonic synapses. The participation of GABA in this process is suggested by the depolarization of afferent terminals demonstrated in amphibian spinal cord preparations (SCHMIDT, 1963; TEBĒCIS and PHILLIS, 1969; DAVIDOFF, 1972a; BARKER and NICOLL, 1973), an effect which is dependent on sodium ions, unaffected by tetrodotoxin (BARKER and NICOLL, 1973), and like PAD is blocked by picrotoxinin, bicuculline and penicillin (SCHMIDT, 1971; DAVIDOFF, 1972a, b; DAVIDOFF, SILVEY, KOBETZ and SPIRA, 1972; BARKER and NICOLL, 1973). Such direct evidence for terminal depolarization by GABA has not been obtained in the cat, in fact electrophoretic GABA depressed terminal excitability (CURTIS and RYALL, 1966b). Nevertheless picrotoxinin, bicuculline and penicillin reduce both PAD and the inhibitory effect on reflexes of incoming volleys (SCHMIDT, 1971; CURTIS et al., 1971a; LEVY and ANDERSON, 1972; DAVIDOFF, 1972c). Additionally, administration of semicarbazide to acute spinal cats, which reduces spinal GABA levels, suppresses dorsal root potentials and reflexes, and "presynaptic" inhibition (BELL and ANDERSON, 1972).

On the other hand prolonged inhibition is considered to be *postsynaptic* in nature, at synapses on dendrites so distant from the soma of motoneurones that changes in membrane potential and conductance are not readily detected by an intracellular microelectrode (GRANIT, 1968; KELLERTH, 1968). Evidence has been provided of inhibitory hyperpolarizations which are reversed by intracellular chloride ion injection and which are reduced by picrotoxin but not by strychnine (KELLERTH, 1968; COOK and CANGIANO, 1972). The participation of GABA as an inhibitory transmitter at such synapses would be entirely consistent with its inhibitory effect when administered electrophoretically near spinal neurones, and for the reduction of this synaptic inhibition by bicuculline. Furthermore, elevation of the extracellular potassium concentration as a consequence of prolonged activation of GABA receptors could account for the depolarization of neighbouring synaptic terminals and hence for PAD (see CURTIS et al., 1971a). Increases in extracellular potassium levels have been demonstrated in spinal tissue to follow afferent volleys which produce "presynaptic" inhibition (Cat: SOMJEN, 1970; KRNJEVIĆ and MORRIS, 1972. Rat: VYKLICKÝ, SYKOVA KŘIŽ, and UJEC, 1972), and the depolarization would be reduced by GABA antagonists and by substances interfering with GABA release. The only observation inconsistent with this explanation of PAD is the apparent dependence on sodium ions of the depolarization of amphibian spinal afferent terminals by GABA (BARKER and NICOLL, 1973), and further investigation is necessary of the ionic mechanism of PAD in mammals.

The effects of tetanus toxin on spinal inhibitions are relevant to the role of GABA as an inhibitory transmitter since the toxin suppresses GABA-mediated inhibition in the cerebellum (Section 4.2.2). Following intramuscular (SVERDLOV and ALEKSEEVA, 1965; SVERDLOV and KOZHECHKIN, 1969) or intraspinal (CURTIS et al., 1973a) administration, short latency glycine-mediated inhibitions of motoneurones are diminished, and subsequently there is a reduction in prolonged inhibitions accompanied by reduced PAD (CURTIS et al., 1973a). Since Renshaw cells affected by the toxin remain sensitive to GABA (CURTIS and DE GROAT, 1968), and spinal levels of GABA are unaffected (JOHNSTON et al., 1969; SEMBA and KANO, 1969; FEDINEC and SHANK, 1971), the toxin presumably interferes with the synaptic release of GABA, an action similar to that occurring at glycine-releasing terminals. The differences in the latency of the action of this toxin on the two types of spinal inhibition may be related to the predominantly somatic location of glycine releasing terminals and the probable association of GABA with axodendritic inhibitory synapses.

Several processes thus appear responsible for the long latency and prolonged inhibition of spinal motoneurones by afferent volleys in segmental afferents and from higher centres (SCHMIDT, 1971). The central effects of these volleys are complex, both presynaptic and postsynaptic factors may be involved, but the release of GABA from the terminals of a particular type of inhibitory interneurone appears to be essential to the prolonged inhibitory process. To what extent this type of inhibition normally regulates the activity of spinal neurones other than motoneurones remains to be elucidated, as does the location and characterization of the interneurones and their synaptic interconnections with different types of afferent fibre, fibres from supraspinal regions and the axons of other spinal interneurones. Neurochemical evidence suggests that most of these GABA releasing interneurones, or their synaptic endings, are located in the dorsal horn. Other features requiring further investigation are the prolonged time course of this type of inhibition and its enhancement by anaesthetics and other agents (SCHMIDT, 1971), an effect which may be common to other synapses operated by GABA where transmitter action is similarly of long duration (NICOLL, 1972).

4.13. Non-Central Neurones

Although possibly not strictly relevant to the central theme of this review, the effects of amino acids upon dorsal root sensory ganglia and autonomic ganglia are of considerable interest, and may be of significance in relation to transmitter functions of amino acids.

4.13.1. Dorsal Root Ganglia

Analyses of lumbar dorsal ganglia indicate relatively high intraganglionic levels of a number of amino acids, including aspartate, glutamate and glycine (Table 8), but extremely low levels of GABA (< 0.06 μmole/g, concentration approx. 0.2 mM, Cat: OTSUKA et al., 1971. Rat: 0.3 μmole/g, GOTTESFELD, KELLY, and

SCHON, 1973b). When incubated with labelled amino acids cat and rat ganglia accumulate GABA by a high affinity system ($K_m$approx. 2.5×10^{-5} M) which is facilitated by amino-oxyacetic acid and occurs predominantly into satellite glial cells (GOTTESFELD et al., 1973b). In contrast, α-alanine and glycine transport is less rapid and occurs into both glial cells and neurones.

When administered by close intra-arterial injection, GABA depolarizes the nodose ganglion of the cat, the threshold dose being 0.05–2 μg (DE GROAT, 1972). No action could be demonstrated for higher doses of glycine, L-glutamate, L-aspartate or DL-homocysteate, and the depolarizing action of GABA was abolished by picrotoxin (100–300 μg) or bicuculline (100–300 μg) but was unaffected by strychnine (200–400 μg). Essentially similar results were obtained using lumbar dorsal root ganglia when GABA was administered intra-arterially or intravenously (Cat: DE GROAT, LALLEY, and SAUM, 1972), threshold concentrations of GABA after intravenous administration being of the order of 10^{-7}–10^{-6} M. Depolarization by 3-aminopropane sulphonic acid, β-alanine and δ-aminovaleric acid were also demonstrated (DE GROAT et al., 1972). Furthermore electrophoretic GABA depolarizes cultured dorsal root ganglion cells, an effect also blocked by picrotoxin (OBATA, 1972a).

This depolarizing action of GABA, which appears not to modify impulse transmission through sensory ganglia (DE GROAT et al., 1972), has been interpreted in terms of enhanced chloride permeability (DE GROAT, 1972), and the relevance of this apparently non-synaptic change in membrane potential to the depolarization of central terminals of afferent fibres during "presynaptic" inhibition (see Section 4.12.3.2) remains to be determined. Neuronal or glial uptake with a subsequent elevation of extracellular potassium levels seems unlikely since neither picrotoxin nor bicuculline interfere with GABA uptake elsewhere in the nervous system.

4.13.2. Autonomic Ganglia

Earlier evidence of the effects of amino acids upon autonomic ganglia has been discussed previously (CURTIS and WATKINS, 1965).

The very low concentrations of GABA (< 0.1 μmole/g, Rat: NAGATA, YOKOI, and TSUKADA, 1966; MASI, PAGGI, POCCHIAR, and TOSCHI, 1969; MCBRIDE and KLINGMAN, 1972) and GAD (NAGATA et al., 1966) within the superior cervical ganglion could be considered to exclude a functional role of this amino acid in such ganglia. Nevertheless, isolated ganglia accumulate GABA by a relatively high affinity sodium dependent system which however differs in substrate specificity from that of brain tissue (Rat: superior cervical ganglion, BOWERY and BROWN, 1972b). The uptake is not modified by preganglionic denervation (BOWERY and BROWN, 1972b) and uptake into nerve terminals also seems excluded in radio-autographic studies, the amino acid being located in neurones and capsular cells (BOWERY and BROWN, 1971). An equivalent or greater accumulation of GABA was demonstrated for the preganglionic cervical sympathetic trunk and the vagus nerve (BOWERY and BROWN, 1972b). Although electrical stimulation of pre- or postganglionic trunks, or treatment with carbamylcholine, did not release labelled GABA from the ganglion, an efflux was demonstrated during high frequency electrical stimulation of the ganglion body or elevation of the extracellular potassium concentration; this potassium-induced efflux not being affected by preganglionic denervation (BOWERY and BROWN, 1972b).

Studies of other amino acids in ganglia have been largely concerned with metabolism (NAGATA et al., 1966; MCBRIDE and KLINGMAN, 1972; MASI et al., 1969). The level of glutamate in the cat inferior cervical and thoracic ganglia (3 μmole/g, JOHNSON and APRISON, 1970) is less than that of dorsal root ganglia (Table 8). Glycine values are relatively high (> 3 μmole/g, Rat: MCBRIDE and KLINGMAN, 1972) and glycine is accumulated by isolated ganglia (BOWERY and BROWN, 1972b).

Although these neurochemical studies with GABA support the non-involvement of this amino acid as a transmitter in autonomic ganglia, there is strong evidence that GABA (and closely related analogues) depresses ganglionic transmission and depolarizes ganglionic neurones (Cat: superior cervical, inferior mesenteric, pelvic ganglia *in vivo*, DE GROAT, 1970a, Rat: superior cervical ganglion *in vitro*, BOWERY and BROWN, 1972a; see also SRIMAL and BHARGAVA, 1966). Threshold concentrations were of the order of 0.02 μg (intra-arterial, DE GROAT, 1970a) and 10^{-6} M (isolated, BOWERY and BROWN, 1972a), and intracellular studies indicate a depolarization and an increased membrane conductance which is most probably due to an enhanced permeability of the membrane to chloride ions (Rat: ADAMS and BROWN, 1973). Ganglia are insensitive to glycine and acidic amino acids (DE GROAT, 1970a; BOWERY and BROWN, 1972a) and the effects of GABA were not altered by strychnine (DE GROAT, 1970a) but were antagonized by picrotoxin and bicuculline (DE GROAT, 1970a; DE GROAT, LALLEY and BLOCK, 1971; BOWERY and BROWN, 1972a) neither of which seriously affected ganglionic transmission.

A picrotoxin-sensitive depolarization by GABA has also been demonstrated using cultured rat sympathetic ganglion cells (OBATA, 1972a).

Chronic preganglionic denervation has no effect on the depolarizing action of GABA on the cat superior cervical ganglion (DE GROAT, 1969) and the relevance of these pharmacological observations to synaptic transmission within ganglia by either GABA or a related amino acid seems rather remote. However further investigation is warranted.

5. Summarizing and Concluding Remarks

5.1. Glycine

There are reasonable grounds for considering glycine as the transmitter mediating "direct" and "recurrent" inhibition of spinal motoneurones, as well as other strychnine-sensitive inhibitions of both segmental and supraspinal origin which influence the firing of medullary and spinal motoneurones and interneurones. There is at present little evidence to indicate that glycine operates as an inhibitory transmitter in other areas of the CNS, and the importance of this amino acid appears to be its involvement in the regulation of spinal and brain stem reflexes.

5.2. GABA

The most fully documented case for an amino acid transmitter, and indeed for any central transmitter, is that for GABA, released at terminals of Purkinje cell axons to inhibit Deiters' neurones. Somewhat less, though adequate, evidence is available that GABA is an inhibitory transmitter in the cerebral, cerebellar and hippocampal cortices released by basket type cells. In addition, GABA seems likely to be an inhibitory transmitter in the substantia nigra (which contains the highest levels of this amino acid in the brain), in the retina, olfactory bulb, thalamic relay nuclei, dorsal column nuclei and the spinal cord. In the two latter regions GABA is associated with prolonged inhibition, a process sensitive to picrotoxinin and bicuculline, and which may be at least partially presynaptic in nature.

5.3. Aspartate and Glutamate

Glutamate, the most abundant amino acid in the adult brain, and aspartate, may be excitatory transmitters of the main afferent and efferent pathways within the CNS. The evidence is far from convincing, the proposal being based largely on negative evidence for the involvement of other substances, the distribution of these amino acids in the brain and spinal cord, and the excitation by both of neurones in all regions.

Neurochemical studies suggest that glutamate could be the excitatory transmitter of spinal primary afferent fibres and aspartate a transmitter released by spinal excitatory interneurones. Further investigations are required of the relative sensitivities of central neurones to these excitants, a study confined so far to the thalamus and spinal cord. Additionally there is a vital need for selective antagonists of both the excitation by, and the uptake of, aspartate and glutamate in order to establish the central pathways which operate by releasing these amino acids.

5.4. Other Amino Acids

It is possible that amino acids other than glycine, GABA, aspartate and glutamate function as synaptic transmitters. These include α- and β-alanine, γ-aminobutyrylcholine, cystathionine, cysteate, cysteine sulphinate, hypotaurine, taurine, imidazole-4-acetate and serine. All influence the firing of neurones when administered electrophoretically, and even the levels of those which occur in "trace" amounts, in relation to GABA, are comparable with the amounts of acetylcholine, noradrenaline, dopamine and 5-hydroxytryptamine in nervous tissue.

5.5. Concluding Remarks

This review has been concerned with the function of amino acids of relatively simple chemical structure as excitatory and inhibitory transmitters regulating the activity of neurones throughout the mammalian central nervous system. The evidence presented suggests that these substances deserve consideration as major central transmitters involved with the relay of afferent information from the spinal to the cortical level, the control of cerebral pyramidal and cerebellar Purkinje cells, and of other cortical, subcortical and spinal neurones. The case could even be argued that other substances such as acetylcholine, noradrenaline, dopamine and 5-hydroxytryptamine, widely acknowledged as central transmitters, are of relatively minor but nonetheless vital importance, being associated with the finer control of functions such as movement, mood and behaviour. Due consideration is thus required of amino acids *and* the "amines" when attempting to understand the "magic loom" of synaptic interconnections that constitutes the central nervous system.

With further elucidation and appreciation of the essential role of amino acids as transmitters, and of disturbances of CNS function resulting from abnormalities at the synaptic level, increasing opportunities will arise for therapeutic measures based on chemical manipulation of the synthesis, storage, release, postsynaptic action or inactivation of particular amino acids.

Acknowledgement. The authors are indebted to Mrs. R. MACLACHLAN, Mrs. P. SEARLE, Mrs. A. STEPHANSON, and particularly Mrs. H. WALSH for invaluable assistance in preparing this manuscript, and are grateful for material provided prior to publication by colleagues in North America, Europe and Japan.

References

ADAMS, P.R., BROWN, D.A.: Action of γ-aminobutyric acid (GABA) on rat sympathetic ganglion cells. Brit. J. Pharmacol. **47**, 639–640P (1973).

ADOLPH, A.R.: Excitation and inhibition of electrical activity in the *limulus* eye by neuropharmacological agents. In: The functional organization of the compound eye, p. 465–482. Oxford: Pergamon Press 1966.

AGHAJANIAN, G.K., HAIGLER, H.J., BLOOM, F.E.: Lysergic acid diethylamide and serotonin: direct actions on serotonin-containing neurons in rat brain. Life Sci. **11**, Part I, 615–622 (1972).

AGRAWAL, H.C., DAVIS, J.M., HIMWICH, W.A.: Postnatal changes in free amino acid pool of rat brain. J. Neurochem. **13**, 607–615 (1966).

AGRAWAL, H.C., DAVISON, A.N., KACZMAREK, L.K.: Subcellular distribution of taurine and cysteinesulphinate decarboxylase in developing rat brain. Biochem. J. **122**, 759–763 (1971).

AJMONE-MARSAN, C.: Acute effects of topical epileptogenic agents. In: Basic mechanisms of the epilepsies, eds. H.H. Jasper, A.A. Ward, Jr., and A. Pope, p. 299–319. Boston: Little, Brown & Co. 1969.

ALBERS, R.W., BRADY, R.O.: Distribution of glutamic decarboxylase in the nervous system of the Rhesus monkey. J. biol. Chem. **234**, 926–928 (1959).

ALBERT, A.: Quantitative studies of the avidity of naturally occurring substances for trace metals. 1. Amino-acids having only two ionizing groups. Biochem. J. **47**, 531–538 (1950).

ALTMANN, H., BRUGGENCATE, G. TEN, SONNHOF, U.: Differential strength of action of glycine and GABA in hypoglossus nucleus. Pflügers Arch. **331**, 90–94 (1972).

ALTMANN, H., STEINBERG, R., BRUGGENCATE, G. TEN, SONNHOF, U.: Actions of amino acids applied iontophoretically in the red nucleus. Pflügers Arch. **335**, Suppl., Abstract 132 (1972).

AMES, A., III, POLLEN, D.A.: Neurotransmission in central nervous tissue; a pharmacologic study of isolated rabbit retina. J. Neurophysiol. **32**, 424–442 (1969).

ANDERSEN, P., CURTIS, D.R.: The excitation of thalamic neurones by acetylcholine. Acta physiol. scand. **61**, 85–99 (1964a).

ANDERSEN, P., CURTIS, D. R.: The pharmacology of the synaptic and acetylcholine-induced excitation of ventrobasal thalamic neurones. Acta physiol. scand. **61**, 100–120 (1964b).

ANDERSEN, P., ECCLES, J.C., LØYNING, Y.: Pathway of postsynaptic inhibition in the hippocampus. J. Neurophysiol. **27**, 608–619 (1964).

ANDERSEN, P., ECCLES, J.C., LØYNING, Y., VOORHOEVE, P.E.: Strychnine-resistant inhibition in the brain. Nature (Lond.) **200**, 843–845 (1963).

ANDERSEN, P., ECCLES, J.C., OSHIMA, T., SCHMIDT, R.F.: Mechanisms of synaptic transmission in the cuneate nucleus. J. Neurophysiol. **27**, 1096–1116 (1964).

ANDERSEN, P., ETHOLM, B., GORDON, G.: Presynaptic and postsynaptic inhibition elicited in the cat's dorsal column nuclei by mechanical stimulation of skin. J. Physiol. (Lond.) **210**, 433–455 (1970).

APRISON, M.H.: Evidence of the release of C^{14}-glycine from hemisectioned toad spinal cord with dorsal root stimulation. Pharmacologist **12**, No. 2, 126 (1970).

APRISON, M.H., DAVIDOFF, R.A., WERMAN, R.: Glycine: its metabolic and possible transmitter roles in nervous tissue. In: Handbook of neurochemistry, vol. 3, Metabolic reactions in the nervous system, ed. A. Lajtha, p. 381–397. New York: Plenum Press 1970.

APRISON, M.H., MCBRIDE, W.J.: Evidence for the net accumulation of glycine into a synaptosomal fraction isolated from the telecephalon and spinal cord of the rat. Life Sci. **12**, Part I, 449–458 (1973).

APRISON, M.H., SHANK, R.P., DAVIDOFF, R.A.: A comparison of the concentration of glycine. a transmitter suspect, in different areas of the brain and spinal cord in seven different vertebrates. Comp. Biochem. Physiol. **28**, 1345–1355 (1969).

APRISON, M.H., WERMAN, R.: A combined neurochemical and neurophysiological approach to identification of central nervous system transmitters. In: Neurosciences research, vol. 1, eds. S. Ehrenpreis and P.C. Solnitzky, p. 143–174. New York: Academic Press 1968.

ARNFRED, T., HERZ, L.: Effects of potassium and glutamate on brain cortex slices: uptake and release of glutamic and other amino acids. J. Neurochem. **18**, 259–265 (1971).

ARREGUI, A., LOGAN, W.J., BENNETT, J.P., SNYDER, S.H.: Specific glycine – accumulating synaptosomes in the spinal cord of rats. Proc. nat. Acad. Sci. (Wash.) **69**, 3485–3489 (1972).

AWAPURA, J.: The taurine concentration of organs from fed and fasted rats. J. biol. Chem. **218**, 571–576 (1956).

BAKER, W.W., KRATKY, M., BENEDICT, F.: Electrographic responses to intrahippocampal injections of convulsant drugs. Exp. Neurol. **12**, 136–145 (1965).

BALÁZS, R., MACHIYAMA, Y., HAMMOND, B.J., JULIAN, T., RICHTER, D.: The operation of the γ-aminobutyrate bypath of the tricarboxylic acid cycle in brain tissue *in vitro*. Biochem. J. **116**, 445–467 (1970).

BALÁZS, R., PATEL, A.J., RICHTER, D.: Metabolic compartments in the brain: their properties and relation to morphological structures. In: Metabolic compartmentation in the brain, eds. R. Balázs and J.E. Cremer, p. 167–184. London: Macmillan 1973.

BALCAR, V.J., JOHNSTON, G.A.R.: Glutamate uptake by brain slices and its relation to the depolarization of neurones by acidic amino acids. J. Neurobiol. **3**, 295–301 (1972a).

BALCAR, V.J., JOHNSTON, G.A.R.: The structural specificity of the high affinity uptake of L-glutamate and L-aspartate by rat brain slices. J. Neurochem. **19**, 2657–2666 (1972b).

BALCAR, V.J., JOHNSTON, G.A.R.: High affinity uptake of transmitters: studies on the uptake of L-aspartate, GABA, L-glutamate and glycine in cat spinal cord. J. Neurochem. **20**, 529–539 (1973).

BANNA, N.R., JABBUR, S.J.: Pharmacological studies on inhibition in the cuneate nucleus of the cat. Int. J. Neuropharmacol. **8**, 299–307 (1969).

BANNA, N.R., JABBUR, S.J.: The effects of depleting GABA on cuneate presynaptic inhibition. Brain Res. **33**, 530–532 (1971).

BANNA, N.R., NACCACHE, A., JABBUR, S.J.: Picrotoxin-like action of bicuculline. Europ. J. Pharmacol. **17**, 301–302 (1972).

BAÑOS, G., DANIEL, P.M., MOORHOUSE, S.R., PRATT, O.E.: The entry of amino acids into the brain of the rat during the postnatal period. J. Physiol. (Lond.) **213**, 45–46P (1971).

BARKER, J.L., CRAYTON, J.W., NICOLL, R.A.: Noradrenaline and acetylcholine responses of supraoptic neurosecretory cells. J. Physiol. (Lond.) **218**, 19–32 (1971).

BARKER, J.L., NICOLL, R.A.: The pharmacology and ionic dependency of amino acid responses in the frog spinal cord. J. Physiol. (Lond.) **228**, 259–277 (1973).

BATTISTIN, L., GRYNBAUM, A., LAJTHA, A.: Distribution and uptake of amino acids in various regions of the cat brain *in vitro*. J. Neurochem. **16**, 1459–1468 (1969a).

BATTISTIN, L., GRYNBAUM, A., LAJTHA, A.: Energy dependence of amino acid uptake in brain slices. Brain Res. **16**, 187–197 (1969b).

BAXTER, C.F.: The nature of γ-aminobutyric acid. In: Handbook of neurochemistry, vol. 3, ed. A. Lajtha, p. 289–353. New York: Plenum Press 1970.

BAXTER, C.F.: ASSAY OF γ-aminobutyric acid and enzymes involved in its metabolism. In: Methods of neurochemistry, vol. 3, ed. R. Fried, p. 1–73. New York: Marcel Dekker 1972.

BAXTER, C.F., TEWARI, S.: Regulation by amino acids of protein synthesis in a cell free system from immature rat brain: stimulatory effect of γ-aminobutyric acid and glycine. In: Protein metabolism of the nervous system, p. 439–456. New York: Plenum Press 1970.

BAXTER, C.F., TEWARI, S., RAEBURN, S.: The possible role of gamma-aminobutyric acid in the synthesis of protein. In: Advances in biochemical psychopharmacology, vol. 4, eds. M.S. Ebadi and E. Costa, p. 195–216. New York: Raven Press (1972).

BEART, P.M., CURTIS, D.R., JOHNSTON, G.A.R.: 4-Aminotetrolic acid: a new conformationally-restricted analogue of γ-aminobutyric acid. Nature (Lond.) New Biol. **234**, 80–81 (1971).

BEART, P.M., JOHNSTON, G.A.R.: Transmination of analogues of γ-aminobutyric acid by extracts of rat brain mitochondria. Brain Res. **49**, 459–462 (1973a).

BEART, P.M., JOHNSTON, G.A.R.: GABA uptake in rat brain slices: inhibition by GABA analogues and by various drugs. J. Neurochem. **20**, 319–324 (1973b).

BEART, P.M., JOHNSTON, G.A.R., UHR, M.L.: Competitive inhibition of GABA uptake in rat brain slices by some GABA analogues of restricted conformation. J. Neurochem. **19**, 1855–1861 (1972).

BELL, J.A., ANDERSON, E.G.: The influence of semicarbazide-induced depletion of γ-aminobutyric acid on presynaptic inhibition. Brain Res. **43**, 161–169 (1972).

BENJAMIN, A.M., QUASTEL, J.H.: Locations of amino acids in brain slices from the rat. Tetrodotoxin-sensitive release of amino acids. Biochem. J. **128**, 631–646 (1972).

BENNETT, J.P., JR., LOGAN, W.J., SNYDER, S.H.: Amino acid neurotransmitter candidates: sodium-dependent high-affinity uptake by unique synaptosomal fractions. Science **178**, 997–999 (1972).

BENUCK, M., STERN, F., LAJTHA, A.: Transamination of amino acids in homogenates of rat brain. J. Neurochem. **18**, 1555–1567 (1971).

BENUCK, M., STERN, F., LAJTHA, A.: Regional and subcellular distribution of aminotransferases in rat brain. J. Neurochem. **19**, 949–957 (1972).

BERGERET, B., CHATAGNE, F.: Sur la présence d'acide cysteinesulfinique dans le cerveau du rat normal. Biochim. biophys. Acta (Amst.) **14**, 297 (1954).

BERL, S., CLARKE, D.D.: Compartmentation of amino acid metabolism. In: Handbook of neurochemistry, vol. 2, ed. A. Lajtha, p. 447–472. New York: Plenum Press 1969.

BERL, S., MCMURTRY, J.G.: Isolated cerebral cortex: changes in levels of glutamic acid, glutamine, aspartic acid, and γ-aminobutyric acid. Arch. Biochem. Biophys. **118**, 645–648 (1967).

BERL, S., NICKLAS, W.J., CLARKE, D.D.: Compartmentation of glutamic acid metabolism in brain slices. J. Neurochem. **15**, 131–140 (1968).

BHAGAVAN, H.N., COURSIN, D.B., STEWART, C.N.: Monosodium glutamate induces convulsive disorders in rats. Nature (Lond.) **232**, 275–276 (1971).

BHARGAVA, K.P., SRIVASTAVA, R.K.: Nonspecific depressant action of γ-aminobutyric acid on somatic reflexes. Brit. J. Pharmacol. **23**, 391–398 (1964).

BISCOE, T.J., CURTIS, D.R.: Noradrenaline and inhibition of Renshaw cells. Science **151**, 1230–1231 (1966).

BISCOE, T.J., CURTIS, D.R.: Strychnine and cortical inhibition. Nature (Lond.) **214**, 914–915 (1967).

BISCOE, T.J., DUGGAN, A.W., LODGE, D.: Antagonism between bicuculline, strychnine and picrotoxin and depressant amino acids in the rat central nervous system. Comp. J. Gen. Pharmac. **3**, 423 (1972).

BISCOE, T.J., STRAUGHAN, D.W.: Micro-electrophoretic studies of neurones in the cat hippocampus. J. Physiol. (Lond.) **183**, 341–359 (1966).

BISTI, S., IOSIF, G., MARCHESI, G.F., STRATA, P.: Pharmacological properties of inhibitions in the cerebellar cortex. Exp. Brain Res. **14**, 24–37 (1971).

BLASBERG, R.G.: Specificity of cerebral amino acid transport: a kinetic analysis. Progr. Brain Res. **29**, 245–256 (1968).

BLOOM, F.E.: Amino acids and polypeptides in neuronal function. A report based on an NRP work session. Neurosci. Res. Prog. Bull. **10**, 121–251 (1972).

BLOOM, F. E., COSTA, E., SALMOIRAGHI, G. C.: Anesthesia and the responsiveness of individual neurons of the caudate nucleus of the cat to acetylcholine, norepinephrine and dopamine administered by microelectrophoresis. J. Pharmacol. exp. Ther. **150**, 244–252 (1965).

BOAKES, R. J., BRADLEY, P. B., BRIGGS, I., DRAY, A.: Antagonism of 5-hydroxytryptamine by LSD 25 in the central nervous system: a possible neuronal basis for the actions of LSD 25. Brit. J. Pharmacol. **40**, 202–218 (1970).

BOEHME, D. H., FORDICE, M. W., MARKS, N., VOGEL, W.: Distribution of glycine in human spinal cord and selected regions of brain. Brain Res. **50**, 353–359 (1973).

BORNSCHEIN, H., HEISS, W. D.: Strychnine-resistant inhibition in the retina. Experientia, (Basel) **22**, 49 (1966).

BORRE, C., GENESER-JENSEN, F. A.: Distribution of aspartate aminotransferase in the hippocampal region of the guinea pig. Z. Zellforsch. **127**, 209–219 (1972).

BOWERY, N. G., BROWN, D. A.: Observations on [^{3}H]-aminobutyric acid accumulation and efflux in isolated sympathetic ganglia. J. Physiol. (Lond.) **218**, 32–33 P (1971).

BOWERY, N. G., BROWN, D. A.: Depolarization of isolated rat ganglia by γ-aminobutyric acid and related compounds. Brit. J. Pharmacol. **45**, 160–161 P (1972a).

BOWERY, N. G., BROWN, D. A.: γ-Aminobutyric acid uptake by sympathetic ganglia. Nature (Lond.) New Biol. **238**, 89–91 (1972b).

BOYD, E. S., MERITT, D. A., GARDNER, C.: The effect of convulsant drugs on transmission through the cuneate nucleus. J. Pharmacol. exp. Ther. **154**, 398–409 (1966).

BRADFORD, H. F.: Metabolic response of synaptosomes to electrical stimulation: release of amino acids. Brain Res. **19**, 239–247 (1970).

BRADFORD, H. F.: Cerebral cortex slices and synaptosomes: *in vitro* approaches to brain metabolism. In: Methods of neurochemistry, vol. 3, ed. R. Fried, p. 155–202. New York: Marcel Dekker 1972.

BRADLEY, K., EASTON, D. M., ECCLES, J. C.: An investigation of primary or direct inhibition. J. Physiol. (Lond.) **122**, 474–488 (1953).

BRIDGERS, W. F.: Serine transhydroxymethylase in developing mouse brain. J. Neurochem. **15**, 1325–1328 (1968).

BRIDGERS, W. F.: Purification of mouse brain phosphoserine phosphohydrolase and phosphotransferase. Arch. Biochem. Biophys. **133**, 201–207 (1969).

BRIEL, G., NEUHOFF, V.: Microanalysis of amino acids and their determination in biological material using dansyl chloride. Hoppe-Seyler's Z. Physiol. Chem. **353**, 540–553 (1972).

BRIGGS, I.: Ergothioneine in the central nervous system. J. Neurochem. **19**, 27–35 (1972).

BRODAL, A., POMPEIANO, O., WALBERG, A.: The vestibular nuclei and their connections. Anatomy and functional correlations. Edinburgh: Oliver & Boyd 1962.

BRODERICK, D. S., CANDLAND, K. L., NORTH, J. A., MANGUM, J. H.: The isolation of serine transhydroxymethylase from bovine brain. Arch. Biochem. Biophys. **148**, 196–198 (1972).

BRODY, J. F., DeFEUDIS, P. A., DEFEUDIS, F. V.: Effects of microinjections of L-glutamate into the hypothalamus on attack and flight behaviour in cats. Nature (Lond.) **224**, 1330 (1969).

BROOKS, V. B., ASANUMA, H.: Pharmacological studies of recurrent cortical inhibition and facilitation. Amer. J. Physiol. **208**, 674–681 (1965).

BROOKS, V. B., CURTIS, D. R., ECCLES, J. C.: The action of tetanus toxin on the inhibition of motoneurones. J. Physiol. (Lond.) **135**, 655–672 (1957).

BROWN, F. C., GORDON, P. H.: A study of L-^{14}C-cystathionine metabolism in the brain, kidney and liver of pyridoxine-deficient rats. Biochim. biophys. Acta (Amst.) **230**, 434–445 (1971).

BRUGGENCATE, G. TEN, ENGBERG, I.: Analysis of glycine actions on spinal interneurones by intracellular recording. Brain Res. **11**, 446–450 (1968).

BRUGGENCATE, G. TEN, ENGBERG, I.: Iontophoretic studies in Deiters' nucleus of the inhibitory actions of GABA and related amino acids and the interactions of strychnine and picrotoxin. Brain Res. **25**, 431–448 (1971).

BRUGGENCATE, G. TEN, SONNHOF, U.: Glycine and GABA actions in hypoglossus nucleus and blocking effects of strychnine and picrotoxin. Experientia (Basel) **27**, 1109 (1971).

BRUGGENCATE, G. TEN, SONNHOF, U.: Effects of glycine and GABA, and blocking actions of strychnine and picrotoxin in the hypoglossus nucleus. Pflügers Arch. **334**, 240–252 (1972).

BRUUN, A., EHINGER, B.: Uptake of the putative neurotransmitter, glycine, into the rabbit retina. Invest. Ophthal. **11**, 191–198 (1972).

BURKE, R.E., FEDINA, L., LUNDBERG, A.: Spatial synaptic distribution of recurrent and group Ia inhibitory systems in cat spinal motoneurones. J. Physiol. (Lond.) **214**, 305–326 (1971).

BURKHARDT, D.A.: Effects of picrotoxin and strychnine upon electrical activity of the proximal retina. Brain Res. **43**, 246–249 (1972).

CACIAPPO, F., PANDOLFO, L., DI CHIARA, G.: Transamination reaction between 4-aminobutyric acid and α-ketoglutaric acid in certain rat tissues. Boll. Soc. Ital. Biol. Sper. **36**, 465–467 (1959).

CASOLA, L., DI MATTEO, G.: Studies on the dansylation reaction by the use of ^{14}C dansyl chloride: application to analysis of free amino acids in the rat optic nerve. Analyt. Biochem. **49**, 416–429 (1972).

CAVANAUGH, J.R.: The rotational isomerism of phenylalanine by nuclear magnetic resonance. J. Amer. chem. Soc., **89**, 1558–1563 (1967).

CHAMBERS, G., ELDRED, E., EGGETT, C.: Anatomical observations on the arterial supply to the lumbosacral spinal cord of the cat. Anat. Rec. **174**, 421–434 (1972).

CHAPMAN, J.B., MCCANCE, I.: Acetylcholine sensitive cells in the intracerebellar nuclei of the cat. Brain Res. **5**, 535–538 (1967).

CHEN, G., HAUCK, F.P.: The central effect of hexahydro-2′-methylspiro[cyclohexane-1,8′(6H)-oxazino(3,4-A)pyrazine]. Fed. Proc. **20**, 323 (1961).

CHRISTENSEN, H.N.: Linked ion and amino acid transport. In: Membrane and ion transport, vol. 1, p. 365–394, ed. H.H. Bittar. London-New York: Wiley-Interscience 1970.

CHRISTENSEN, H.N.: Nature and role of receptor sites for amino acid transport. In: Advances in biochemical psychopharmacology vol. 4, Role of vitamin B6 in neurobiology, eds. M.S. Ebadi and E. Costa, p. 39–62. New York: Raven Press 1972.

CHU, S.: Strychnine-sensitive and -insensitive inhibitions in cats retina. Tohoku J. exp. Med. **96**, 37–43 (1968).

CLARKE, G., HILL, R.G.: Effects of a focal penicillin lesion on responses of rabbit cortical neurones to putative neurotransmitters. Brit. J. Pharmacol. **44**, 435–441 (1972).

COHEN, S.R., LAJTHA, A.: Amino acid transport. In: Handbook of neurochemistry, vol. 7, ed. A. Lajtha, p. 543–572. New York: Plenum Press 1972.

COLLINS, G.D.S.: Effect of aminooxyacetic acid, thiosemicarbazide and haloperidol on the metabolism and half-lives of glutamate and GABA in rat brain. Biochem. Pharmacol. **22**, 101–111 (1973).

CONNOR, J.D.: Caudate nucleus neurones: correlation of the effects of substantia nigra stimulation with iontophoretic dopamine. J. Physiol. (Lond.) **208**, 691–703 (1970).

CONSTANTINESCU, E., FLORESCU, D., HATEGANU, D.: Level of some aminoacids in the cerebral cortex of the cat following tetanization of the ventralis lateralis nucleus. Rev. roum. Neurol. **7**, 115–121 (1970).

COOK, W.A., JR., CANGIANO, A.: Presynaptic and postsynaptic inhibition of spinal motoneurones. J. Neurophysiol. **35**, 389–403 (1972).

COOMBS, J.S., ECCLES, J.C., FATT, P.: The specific ionic conductances and the ionic movements across the motoneuronal membrane that produce the inhibitory post-synaptic potential. J. Physiol. (Lond.) **130**, 326–373 (1955a).

COOMBS, J.S., ECCLES, J.C., FATT, P.: The inhibitory suppression of reflex discharges from motoneurones. J. Physiol. (Lond.) **130**, 396–413 (1955b).

CORRIGAN, J.J.: D-Amino acids in animals. Science **164**, 142–149 (1969).

COTMAN, C.W.: Principles for the optimization of centrifugation conditions for fractionation of brain tissue. In: Research methods in neurochemistry, vol. 1, eds. N. Marks and R. Rodnight, p. 45–93. New York: Plenum Press 1972.

COTMAN, C.W., HERSCHMAN, H., TAYLOR, D.: Subcellular fractionation of cultured glial cells. J. Neurobiol. **2**, 169–180 (1971).

CRAWFORD, J.M.: The effect upon mice of intra-ventricular injection of excitant and depressant amino acids. Biochem. Pharmacol. **12**, 1443–1444 (1963).

CRAWFORD, J.M.: The sensitivity of cortical neurones to acidic amino acids and acetylcholine. Brain Res. **17**, 287–296 (1970).

CRAWFORD, J.M., CURTIS, D.R.: The excitation and depression of mammalian cortical neurones by amino acids. Brit. J. Pharmacol. **23**, 313–329 (1964).

CRAWFORD, J.M., CURTIS, D.R., VOORHOEVE, P.E., WILSON, V.J.: Strychnine and cortical inhibition. Nature (Lond.) **200**, 845–846 (1963).

CRAWFORD, J. M., CURTIS, D. R., VOORHOEVE, P. E., WILSON, V. J.: Acetylcholine sensitivity of cerebellar neurones in the cat. J. Physiol. (Lond.) **186**, 139–165 (1966).
CROWSHAW, K., JESSUP, S. J., RAMWELL, P. W.: Thin-layer chromatography of 1-dimethylaminonaphthalen-5-sulphonyl derivatives of amino acids present in superfusates of cat cerebral cortex. Biochem. J. **103**, 79–85 (1967).
CSILLIK, B., GEREBTZOFF, A. M., KISS, J., KNYIHÁR, E.: Zur Histochemie der limbischen Hemmung. Zytochemische und autoradiographische Untersuchungen über die Lokalisation der Enzyme des Gamma-aminoButtersäure-Stoffwechsels im Hippocampus der Ratte. Histochemie **28**, 38–54 (1971).
CURTIS, D. R.: The depression of spinal inhibition by electrophoretically administered strychnine. Int. J. Neuropharmacol. **1**, 239–250 (1962).
CURTIS, D. R.: The pharmacology of central and peripheral inhibition. Pharmacol. Rev. **15**, 333–364 (1963).
CURTIS, D. R.: Microelectrophoresis. In: Physical techniques in biological research, vol. 5, ed. W. L. Nastuk, p. 144–190. New York: Academic Press 1964.
CURTIS, D. R.: The actions of amino acids upon mammalian neurones. In: Studies in physiology, presented to J. C. Eccles, eds. D. R. Curtis and A. K. McIntyre, p. 34–42. Berlin-Heidelberg-New York: Springer 1965.
CURTIS, D. R.: The pharmacology of spinal postsynaptic inhibition. Progr. Brain Res. **31**, 171–189 (1969).
CURTIS, D. R.: Amino acid transmitters in the mammalian central nervous system. Proc. IV Int. Cong. Pharmac. **1**, 9–31 (1970).
CURTIS, D. R.: Neutral amino acids as central transmitters. The use of amino acid antagonists. In: Research in physiology, eds. F. F. Kao, K. Koizumi and M. Vassalle, p. 403–411. Bologna: Aulo Gaggi 1971.
CURTIS, D. R., CRAWFORD, J. M.: Central synaptic transmission – microelectrophoretic studies. Ann. Rev. Pharmacol. **9**, 209–240 (1969).
CURTIS, D. R., DAVIS, R.: Pharmacological studies upon neurones of the lateral geniculate nucleus of the cat. Brit. J. Pharmacol. **18**, 217–246 (1962).
CURTIS, D. R., DE GROAT, W. C.: Tetanus toxin and spinal inhibition. Brain Res. **10**, 208–212 (1968).
CURTIS, D. R., DUGGAN, A. W.: The depression of spinal inhibition by morphine. Agents and Action **1**, 14–19 (1969).
CURTIS, D. R., DUGGAN, A. W., FELIX, D.: GABA and inhibition of Deiters' neurones. Brain Res. **23**, 117–120 (1970a).
CURTIS, D. R., DUGGAN, A. W., FELIX, D., JOHNSTON, G. A. R.: Bicuculline, an antagonist of GABA and synaptic inhibition in the spinal cord. Brain Res. **32**, 69–96 (1971a).
CURTIS, D. R., DUGGAN, A. W., FELIX, D., JOHNSTON, G. A. R., MCLENNAN, H.: Antagonism between bicuculline and GABA in the cat brain. Brain Res. **33**, 57–73 (1971b).
CURTIS, D. R., DUGGAN, A. W., FELIX, D., JOHNSTON, G. A. R., TEBĒCIS, A. K., WATKINS, J. C.: Excitation of mammalian central neurones by acidic amino acids. Brain Res. **41**, 283–301 (1972a).
CURTIS, D. R., DUGGAN, A. W., JOHNSTON, G. A. R.: Glycine, strychnine, picrotoxin and spinal inhibition. Brain Res. **14**, 759–762 (1969).
CURTIS, D. R., DUGGAN, A. W., JOHNSTON, G. A. R.: The inactivation of extracellularly administered amino acids in the feline spinal cord. Exp. Brain Res. **10**, 447–462 (1970b).
CURTIS, D. R., DUGGAN, A. W., JOHNSTON, G. A. R.: The specificity of strychnine as a glycine antagonist in the mammalian spinal cord. Exp. Brain Res. **12**, 547–565 (1971c).
CURTIS, D. R., FELIX, D.: The effect of bicuculline upon synaptic inhibition in the cerebral and cerebellar cortices of the cat. Brain Res. **34**, 301–321 (1971).
CURTIS, D. R., FELIX, D., GAME, C. J. A., MCCULLOCH, R. M.: Tetanus toxin and the synaptic release of GABA. Brain Res. **51**, 358–362 (1973a).
CURTIS, D. R., FELIX, D., MCLENNAN, H.: GABA and hippocampal inhibition. Brit. J. Pharmacol. **40**, 881–883 (1970c).
CURTIS, D. R., GAME, C. J. A., JOHNSTON, G. A. R., MCCULLOCH, R. M., MACLACHLAN, R. M.: Convulsive action of penicillin. Brain Res. **43**, 242–245 (1972b).
CURTIS, D. R., HÖSLI, L., JOHNSTON, G. A. R.: A pharmacological study of the depression of spinal neurones by glycine and related amino acids. Exp. Brain Res. **6**, 1–18 (1968a).

CURTIS, D. R., HÖSLI, L., JOHNSTON, G. A. R., JOHNSTON, I. H.: The hyperpolarization of spinal motoneurones by glycine and related amino acids. Exp. Brain Res. **5**, 235–258 (1968b).

CURTIS, D. R., JOHNSTON, G. A. R.: Amino acid transmitters. In: Handbook of neurochemistry, vol. 4, ed. A. Lajtha, p. 115–135. New York: Plenum Press 1970.

CURTIS, D. R., JOHNSTON, G. A. R., GAME, C. J. A., MCCULLOCH, R. M.: Antagonism of neuronal excitation by 1-hydroxy-3-amino-pyrrolidone-2. Brain Res. **49**, 467–470 (1973b).

CURTIS, D. R., PHILLIS, J. W., WATKINS, J. C.: The depression of spinal neurones by γ-amino-n-butyric acid and β-alanine. J. Physiol. (Lond.) **146**, 185–203 (1959).

CURTIS, D. R., PHILLIS, J. W., WATKINS, J. C.: The chemical excitation of spinal neurones by certain acidic amino acids. J. Physiol. (Lond.) **150**, 656–682 (1960).

CURTIS, D. R., RYALL, R. W.: The synaptic excitation of Renshaw cells. Exp. Brain Res. **2**, 81–96 (1966a).

CURTIS, D. R., RYALL, R. W.: Pharmacological studies upon spinal presynaptic fibres. Exp. Brain Res. **1**, 195–204 (1966b).

CURTIS, D. R., TEBĒCIS, A. K.: Bicuculline and thalamic inhibition. Exp. Brain Res. **16**, 210–218 (1972).

CURTIS, D. R., WATKINS, J. C.: The excitation and depression of spinal neurones by structurally related amino acids. J. Neurochem. **6**, 117–141 (1960).

CURTIS, D. R., WATKINS, J. C.: Acidic amino acids with strong excitatory actions on mammalian neurones. J. Physiol. (Lond.) **166**, 1–14 (1963).

CURTIS, D. R., WATKINS, J. C.: The pharmacology of amino acids related to gamma-aminobutyric acid. Pharmacol. Rev. **17**, 347–392 (1965).

CUTLER, R. W. P., HAMMERSTAD, J. P., CORNICK, L. R., MURRAY, J. E.: Efflux of amino acid neurotransmitters from rat spinal cord slices. I. Factors influencing the spontaneous efflux of [^{14}C]glycine and ^{3}H-GABA. Brain Res. **35**, 337–355 (1971).

DALE, H. H.: Nomenclature of fibres in the autonomic system and their effects. J. Physiol. (Lond.) **80**, 10–11P (1933).

DALE, H. H.: Pharmacology and nerve-endings. Proc. roy. Soc. Med. **28**, 319–332 (1935).

DAVIDOFF, R. A.: The effects of bicuculline on the isolated spinal cord of the frog. Exp. Neurol. **35**, 179–193 (1972a).

DAVIDOFF, R. A.: Penicillin and presynaptic inhibition in the amphibian spinal cord. Brain Res. **36**, 218–222 (1972b).

DAVIDOFF, R. A.: Penicillin and inhibition in the cat spinal cord. Brain Res. **45**, 638–642 (1972c).

DAVIDOFF, R. A.: Gamma-aminobutyric acid antagonism and presynaptic inhibition in the frog spinal cord. Science **175**, 331–333 (1972d).

DAVIDOFF, R. A., APRISON, M. H.: Picrotoxin antagonism of the inhibition of interneurones by glycine. Life Sci. **8**, 107–112 (1969).

DAVIDOFF, R. A., APRISON, M. H., WERMAN, R.: The effects of strychnine on the inhibition of interneurons by glycine and γ-aminobutyric acid. Int. J. Neuropharmacol. **8**. 191–194 (1969).

DAVIDOFF, R. A., GRAHAM, L. T. JR., SHANK, R. P., WERMAN, R., APRISON, M. H.: Changes in amino acid concentrations associated with loss of spinal interneurons. J. Neurochem. **14**, 1025–1031 (1967).

DAVIDOFF, R. A., SILVEY, G. E., KOBETZ, S. A., SPIRA, H. M.: N-Methyl bicuculline and primary afferent depolarization. Exp. Neurol. **38**, 525–528 (1973).

DAVIDSON, N., REISINE, H.: Presynaptic inhibition in cuneate blocked by GABA antagonists. Nature (Lond.) New Biol. **234**, 223–224 (1971).

DAVIDSON, N., SOUTHWICK, C. A. P.: Amino acids and presynaptic inhibition in the rat cuneate nucleus. J. Physiol. (Lond.) **219**, 689–708 (1971).

DAVIES, J., WATKINS, J. C.: Is 1-hydroxy-3-aminopyrrolidone-2 (HA-966) a selective excitatory amino acid antagonist? Nature (Lond.) New Biol. **238**, 61–63 (1972).

DAVIES, J., WATKINS, J. C.: Microelectrophoretic studies on the depressant action of HA-966 on chemically and synaptically-excited neurones in the cat cerebral cortex and cuneate nucleus. Brain Res. **59**, 311–322 (1973).

DAVIES, L. P., JOHNSTON, G. A. R.: Serine hydroxymethyltransferase in the central nervous system: regional and subcellular distribution studies. Brain Res. **54**, 149–156 (1973).

DAVIS, R., HUFFMAN, R. D.: Pharmacology of the brachium conjunctivum-red nucleus synaptic system in the baboon. Fed. Proc. **28**, 775 (1969).

DAVIS, R., VAUGHAN, P. C.: Pharmacological properties of feline red nucleus. Int. J. Neuropharmacol. **8**, 475–488 (1969).

DAVSON, H.: The cerebrospinal fluid. In: Handbook of neurochemistry, vol. II, chap. 3, ed. A. Lajtha, p. 23–48. New York: Plenum Press 1969.

DEFEUDIS, F. V.: Effects of electrical stimulation on the efflux of L-glutamate from peripheral nerve *in vitro*. Exp. Neurol. **30**, 291–296 (1971).

DEFEUDIS, F. V., DELGADO, J. M. R., ROTH, R. H.: Content, synthesis and collectability of amino acids in various structures of the brains of rhesus monkeys. Brain Res. **18**, 15–23 (1970).

DE GROAT, W. C.: Action of gamma-amino butyric acid (GABA) and related amino acids on mammalian autonomic ganglia. Pharmacologist **11**, 286 (1969).

DE GROAT, W. C.: The actions of γ-aminobutyric acid and related amino acids on mammalian autonomic ganglia. J. Pharmacol. exp. Ther. **172**, 384–396 (1970a).

DE GROAT, W. C.: The effects of glycine, GABA and strychnine on sacral parasympathetic preganglionic neurones. Brain Res. **18**, 542–544 (1970b).

DE GROAT, W. C.: GABA-depolarization of a sensory ganglion: antagonism by picrotoxin and bicuculline. Brain Res. **38**, 429–432 (1972).

DE GROAT, W. C., LALLEY, P. M., BLOCK, M.: The effects of bicuculline and GABA on the superior cervical ganglion of the cat. Brain Res. **25**, 665–668 (1971).

DE GROAT, W. C., LALLEY, P. M., SAUM, W. R.: Depolarization of dorsal root ganglia in the cat by GABA and related amino acids: antagonism by picrotoxin and bicuculline. Brain Res. **44**, 273–277 (1972).

DE GROAT, W. C., RYALL, R. W.: The identification and characteristics of sacral parasympathetic preganglionic neurones. J. Physiol. (Lond.) **196**, 563–577 (1968).

DE MARCHI, W. J., JOHNSTON, G. A. R.: The oxidation of glycine by D-amino acid oxidase in extracts of mammalian central nervous system. J. Neurochem. **16**, 355–361 (1969).

DERISSEN, J. L., ENDEMAN, H. J., PEERDEMAN, A. F.: The crystal structure of L-aspartic acid. Acta Cryst., Sect. B. **24**, 1349–1354 (1968).

DE ROBERTIS, E.: Isolation of inhibitory nerve endings from brain. In: Structure and function of inhibitory neuronal mechanisms, vol. 10, eds. C. von Euler, S. Skoglund and U. Söderberg, p. 511–522. Oxford: Pergamon Press 1968.

DHAWAN, B. N., SHARMA, J. N., SRIMAL, R. C.: Selective inhibition by glycine of some somatic reflexes in the cat. Brit. J. Pharmacol. **44**, 404–412 (1972).

DIAMOND, J.: The activation and distribution of GABA and L-glutamate receptors on goldfish Mauthner neurones: an analysis of dendritic remote inhibition. J. Physiol. (Lond.) **194**, 669–723 (1968).

DO CARMO, R. J., LEÃO, A. A. P.: On the relation of glutamic acid and some allied compounds to cortical spreading depression. Brain Res. **39**, 515–518 (1972).

DOWLING, J. E.: Organization of vertebrate retinas. Invest. Opthal. **9**, 655–680 (1970).

DREIFUSS, J. J., KELLY, J. S.: The activity of identified supraoptic neurones and their response to acetylcholine applied by iontophoresis. J. Physiol. (Lond.) **220**, 105–118 (1972).

DREIFUSS, J. J., KELLY, J. S., KRNJEVIĆ, K.: Cortical inhibition and γ-aminobutyric acid. Exp. Brain Res. **9**, 137–154 (1969).

DREIFUSS, J. J., MATTHEWS, E. K.: Antagonism between strychnine and glycine, and bicuculline and GABA, in the ventromedial hypothalamus. Brain Res. **45**, 599–603 (1972).

DUDAR, J. D.: Glutamic acid sensitivity of hippocampal pyramidal cell dendrites. Acta physiol. scand. **84**, 28 A (1972).

DUGGAN, A. W.: The differential sensitivity to L-glutamate and L-aspartate of spinal interneurones and Renshaw cells. Exp. Brain Res., in press (1974).

DUGGAN, A. W., JOHNSTON, G. A. R.: Glutamate and related amino acids in cat spinal roots, dorsal root ganglia and peripheral nerves. J. Neurochem. **17**, 1205–1208 (1970a).

DUGGAN, A. W., JOHNSTON, G. A. R.: Glutamate and related amino acids in cat, dog, and rat spinal roots. Comp. Gen. Pharmac. **1**, 127–128 (1970b).

DUGGAN, A. W., LODGE, D., BISCOE, T. J.: The inhibition of hypoglossal motoneurones by impulses in the glossopharyngeal nerve of the rat. Exp. Brain Res. **17**, 261–270 (1973).

DUGGAN, A. W., MCLENNAN, H.: Bicuculline and inhibition in the thalamus. Brain Res. **25**, 188–191 (1971).

ECCLES, J. C.: The ionic mechanisms of excitatory and inhibitory synaptic action. Ann. N. Y. Acad. Sci. **137**, 473–494 (1966).

ECCLES, J. C., FATT, P., KOKETSU, K.: Cholinergic and inhibitory synapses in a pathway from motor-axon collaterals to motoneurones. J. Physiol. (Lond.) **126**, 524–562 (1954).

ECCLES, J. C., ITO, M., SZENTAGOTHAI, J.: The cerebellum as a neuronal machine. Berlin-Heidelberg-New York: Springer 1967.

ECCLES, J. C., SCHMIDT, R. F., WILLIS, W. D.: Pharmacological studies on presynaptic inhibition. J. Physiol. (Lond.) **168**, 500–530 (1963).

EDWARDSON, J. A., BENNETT, G. W., BRADFORD, H. F.: Release of amino-acids and neurosecretory substances after stimulation of nerve endings (synaptosomes). Nature (Lond.) **240**, 554–556 (1972).

EHINGER, B.: Cellular location of the uptake of some amino acids into the rabbit retina. Brain Res. **46**, 297–311 (1972a).

EHINGER, B.: Uptake of tritiated glycine into neurons of the human retina. Experientia (Basel) **28**, 1042–1043 (1972b).

EHINGER, B., FALCK, B.: Autoradiography of some suspected neurotransmitter substances: GABA, glycine, glutamic acid, histamine, dopamine, and L-Dopa. Brain Res. **33**, 157–172 (1971).

ELLAWAY, P. H.: Recurrent inhibition of fusimotor neurones exhibiting background discharges in the decerebrate and the spinal cat. J. Physiol. (Lond.) **216**, 419–439 (1971).

ELLIOTT, K. A. C., JASPER, H. H.: Gamma-aminobutyric acid. Physiol. Rev. **39**, 383–406 (1959).

ELLIOTT, K. A. C., PAGE, I. H., QUASTEL, J. H.: Neurochemistry, 2nd edn. Springfield: Charles C. Thomas 1962.

ELLIOTT, K. A. C., VAN GELDER, N. M.: Occlusion and metabolism of γ-aminobutyric acid by brain tissue. J. Neurochem. **3**, 28–40 (1958).

ENDO, K., ARAKI, T.: Study of glycine induced potentials in motoneurones. J. Physiol. Soc. Japan **31**, 118 (1969).

ENGBERG, I., THALLER, A.: On the interaction of picrotoxin with GABA and glycine in the spinal cord. Brain Res. **19**, 151–154 (1970).

FAHN, S., CÔTÉ, L. J.: Regional distribution of γ-aminobutyric acid (GABA) in brain of the Rhesus monkey. J. Neurochem. **15**, 209–213 (1968).

FASTIER, F. N.: Tutu poisoning. In: Research in physiology, a liber memorialis in honor of Professor C. McC. Brooks, eds. F. F. Kao, K. Koizumi and M. Vassalle, p. 653–659. Bologna: Aulo Gaggi Publ. 1971.

FATT, P.: Biophysics of junctional transmission. Physiol. Rev. **34**, 674–710 (1954).

FEDINEC, A. A., SHANK, R. P.: Effect of tetanus toxin on the content of glycine, gamma-aminobutyric acid, glutamate, glutamine and aspartate in the rat spinal cord. J. Neurochem. **18**, 2229–2234 (1971).

FELIX, D., MCLENNAN, H.: The effect of bicuculline on the inhibition of mitral cells of the olfactory bulb. Brain Res. **25**, 661–664 (1971).

FELPEL, L. P.: Effects of strychnine, bicuculline and picrotoxin on labyrinthine-evoked inhibition in neck motoneurons of the cat. Exp. Brain Res. **14**, 494–502 (1972).

FELTZ, P.: γ-Aminobutyric acid and a caudato-nigral inhibition. Can. J. Physiol. Pharmac. **49**, 1114–1115 (1971).

FLOREY, E.: Amino acids as transmitter substances. In: Major problems in neuroendocrinology, eds. E. Bajusz and G. Jasmin, p. 17–41. Basel: S. Karger 1964.

FONNUM, F.: The distribution of glutamate decarboxylase and aspartate transminase in subcellular fractions of rat and guinea-pig brain. Biochem. J., **106**, 401–412 (1968).

FONNUM, F.: Localization of cholinergic and γ-aminobutyric acid containing pathways in brain. In: Metabolic compartmentation in the brain, eds. J. Cremer and R. Balázs, p. 243–255. London: MacMillan 1972.

FONNUM, F., STORM-MATHISEN, J., WALBERG, F.: Glutamate decarboxylase in inhibitory neurons. A study of the enzyme in Purkinje cell axons and boutons in the cat. Brain Res. **20**, 259–275 (1970).

FONNUM, F., WALBERG, F.: An estimation of the concentration of γ-aminobutyric acid and glutamate decarboxylase in the inhibitory Purkinje axon terminals in the cat. Brain Res. **54**, 115–127 (1973).

FUKUDA, J., HIGHSTEIN, S. M., ITO, M.: Cerebellar inhibitory control of the vestibulo-ocular reflex investigated in rabbit IIIrd nucleus. Exp. Brain Res. **14**, 511–526 (1972).

GAITONDE, M. K.: Sulfur amino acids. In: Handbook of neurochemistry, vol. 3, metabolic reactions in the nervous system, ed. A. Lajtha, p. 225–287. New York: Plenum Press 1970.

GALINDO, A.: GABA-picrotoxin interaction in the mammalian central nervous system. Brain Res. **14**, 763–767 (1969).

GALINDO, A., KRNJEVIĆ, K., SCHWARTZ, S.: Micro-iontophoretic studies on neurones in the cuneate nucleus. J. Physiol. (Lond.) **192**, 359–377 (1967).
GALINDO, A., KRNJEVIĆ, K., SCHWARTZ, S.: Patterns of firing in cuneate neurones and some effects of flaxedil. Exp. Brain Res. **5**, 87–101 (1968).
GESSI, T., RABINI, C., VOLTA, F.: Effects of strychnine, picrotoxin and γ-aminobutyric acid on synaptic and antidromic responses of hippocampal pyramidal neurons and granule cells. Arch. Sci. biol. **51**, 1–23 (1967).
GFELLER, E., KUHAR, M. J., SNYDER, S. H.: Neurotransmitter-specific synaptosomes in rat corpus striatum: morphological variations. Proc. nat. Acad. Sci. (Wash.) **68**, 155–159 (1971).
GINSBORG, B. L.: Ion movements junctional transmission. Pharmacol. Rev. **19**, 289–316 (1967).
GJESSING, L. R., GJESDAHL, P., SHAASTAD, O.: The free amino acids in human cerebrospinal fluid. J. Neurochem. **19**, 1807–1808 (1972).
GJESSING, L. R., TORVICK, A.: Distribution of cystathionine in human brain. Scand J. clin. Lab. Invest. **18**, 565 (1966).
GLAGOLEVA, A. A.: Aqueous solutions of aspartic and glutamic acids. Zh.Obshch. Khim. **37**, 43–47 (1967).
GLOBUS, A., LUX, H. D., SCHUBERT, P.: Somadendritic spread of intracellulary injected tritiated glycine in cat spinal motoneurons. Brain Res. **11**, 440–445 (1968).
GOLDBERG, L. J.: Excitatory and inhibitory effects of lingual nerve stimulation on reflexes controlling the activity of masseteric motoneurons. Brain Res. **39**, 95–108 (1972).
GOODCHILD, M., NEAL, M. J.: The uptake of ^{3}H-γ-aminobutyric acid by the retina. Brit. J. Pharmacol. **47**, 529–542 (1973).
GOTTESFELD, Z., ELLIOTT, K. A. C.: Factors that affect the binding and uptake of GABA by brain tissue. J. Neurochem. **18**, 683–690 (1971).
GOTTESFELD, Z., KELLY, J. S., RAYNER, C. N.: Depletion of glutamic acid decarboxylase (GAD I) activity from the spinal cord following dorsal-root section. J. Physiol. (Lond.) **231**, 25–26P (1973a).
GOTTESFELD, Z., KELLY, J. S., RENAUD, L. P.: The *in vivo* neuropharmacology of amino-oxyacetic acid in the cerebral cortex of the cat. Brain Res. **42**, 311–318 (1972).
GOTTESFELD, Z., KELLY, J. S., SCHON, F.: Uptake of γ-aminobutyric acid (GABA) by sensory root ganglia. Brit. J. Pharmacol. **47**, 640P (1973b).
GRAHAM, L. T., JR.: Intraretinal distribution of GABA content and GAD activity. Brain Res. **36**, 476–479 (1972).
GRAHAM, L. T., JR., APRISON, M. H.: Distribution of some enzymes associated with the metabolism of glutamate, aspartate, γ-aminobutyrate and glutamine in cat spinal cord. J. Neurochem. **16**, 559–566 (1969).
GRAHAM, L. T., JR., BAXTER, C. F., LOLLEY, R. N.: *In vivo* influence of light or darkness on the GABA system in the retina of the frog (*Rana pipiens*). Brain Res. **20**, 379–388 (1970).
GRAHAM, L. T., JR., PONG, S. F.: Evidence for a function of γ-aminobutyric acid (GABA) in the neural mechanism of light adaptation in rat retina. Trans. Amer. Soc. Neurochem. **3**, 82 (1972a).
GRAHAM, L. T., JR., PONG, S. F.: Rhythmic potentials originating in rat retina. Exp. Neurol. **36**, 399–403 (1972b).
GRAHAM, L. T., JR., SHANK, R. P., WERMAN, R., APRISON, M. H.: Distribution of some synaptic transmitter suspects in cat spinal cord: glutamic acid, aspartic acid, γ-aminobutyric acid, glycine and glutamine. J. Neurochem. **14**, 465–472 (1967).
GRANIT, R.: The case for presynaptic inhibition by synapses on the terminals of motoneurons. In: Structure and function of inhibitory neuronal mechanisms, eds. C. von Euler, S. Skoglund and U. Söderberg, p. 183–195. Oxford: Pergamon Press 1968.
GREEN, J. D., MANCIA, M., VON BAUMGARTEN, R.: Recurrent inhibition in the olfactory bulb. I. Effects of antidromic stimulation of the lateral olfactory tract. J. Neurophysiol. **25**, 467–488 (1962).
GRUNDFEST, H.: The role of GABA in modern concepts of neurophysiology and pharmacology of the central nervous system. In: Recent advances in biological psychiatry, p. 76–93. U.S.A.: Grune & Stratton, Inc. 1960.
GUIDOTTI, A., BADIANI, G., PEPEU, G.: Taurine distibution in cat brain. J. Neurochem. **19**, 431–435 (1972),
GUSHCHIN, I. S., KOZHECHKIN, S. N., SVERDLOV, Y. S.: On the presynaptic nature of the suppression of postsynaptic inhibition by tetanus toxin. Dokl. Akad. Nauk SSSR **187**, No. 3, 685–688 (1969).

HAAS, H. L., ANDERSON, E. G., HÖSLI, L.: Histamine and metabolites: their effects and interactions with convulsants on brain stem neurones. Brain Res. **51**, 269–278 (1972).

HAAS, H. L., HÖSLI, L.: The depression of brain stem neurones by taurine and its interaction with strychnine and bicuculline. Brain Res. **52**, 399–402 (1973a).

HAAS, H. L., HÖSLI, L.: Strychnine and inhibition of bulbar reticular neurones. Experientia **29**, 542–544 (1973b).

HABER, B., KURIYAMA, K., ROBERTS, E.: L-Glutamic acid decarboxylase: a new type in glial cells and human brain gliomas. Science **168**, 598–599 (1970).

HAGEN, J.: Schwere Vergiftungen in einer Polstermöbelfabrik durch einen neuartigen hochtoxischen Giftstoff (Tetramethylendisulfotetramin). Dtsch. med. Wschr. **75**, 183–184 (1950).

HALDEMAN, S., HUFFMAN, R. D., MARSHALL, K. C., MCLENNAN, H.: The antagonism of the glutamate-induced and synaptic excitation of thalamic neurones. Brain Res. **39**, 419–425 (1972).

HALDEMAN, S., MCLENNAN, H.: The antagonistic action of glutamic acid diethylester towards amino acid-induced and synaptic excitations of central neurones. Brain Res. **45**, 393–400 (1972).

HAMMERSCHLAG, R., WEINREICH, D.: Glutamic acid and primary afferent transmission. Advanc. Biochem. Psychopharm. **6**, 165–180 (1972).

HAMMERSTAD, J. P., MURRAY, J. E., CUTLER, R. W. P.: Efflux of amino acid neurotransmitters from rat spinal cord slices. II. Factors influencing the electrically induced efflux of [^{14}C]-glycine and [^{3}H]GABA. Brain Res. **35**, 357–367 (1971).

HARRIS, M., HOPKIN, J. M., NEAL, M. J.: Effect of centrally acting drugs on the uptake of γ-aminobutyric acid (GABA) by slices of rat cerebral cortex. Brit. J. Pharmacol. **47**, 229–239 (1973).

HARVEY, J. A., MCILWAIN, H.: Electrical phenomena and isolated tissues from the brain. In: Handbook of neurochemistry, vol. 2, ed. A. Lajtha, p. 115–136. New York: Plenum Press 1969.

HAULICĂ, I., ČAPÂLNĂ, S., NESTIANU, V., BORDEIANU, A., BĂDESCU, A.: Contributions to the study of gamma-aminobutyric acid (GABA) in the hypothalamo-hypophyseal pathways. Int. J. Neuropharmacol. **3**, 465–472 (1964).

HAYASHI, T.: A physiological study of epileptic seizures following cortical stimulation in animals and its application to human clinics. Jap. J. Physiol. **3**, 46–64 (1952).

HAYASHI, T.: Neurophysiology and neurochemistry of convulsion. Tokyo: Dainihon-Tosho Co.Ltd. 1959.

HAYASHI, T.: Gamma-aminobutyric acid and its derivatives in mental health. In: Enzymes in mental health, eds. G. J. Martin and B. Kisch, p. 160–170. Philadelphia: J. B. Lippincott Co. 1966.

HENDRICK, J. L., SALLACH, H. J.: The non-oxidative decarboxylation of hydroxypyruvate in mammalian systems. Arch. Biochem. Biophys. **105**, 261–269 (1964).

HENN, F. A., HAMBERGER, A.: Glial cell function: uptake of transmitter substances. Proc. nat. Acad. Sci. (Wash.) **68**, 2686–2690 (1971).

HENNECKE, H., WIECHERT, P.: Seizures and the dose of L-glutamic acid in rats. Epilepsia **11**, 327–331 (1970).

HENSCHEN, A., SÖDERBERG, U.: Net exchange of free amino acids between blood and brain and its significance for the electrical activity. In: Structure and function of inhibitory neuronal mechanisms, eds. C. von Euler, S. Skoglund and U. Söderberg, p. 505–509. London: Pergamon Press 1968.

HERZ, A., GOGOLAK, G.: Mikroelektrophoretische Untersuchungen am Septum des Kaninchens. Pflügers Arch. ges. Physiol. **285**, 317–330 (1965).

HERZ, A., NACIMIENTO, A.: Über die Wirkung von Pharmaka auf Neurone des Hippocampus nach mikroelektrophoretischer Verabfolgung. Naunyn-Schmiedebergs Arch. exp. Path. Pharmak. **250**, 258–259 (1965).

HERZ, A., VON FREYTAG-LORINGHOVEN, HJ.: Über die synaptische Erregung im Corpus striatum und deren antagonistische Beeinflussung durch mikroelektrophoretisch verabfolgte Glutaminsäure und Gamma-Aminobuttersäure. Pflügers Arch. ges. Physiol. **299**, 167–184 (1968).

HIGHSTEIN, S. M., ITO, M., TSUCHIYA, T.: Synaptic linkage in the vestibulo-ocular reflex pathway of rabbit. Exp. Brain Res. **13**, 306–326 (1971).

HILL, R. G., SIMMONDS, M. A., STAUGHAN, D. W.: Convulsive properties of d-tubocurarine and cortical inhibition. Nature (Lond.) **240**, 51–52 (1972a).

HILL, R. G., SIMMONDS, M. A., STRAUGHAN, D. W.: Antagonism of GABA by picrotoxin in the feline cerebral cortex. Brit. J. Pharmacol. **44**, 807–809 (1972b).

HIRSCH, H., GREENBERG, D. M.: Studies on phosphoserine aminotransferase of sheep brain. J. biol. Chem. **242**, 2283–2287 (1967).

HIRSCH, H. E., ROBINS, E.: Distribution of γ-aminobutyric acid in the layers of the cerebral and cerebellar cortex. Implications for its physiological role. J. Neurochem. **9**, 63–70 (1962).

HOCKMAN, C. H., LLOYD, K. G., FARLEY, I. J., HORNYKIEWICZ, O.: Experimental midbrain lesions: neurochemical comparison between the animal model and Parkinson's disease. Brain Res. **35**, 613–618 (1971).

HOFFMANN, K.-P., STONE, J., SHERMAN, S. M.: Relay of receptive field properties in dorsal lateral geniculate nucleus of the cat. J. Neurophysiol. **35**, 518–531 (1972).

HÖKFELT, T., JONSSON, G., LJUNGDAHL, A.: Regional uptake and subcellular localization of (^{3}H)-gamma-aminobutyric acid (GABA) in rat brain slices. Life Sci. **9**, Part 1, 203–212 (1970).

HÖKFELT, T., LJUNGDAHL, A.: Light and electron microscopic autoradiography on spinal cord slices after incubation with labeled glycine. Brain Res. **32**, 189–194 (1971a).

HÖKFELT, T., LJUNGDAHL, A.: Uptake of [^{3}H]noradrenaline and γ-[^{3}H]aminobutyric acid in isolated tissues of rat: an autoradiographic and fluorescence microscopic study. Progr. Brain Res. **34**, 87–102 (1971b).

HÖKFELT, T., LJUNGDAHL, A.: Autoradiographic identification of cerebral and cerebellar cortical neurons accumulating labeled gamma-aminobutyric acid (^{3}H-GABA). Exp. Brain Res. **14**, 354–362 (1972a).

HÖKFELT, T., LJUNGDAHL, A.: Histochemical determination of neurotransmitter distribution. Res. Publ. Ass. nerv. Ment. Dis. **50**, 1–24 (1972b).

HÖKFELT, T., LJUNGDAHL, A.: Application of cytochemical techniques to the study of suspected transmitter substances in the nervous system. Advanc. Biochem. Psychopharmac. **6**, 1–36 (1972c).

HONGO, T., RYALL, R. W.: Electrophysiological and micro-electophoretic studies on sympathetic preganglionic neurones in the spinal cord. Acta physiol. scand. **68**, 96–104 (1966).

HOPKIN, J., NEAL, M. J.: Effect of electrical stimulation and high potassium concentrations on the efflux of ^{14}C glycine from slices of spinal cord. Brit. J. Pharmacol. **42**, 215–223 (1971).

HÖSLI, E., LJUNGDAHL, A., HÖKFELT, T., HÖSLI, L.: Spinal cord tissue cultures—a model for autoradiographic studies on uptake of putative neurotransmitters like glycine and GABA. Experientia (Basel) **28**, 1342–1344 (1972).

HÖSLI, L., HAAS, H. L.: The hyperpolarization of neurones of the medulla oblongata by glycine. Experientia (Basel) **28**, 1057–1058 (1972).

HÖSLI, L., TEBĒCIS, A. K.: Actions of amino acids and convulsants on bulbar reticular neurones. Exp. Brain Res. **11**, 111–127 (1970).

HÖSLI, L., TEBĒCIS, A. K., HAAS, H. L.: Depressant amino acids and possible antagonists on medullary reticular neurones. Experientia (Basel) **27**, 732–733 (1971).

HUFFMAN, R. D., MCFADIN, L. S.: Effects of bicuculline on central inhibition. Neuropharmacology **11**, 789–799 (1972).

HUFFMAN, R. D., YIM, G. K. W.: Effects of diphenylaminoethanol and lidocaine on central inhibition. Int. J. Neuropharmacol. **8**, 217–225 (1969).

HULTBORN, H., JANKOWSKA, E., LINDSTRÖM, S.: Recurrent inhibition of interneurones monosynaptically activated from Group Ia afferents. J. Physiol. (Lond.) **215**, 613–656 (1971).

ITO, M., UDO, M., MANO, N.: Long inhibitory and excitatory pathways converging onto cat reticular and Deiters' neurons and their relevance to reticulofugal axons. J. Neurophysiol. **32**, 210–226 (1970).

ITO, M., YOSHIDA, M.: The origin of cerebellar-induced inhibition of Deiters neurones. I. Monosynaptic initation of the inhibitory postsynaptic potentials. Exp. Brain Res. **2**, 330–349 (1966).

ITO, M., YOSHIDA, M., OBATA, K., KAWAI, N., UDO, M.: Inhibitory control of intracerebellar nuclei by the Purkinje cell axons. Exp. Brain Res. **10**, 64–80 (1970).

ITO, S., TANI, M., SATO, S., ODA, M.: Chemical stimulation due to sodium glutamate and aspartate on the motor cortex and subcortical nuclei. J. Physiol. Soc. Japan, **14**, 392–394 (1952).

IVERSEN, L. L.: The uptake, storage, release and metabolism of GABA in inhibitory nerves. In: Perspectives in neuropharmacology, a tribute to Julius Axelrod, ed. S. H. Snyder, p. 75–111. Oxford: Univ. Press 1972.

IVERSEN, L. L., BLOOM, F. E.: Studies of the uptake of [^{3}H]GABA and [^{3}H]glycine in slices and homogenates of rat brain and spinal cord by electron microscopic autoradiography. Brain Res. **41**, 131–143 (1972).

IVERSEN, L. L., JOHNSTON, G. A. R.: GABA uptake in rat central nervous system: comparison of uptake in slices and homogenates and the effects of some inhibitors. J. Neurochem. **18**, 1939–1950 (1971).

IVERSEN, L. L., MITCHELL, J. F., NEAL, M. J., SRINIVASAN, V.: The effect of stimulation of inhibitory pathways on the release of endogenous γ-aminobutyric acid from the cat cerebral cortex. Brit. J. Pharmacol. **38**, 452P (1970).

IVERSEN, L. L., MITCHELL, J. F., SRINIVASAN, V.: The release of γ-aminobutyric acid during inhibition in the cat visual cortex. J. Physiol. (Lond.) **212**, 519–534 (1971).

IVERSEN, L. L., NEAL, M. J.: The uptake of [^{3}H]GABA by slices of rat cerebral cortex. J. Neurochem. **15**, 1141–1149 (1968).

JANKOWSKA, E., LINDSTRÖM, S.: Morphological identification of Renshaw cells. Acta physiol. scand. **81**, 428–430 (1971).

JANKOWSKA, E., LINDSTRÖM, S.: Morphology of interneurones mediating Ia reciprocal inhibition of motoneurones in the spinal cord of the cat. J. Physiol. (Lond.) **226**, 805–823 (1972).

JANKOWSKA, E., ROBERTS, W. J.: Synaptic actions of single interneurones mediating reciprocal Ia inhibition of motoneurones. J. Physiol. (Lond.) **222**, 623–642 (1972).

JARBOE, C. H., PORTER, L. A., BUCKLER, R. T.: Structural aspects of picrotoxinin action. J. Med. Chem. **11**, 729–731 (1968).

JASPER, H. H., KHAN, R. T., ELLIOTT, K. A. C.: Amino acids released from the cerebral cortex in relation to its state of activation. Science **147**, 1448–1449 (1965).

JASPER, H. H., KOYAMA, I.: Rate of release of amino acids from the cerebral cortex in the cat as affected by brainstem and thalamic stimulation. Canad. J. Physiol. Pharmac. **47**, 889–905 (1969).

JOHNSON, E. S., ROBERTS, M. H. T., STRAUGHAN, D. W.: Amino-acid induced depression of cortical neurones. Brit. J. Pharmacol. **38**, 659–666 (1970).

JOHNSON, J. L.: Glutamic acid as a synaptic transmitter in the nervous system. A review. Brain Res. **37**, 1–19 (1972a).

JOHNSON, J. L.: An analysis of the activities of 3 key enzymes concerned with the interconversion of α-ketoglutarate and glutamate: correlations with free glutamate levels in 20 specific regions of the nervous system. Brain Res. **45**, 205–215 (1972b).

JOHNSON, J. L., APRISON, M. H.: The distribution of glutamic acid, a transmitter candidate, and other amino acids in the dorsal sensory neuron of the cat. Brain Res. **24**, 285–292 (1970).

JOHNSON, J. L., APRISON, M. H.: The distribution of glutamate and total free amino acids in thirteen specific regions of the cat central nervous system. Brain Res. **26**, 141–148 (1971).

JOHNSTON, G. A. R.: The intraspinal distribution of some depressant amino acids. J. Neurochem. **15**, 1013–1017 (1968).

JOHNSTON, G. A. R.: Convulsions induced in 10-day old rats by intraperitoneal injection of monosodium glutamate and related excitant amino acids. Biochem. Pharmacol. **22**, 137–140 (1973).

JOHNSTON, G. A. R., BEART, P. M., CURTIS, D. R., GAME, C. J. A., MCCULLOCH, R. M., MACLACHLAN, R. M.: Bicuculline methochloride as a GABA antagonist. Nature (Lond.) New Biol. **240**, 219–220 (1972).

JOHNSTON, G. A. R., CURTIS, D. R.: γ-Aminobutyrylcholine and central inhibition. J. Pharm. Pharmacol. **24**, 251 (1972).

JOHNSTON, G. A. R., DEGROAT, W. C., CURTIS, D. R.: Tetanus toxin and amino acid levels in cat spinal cord. J. Neurochem. **16**, 797–800 (1969).

JOHNSTON, G. A. R., IVERSEN, L. L.: Glycine uptake in rat central nervous system slices and homogenates: evidence for different uptake systems in spinal cord and cerebral cortex. J. Neurochem. **18**, 1951–1961 (1971).

JOHNSTON, G. A. R., VITALI, M. V.: Glycine-producing transminase activity in extracts of spinal cord. Brain Res. **12**, 471–472 (1969a).

JOHNSTON, G. A. R., VITALI, M. V.: Glycine: 2-oxoglutarate transaminase in rat cerebral cortex. Brain Res. **15**, 201–208 (1969b).

JOHNSTON, G. A. R., VITALI, M. V., ALEXANDER, H. M.: Regional and subcellular distribution studies on glycine: 2-oxoglutarate transaminase activity in cat spinal cord. Brain Res. **20**, 361–367 (1970).

JORDAN, C. C., WEBSTER, R. A.: Release of acetylcholine and ^{14}C-glycine from the cat spinal cord *in vivo*. Brit. J. Pharmacol. **43**, 441P (1971).

KACZMAREK, L. K., AGRAWAL, H. C., DAVISON, A. N.: Biochemical studies of taurine in the developing rat brain. Biochem. J. **119**, 45–46P (1970).

KACZMAREK, L. K., AGRAWAL, H. C., DAVISON, A. N.: The biochemistry of taurine in developing rat brain. In: Inherited disorders of sulphur metabolism, eds. N. A. J. Carson and D. N. Raine, p. 63–68. London: Churchill-Livingston 1971.

Kaczmarek, L. K., Davison, A. N.: Uptake and release of taurine from rat brain slices. J. Neurochem. **19**, 2355–2362 (1972).

Kahlson, G., Rosengren, E.: Biogenesis and physiology of histamine. London: Edward Arnold Ltd. 1971.

Kajimoto, Y., Fujimori, H., Hirota, K.: Study on the toxic components of "Shikimi". I. Chemical study. Tokushima J. exp. Med. **2**, 104–107 (1955).

Kanazawa, I., Miyata, Y., Toyokura, Y., Otsuka, M.: The distribution of γ-aminobutyric acid (GABA) in the human substantia nigra. Brain Res. **51**, 363–365 (1973).

Kandera, J., Levi, G., Lajtha, A.: Control of cerebral metabolite levels. II. Amino acid uptake and levels in various areas of the rat brain. Arch. Biochem. Biophys. **126**, 249–260 (1968).

Kawamura, H., Provini, L.: Depression of cerebellar Purkinje cells by microiontophoretic application of GABA and related amino acids. Brain Res. **24**, 293–304 (1970).

Kee, R. D., Wells, J. N., Yim, G. K. M.: Antagonism by diphenylaminoethanol of GABA inhibition of Deiters neurones. Fed. Proc. **30**, 317 A (1971).

Kellerth, J.-O.: Aspects on the relative significance of pre- and postsynaptic inhibition in the spinal cord. In: Structure and function of inhibitory neuronal mechanisms, eds. C. von Euler, S. Skoglund and U. Söderberg, p. 197–212. Oxford: Pergamon Press 1968.

Kellerth, J.-O., Szumski, A. J.: Effects of picrotoxin on stretch activated postsynaptic inhibitions in spinal motoneurones. Acta physiol. scand. **66**, 146–156 (1966).

Kelly, J. S., Krnjević, K.: The action of glycine on cortical neurones. Exp. Brain Res. **9**, 155–163 (1969).

Kelly, J. S., Krnjević, K., Morris, M. E., Yim, G. K. W.: Anionic permeability of cortical neurones. Exp. Brain Res. **7**, 11–31 (1969).

Kelly, J. S., Renaud, L. P.: Post-synaptic inhibition in the cuneate blockade by GABA antagonist. Nature (Lond.) New Biol. **232**, 25–26 (1971).

Kidokoro, J., Kubota, K., Shuto, S., Sumino, R.: Reflex organization of cat masticatory muscles. J. Neurophysiol. **31**, 695–708 (1968).

Kim, J. S., Bak, L. J., Hassler, R., Okada, Y.: Role of γ-aminobutyric acid (GABA) in the extrapyramidal motor system. 2. Evidence for the existence of type of GABA-rich strio-nigral neurons. Exp. Brain Res. **14**, 95–104 (1971).

King, E. J.: The ionization constants of glycine and the effect of sodium chloride upon its second ionization. J. Amer. chem. Soc. **73**, 155–159 (1951).

King, E. J.: The ionization constants of taurine and its activity coefficient in hydrochloric acid solutions from electromotive force measurements. J. Amer. chem. Soc. **75**, 2204–2209 (1953).

King, E. J.: The thermodynamics of ionization of amino acids. I. The ionization constants of γ-aminobutyric acid. J. Amer. chem. Soc. **76**, 1006–1008 (1954).

Kishida, K., Naka, K. I.: Amino acids and the spikes from the retinal ganglion cells. Science **156**, 648–650 (1967).

Konishi, S., Otsuka, M.: Actions of certain polypeptides on frog spinal neurons. Jap. J. Pharmacol. **21**, 685–687 (1971).

Koyama, I.: Amino acids in the cobalt-induced epileptogenic and nonepileptogenic cat's cortex. Canad. J. Physiol. Pharmac. **50**, 740–752 (1972).

Koyama, I., Jasper, H. H.: The release of GABA from cerebral cortex in the cat. Proc. Canad. Fed. Biol. Soc. **15**, Abstr. 315 (1972).

Kravitz, E. A., Potter, D. D.: A further study of the distribution of γ-aminobutyric acid between excitatory and inhibitory axons of the lobster. J. Neurochem. **12**, 323–328 (1965).

Krnjević, K.: Micro-iontophoretic studies on cortical neurones. Int. Rev. Neurobiol. **7**, 41–98 (1964).

Krnjević, K.: Glutamate and γ-aminobutyric acid in brain. Nature (Lond.) **228**, 119–124 (1970).

Krnjević, K.: Microiontophoresis. In: Methods of neurochemistry vol. 1, ed. R. Fried, p. 129–172. New York: Marcel Dekker Inc. 1971.

Krnjević, K., Morris, M. E.: Extracellular K^+ activity and slow potential changes in spinal cord and medulla. Canad. J. Physiol. Pharmac. **50**, 1214–1217 (1972).

Krnjević, K., Phillis, J. W.: Iontophoretic studies of neurones in the mammalian cerebral cortex. J. Physiol. (Lond.) **165**, 274–304 (1963).

Krnjević, K., Randić, M., Straughan, D. W.: Pharmacology of cortical inhibition. J. Physiol. (Lond.) **184**, 78–105 (1966).

Krnjević, K., Schwartz, S.: Some properties of unresponsive cells in the cerebral cortex. Exp. Brain Res. **3**, 306–319 (1967a).

KRNJEVIĆ, K., SCHWARTZ, S.: The action of γ-aminobutyric acid on cortical neurones. Exp. Brain Res. **3**, 320–336 (1967b).
KRNJEVIĆ, K., WHITTAKER, V.P.: Excitation and depression of cortical neurones by brain fractions released from micropipettes. J. Physiol. (Lond.) **179**, 298–322 (1965).
KRŽALIĆ, L., MANDIĆ, V., MIHAILOVIC, L.: On the glutamine and γ-aminobutyric acid contents of various regions of the cat brain. Experientia (Basel) **18**, 368–369 (1962).
KUBIČEK, R., DOLÉNEK, A.: Taurine et acides aminés dans la rétine des animaux. J. Chromat. **1**, 266–268 (1958).
KUNO, M., MUNEOKA, A.: Further studies on site of action of systematic omega-amino acids in the spinal cord. Jap. J. Physiol. **12**, 397–410 (1962).
KURIYAMA, K., HABER, B., ROBERTS, E.: Occurrence of a new L-glutamic acid decarboxylase in several blood vessels of the rabbit. Brain Res. **23**, 121–123 (1970).
KURIYAMA, K., HABER, B., SISKEN, B., ROBERTS, E.: The γ-aminobutyric acid system in rabbit cerebellum. Proc. nat. Acad. Sci. (Wash.) **55**, 846–852 (1966).
KURIYAMA, K., ROBERTS, E., KAKEFUDA, T.: Association of the γ-aminobutyric acid system with a synaptic vesicle fraction from mouse brain. Brain Res. **8**, 132–152 (1968).
KURIYAMA, K., SISKEN, B., HABER, B., ROBERTS, E.: The γ-aminobutyric acid system in rabbit retina. Brain Res. **9**, 165–168 (1968).
LABELLA, F., VIVIAN, S., QUEEN, G.: Abundance of cystathionine in the pineal body. Free amino acids and related compounds of bovine pineal, anterior and posterior pituitary and brain. Biochim. biophys. Acta (Amst.) **158**, 286–288 (1968).
LÄHDESMÄKI, P., OJA, S.S.: Effect of electrical stimulation on the influx and efflux of taurine in brain slices of newborn and adult rats. Exp. Brain Res. **15**, 430–438 (1972).
LAM, D.M.K.: The biosynthesis and content of gamma-aminobutyric acid in the goldfish retina. J. Cell Biol. **54**, 225–231 (1972).
LAM, D.M.K., STEINMAN, L.: The uptake of [γ-^{3}H] aminobutyric acid in the goldfish retina. Proc. nat. Acad. Sci. (Wash.) **68**, 2777–2781 (1971).
LARSON, M.D.: An analysis of the action of strychnine on the recurrent IPSP and amino acid induced inhibitions in the cat spinal cord. Brain Res. **15**, 185–200 (1969).
LEGGE, K.F., RANDIĆ, M., STRAUGHAN, D.W.: The pharmacology of neurones in the pyriform cortex. Brit. J. Pharmacol. **26**, 87–107 (1966).
LEMBECK, F., ZETLER, G.: Substance P, a polypeptide of possible physiological significance, especially within the nervous system. Int. Rev. Neurobiol. **4**, 159–215 (1962).
LEVI, G.: Regional differences in cerebral amino acid transport. Progr. Brain Res. **29**, 219–228 (1968).
LEVI, G., AMALDI, P., MORISI, G.: Gamma-aminobutyric acid (GABA) uptake by the developing mouse brain *in vivo*. Brain Res. **41**, 435–451 (1972).
LEVI, G., RAITERI, M.: Detectability of high and low affinity uptake systems for GABA and glutamate in rat brain slices and synaptosomes. Life Sci. **12**, Part I, 81–88 (1973).
LEVY, R.A., ANDERSON, E.G.: The effect of the GABA antagonists bicuculline and picrotoxin on primary afferent terminal excitability. Brain Res. **43**, 171–180 (1972).
LJUNGDAHL, A., HÖKFELT, T.: Accumulation of ^{3}H-glycine in interneurons of the cat spinal cord. Histochemie **33**, 277 (1973).
LLINAS, B., BAKER, R.: A chloride-dependent inhibitory postsynaptic potential in cat trochlear motoneurons. J. Neurophysiol. **35**, 484–492 (1972).
LOGAN, W.J., SNYDER, S.H.: High affinity uptake systems for glycine, glutamic and aspartic acids in synaptosomes of rat central nervous tissue. Brain Res. **42**, 413–431 (1972).
LOLLEY, R.N.: Metabolic and anatomical specialization within the retina. In: Handbook of neurochemistry, vol. 2, ed. A. Lajtha, p. 473–504. New York: Plenum Press 1969.
LONGO, V.G., PINTO CORRADO, A.: A neuropharmacological investigation of the convulsant action of 4-phenyl-4-formyl-N-methylpiperidine (1762 I.S.). Proc. Soc. exp. Biol. (N.Y.) **107**, 272–274 (1961).
LONGO, V.G., SILVESTRINI, B., BOVET, D.: An investigation of convulsant properties of the 5-7-diphenyl-1-3-diazadamantan-6-ol (1757 I.S.). J. Pharmacol. exp. Ther. **126**, 41–49 (1959).
LOVELL, R.A., ELLIOTT, K.A.C.: The γ-aminobutyric acid and Factor I content of brain. J. Neurochem. **10**, 479–488 (1963).
LOWE, I.P., ROBINS, E., EYERMAN, G.S.: The fluorometric measurement of glutamic decarboxylase and its distribution in brain. J. Neurochem. **3**, 8–18 (1958).

MANDEL, P., MARK, J.: The influence of nitrogen deprivation on free amino acids in rat brain. J. Neurochem. **12**, 987–992 (1965).

MANGAN, J. L., WHITTAKER, V. P.: The distribution of free amino acids in subcellular fractions of guinea-pig brain. Biochem. J. **98**, 128–138 (1966).

MARGOLIS, R. K., HELLER, A., MOORE, R. Y.: Effects of changes in cellular composition following neuronal degeneration on amino acids in brain. Brain Res. **11**, 19–31 (1968).

MARKS, N.: Peptide hydrolases. In: Handbook of neurochemistry, vol. 3, ed. A. Lajtha, p. 133–171. New York: Plenum Press 1970.

MARKS, N., DATTA, R. K., LAJTHA, A.: Distribution of amino acids and of exo- and endopeptidases along vertebrate nerves. J. Neurochem. **17**, 53–63 (1970).

MASI, I., PAGGI, P., POCCHIAR, F., TOSCHI, G.: Metabolism of labelled glucose in sympathetic ganglia *in vitro*, at rest and during sustained activity. Brain Res. **12**, 467–470 (1969).

MATSUSHITA, A., SMITH, C. M.: Spinal cord function in postischemic rigidity in the rat. Brain Res. **19**, 395–410 (1970).

MATUS, A. I., DENNISON, M. E.: An autoradiographic study of uptake of exogenous glycine by vertebrate spinal cord slices *in vitro*. J. Neurocytol. **1**, 27–34 (1972).

MCBRIDE, W. J., KLINGMAN, J. D.: The effects of electrical stimulation and ionic alterations on the metabolism of amino acids and proteins in excised superior cervical ganglia of the rat. J. Neurochem. **19**, 865–880 (1972).

MCCANCE, I., PHILLIS, J. W.: Cholinergic mechanisms in the cerebellar cortex. Int. J. Neuropharmacol. **7**, 447–462 (1968).

MCCANCE, I., PHILLIS, J. W., TEBĒCIS, A. K., Westerman, R. A.: The pharmacology of acetylcholine-excitation of thalamic neurones. Brit. J. Pharmacol. **32**, 652–662 (1968).

MCGEER, P. L., MCGEER, E. G., WADA, J. A., JUNG, E.: Effects of globus pallidus lesions and Parkinson's disease on brain glutamic acid decarboxylase. Brain Res. **32**, 425–431 (1971).

MCLENNAN, H.: Inhibition of long duration in the cerebral cortex. A quantitative difference between excitatory amino acids. Exp. Brain Res. **10**, 417–426 (1970).

MCLENNAN, H.: The pharmacology of inhibition of mitral cells in the olfactory bulb. Brain Res. **29**, 177–184 (1971).

MCLENNAN, H., HUFFMAN, R. D., MARSHALL, K. C.: Patterns of excitation of thalamic neurones by amino-acids and by acetylcholine. Nature (Lond.) **219**, 387–388 (1968).

MCLENNAN, H., YORK, D. H.: The action of dopamine on neurones of the caudate nucleus. J. Physiol. (Lond.) **189**, 393–402 (1967).

MEISTER, A.: The enzymology of amino acid transport. Science **180**, 33–39 (1973).

MELDRUM, B. S., HORTON, R. W.: Consulsive effects of 4-deoxypyridoxine and of bicuculline in photosensitive baboons (*Papio papio*) and in rhesus monkeys (*Macaca mulatta*). Brain Res. **35**, 419–436 (1971).

MEYER, M.: Die Wirkung von Acetylcholin, L-Glutaminsäure und Dopamin auf Neurone im Gebiet der Nuclei cuneatus und gracilis der Katze. Helv. physiol. pharmacol. Acta. **23**, 325–340 (1965).

MITCHELL, J. F., NEAL, M. J., SRINIVASAN, V.: The release of amino-acids from electrically stimulated rat cerebral cortex slices. Brit. J. Pharmacol. **36**, 201–202 P (1969).

MIYATA, Y., OBATA, K., TANAKA, Y., OTSUKA, M.: Determination of γ-aminobutyric acid in isolated nerve cells of the brain stem of the cat. J. Physiol. Soc. Japan, **32**, 377–378 (1970).

MIYATA, Y., OTSUKA, M.: Distribution of γ-aminobutyric acid in cat spinal cord and the alteration produced by local ischaemia. J. Neurochem. **19**, 1833–1834 (1972).

MORGAN, R., VRBOVA, G., WOLSTENCROFT, J. H.: Correlation between the retinal input to lateral geniculate neurones and their relative response to glutamate and aspartate. J. Physiol. (Lond.) **224**, 41–42 P (1972).

MORIMOTO, T., KAWAMURA, Y.: Inhibitory postsynaptic potentials of hypoglossal motoneurons of the cat. Exp. Neurol. **37**, 188–198 (1972).

MORIMOTO, T., TAKATA, M., KAWAMURA, Y.: Effect of lingual nerve stimulation on hypoglossal motoneurons. Exp. Neurol. **22**, 174–190 (1968).

MOSS, R. L., DYBALL, R. E. J., CROSS, B. A.: Responses of antidromically identified supraoptic and paraventricular units to acetylcholine, noradrenaline and glutamate applied iontophoretically. Brain Res. **35**, 573–575 (1971).

MOSS, R. L., URBAN, I., CROSS, B. A.: Microelectrophoresis of cholinergic and aminergic drugs on paraventricular neurons. Amer. J. Physiol. **223**, 310–318 (1972).

MUGNAINI, E., WALBERG, F.: An experimental electron microscopical study on the mode of termination of cerebellar cortico-vestibular fibres in the cat lateral vestibular nucleus (Deiters' nucleus). Exp. Brain Res. **4**, 212–236 (1967).

MÜLLER, P. B., LANGEMANN, H.: Distribution of glutamic acid decarboxylase activity in human brain. J. Neurochem. **9**, 399–401 (1962).

MUNEOKA, A.: Depression and facilitation of spinal reflexes by systemic omega-amino acids. Jap. J. Physiol. **11**, 555–563 (1961).

MURAKAMI, M., OHTSU, K., OHTSUKA, T.: Effects of chemicals on receptors and horizontal cells in the retina. J. Physiol. (Lond.) **227**, 899–913 (1972).

MURAYAMA, S., SMITH, C. M.: Rigidity of hind limbs of cats produced by occlusion of spinal cord blood supply. Neurology (Minneap.) **15**, 565–577 (1965).

MURRAY, J. E., CUTLER, R. W. P.: Transport of glycine from the cerebrospinal fluid. Factors regulating amino acid concentration in feline cerebrospinal fluid. Arch. Neurol. Chic. **23**, 23–31 (1970a).

MURRAY, J. E., CUTLER, R. W. P.: Clearance of glycine from cat cerebrospinal fluid: faster clearance from spinal subarachnoid than from ventricular compartment. J. Neurochem. **17**, 703–704 (1970b).

MUSHAHWAR, I. K., KOEPPE, R. E.: The toxicity of monosodium glutamate in young rats. Biochim. biophys. Acta (Amst.) **244**, 318–321 (1971).

MUSSINI, E., MARCUCCI, F.: Free amino acids in brain after treatment with psychotropic drugs. In: Amino acid pools, ed. J. T. Holden, p. 486–492. Amsterdam: Elsevier 1962.

NADLER, J. V., COOPER, J. R.: Metabolism of the aspartyl moiety of N-acetyl-L-aspartic acid in the rat brain. J. Neurochem. **19**, 2091–2105 (1972).

NAGATA, Y., YOKOI, Y., TSUKADA, Y.: Studies on free amino acid metabolism in excised cervical sympathetic ganglia from the rat. J. Neurochem. **13**, 1421–1431 (1966).

NAKAJIMA, T., WOLFGRAM, F., CLARK, W. G.: Identification of 1,4-methylhistamine, 1,3-diaminopropane and 2,4-diaminobutyric acid in bovine brain. J. Neurochem. **14**, 1113–1118 (1967).

NAKAMURA, Y., WU, C. Y.: Presynaptic inhibition of jaw-opening reflex by high threshold afferents from the masseter muscle of the cat. Brain Res. **23**, 193–211 (1970).

NATHAN, P. W., SMITH, M. C.: Fasciculi proprii of the spinal cord in man. Brain, **82**, 610–668 (1959).

NEAL, M. J.: The uptake of [^{14}C] glycine by slices of mammalian spinal cord. J. Physiol. (Lond.) **215**, 103–117 (1971).

NEAL, M. J., IVERSEN, L. L.: Subcellular distribution of endogenous and ^{3}H-GABA in rat cerebral cortex. J. Neurochem. **16**, 1245–1252 (1969).

NEAL, M. J., IVERSEN, L. L.: Autoradiographic localization of ^{3}H-GABA in rat retina. Nature (Lond.) New Biol. **235**, 217–218 (1972).

NEAL, M. J., STARR, M. S.: Effect of inhibitors of γ-aminobutyrate aminotransferase on the accumulation of ^{3}H-γ-aminobutyric acid by the retina. Brit. J. Pharmacol. **47**, 543–555 (1973).

NEAL, M. J., WHITE, R. D.: Uptake of ^{14}C-L-glutamate by rat retina. Brit. J. Pharmacol. **43**, 442–443P (1971).

NEUBERGER, A.: Dissociation constants and structures of glutamic acid and its esters. Biochem. J. **30**, 2085–2094 (1936).

NICOLL, R. A.: Pharmacological evidence for GABA as the transmitter in granule cell inhibition in the olfactory bulb. Brain Res. **35**, 137–149 (1971).

NICOLL, R. A.: The effects of anesthetics on synaptic excitation and inhibition in the olfactory bulb. J.Physiol. (Lond.) **223**, 803–814 (1972).

NICOLL, R. A., BARKER, J. L.: The pharmacology of recurrent inhibition in the supraoptic neurosecretory system. Brain Res. **35**, 501–511 (1971).

NOELL, W. K.: The visual cell: electric and metabolic manifestations of its life processes. Amer. J. Ophthal. **48**, 347–370 (1959).

OBATA, K.: Acetylcholine and gamma-aminobutyric acid action on tissue-cultured cells from sympathetic ganglion, dorsal-root ganglion and diaphragm muscle. Fed. Proc. **31**, 231 (1972a).

OBATA, K.: The inhibitory action of γ-aminobutyric acid, a probable synaptic transmitter. Int. Rev. Neurobiol. **15**, 167–187 (1972b).

OBATA, K., HIGHSTEIN, S. M.: Blocking by picrotoxin of both vestibular inhibition and GABA action of rabbit oculomotor neurones. Brain Res. **18**, 538–541 (1970).

OBATA, K., ITO, M., OCHI, R., SATO, N.: Pharmacological properties of the postsynaptic inhibition by Purkinje cell axons and the action of γ-aminobutyric acid on Deiters neurones. Exp. Brain Res. **4**, 43–57 (1967).

OBATA, K., TAKEDA, K.: Release of γ-aminobutyric acid into the fourth ventricle induced by stimulation of the cat's cerebellum. J. Neurochem. **16**, 1043–1047 (1969).

OBATA, K., TAKEDA, K., SHINOZAKI, H.: Further study on pharmacological properties of the cerebellar-induced inhibition of Deiters neurones. Exp. Brain Res. **11**, 327–342 (1970).

OKADA, Y., HASSLER, R.: Uptake and release of γ-aminobutyric acid (GABA) in slices of substantia nigra of rat. Brain Res. **49**, 214–217 (1973).

OKADA, Y., NITSCH-HASSLER, C., KIM, J.S., BAK, I.J., HASSLER, R.: Role of γ-aminobutyric acid (GABA) in the extrapyramidal motor system. 1. Regional distribution of GABA in rabbit, rat, guinea pig and baboon CNS. Exp. Brain Res. **13**, 514–518 (1971).

OKAYA, Y.: Refinement of the crystal structure of taurine. An example of computer-controlled experimentation. Acta Cryst. **21**, 726–735 (1966).

OKOMOTO, S.: Epileptogenic action of glutamate directly applied into the brain of animals and inhibitory effect of proteins and tissue emulsions on its action. J. Physiol. Soc. Japan, **13**, 555–562 (1951).

OLDENDORF, W.H.: Brain uptake of radiolabeled amino acids, amines and hexoses after arterial injection. Amer. J. Physiol. **221**, 1629–1639 (1971).

OLNEY, J.W., HO, O.L., RHEE, V.: Cytotoxic effects of acidic and sulphur containing amino acids on the infant mouse central nervous system. Exp. Brain Res. **14**, 61–76 (1971).

OLNEY, J.W., SHARPE, L.G., FEIGIN, R.D.: Glutamate-induced brain damage in infant primates. J. Neuropath. exp. Neurol. **31**, 464–488 (1972).

OOMURA, Y., OOYAMA, H., YAMAMOTO, T., ONO, T., KOBAYASHI, N.: Behavior of hypothalamic unit activity during electrophoretic application of drugs. Ann. N. Y. Acad. Sci. **157**, 642–665 (1969).

ORREGO, F., LIPMANN, F.: Protein synthesis in brain slices. Effects of electrical stimulation and acidic amino acids. J. biol. Chem. **242**, 665–671 (1967).

OTSUKA, M.: γ-Aminobutyric acid in the nervous system. In: The structure and function of nervous tissue, vol. 4, ed. G.H. Bourne, p. 249–289. New York: Academic press 1972.

OTSUKA, M., KONISHI, S., TAKAHASHI, T.: The presence of a motoneuron-depolarizing peptide in bovine dorsal roots of spinal nerves. Proc. Jap. Acad. **48**, 342–346 (1972).

OTSUKA, M., MIYATA, Y.: Application of enzymatic cycling to the measurement of gamma-aminobutyric acid in single neurons of the mammalian central nervous system. Advanc. Biochem. Psychopharmac. **6**, 61–74 (1972).

OTSUKA, M. OBATA, K., MIYATA, Y., TANAKA, Y.: Measurement of γ-aminobutyric acid in isolated nerve cells of cat central nervous system. J. Neurochem. **18**, 287–295 (1971).

PARRY JONES, G., ROBERTS, R.T., AHMED, A.I.: A nuclear magnetic resonance investigation of γ-aminobutyric acid (GABA). Molec. Physiol. **22**, 547–549 (1971).

PASANTES-MORALES, H., KLETHI, J., URBAN, P.F., MANDEL, P.: The physiological role of taurine in retina: uptake and effect on electroretinogram (ERG). Physiol. Chem. Phys. **4**, 339–348 (1972).

PECK, E.J., AWAPARA, J.: Formation of taurine and isethionic acid in rat brain. Biochim. biophys. Acta (Amst.) **141**, 499–506 (1967).

PEREZ, V.J., OLNEY, J.W.: Accumulation of glutamic acid in the arcuate nucleus of the hypothalamus of the infant mouse following subcutaneous administration of monosodium glutamate. J. Neurochem. **19**, 1777–1782 (1972).

PERRY, T.L., BERRY, K., DIAMOND, S., MOK, C.: Regional distribution of amino acids in human brain obtained at autopsy. J. Neurochem. **18**, 513–519 (1971a).

PERRY, T.L., HANSEN, S., BERRY, K., MOK, C., LESK, D.: Free amino acids and related compounds in biopsies of human brain. J. Neurochem. **18**, 521–528 (1971b).

PERRY, T.L., SANDERS, H.D., HANSEN, S., LESK, D., KLOSTER, M., GRAVLIN, L.: Free amino acids and related compounds in five regions of biopsied cat brain. J. Neurochem. **19**, 2651–2656 (1972).

PHILLIS, J.W., OCHS, S.: Excitation and depression of cortical neurones during spreading depression. Exp. Brain Res. **12**, 132–149 (1971).

PHILLIS, J.W., TEBĒCIS, A.K., YORK, D.H.: Histamine and some antihistamines: their actions on cerebral cortical neurones. Brit. J. Pharmacol. **33**, 426–440 (1968).

PINTO CORRADO, A., LONGO, V.G.: An electrophysiological analysis of the convulsant action of morphine, codeine, and thebaine. Arch. int. Pharmacodyn. **132**, 235–269 (1961).

PONG, S.F., GRAHAM, L.T., JR.: N-Methyl bicuculline, a convulsant more potent than bicuculline. Brain Res. **42**, 486–490 (1972).

PONNUSWAMY, P. K., SASISEKLARAN, V.: Conformation of amino acids. II. Potential energy of conformations of glycine, alanine and aminobutyric acid. Int. J. Protein Res. **2**, 37–45 (1970).

POPOV, N., POHLE, W., RÖSLER, V., MATTHIES, H.: Regional distribution of γ-aminobutyric acid, glutamic acid, aspartic acid, dopamine, noradrenaline and serotonin in rat brain. Acta biol. med. germ. **18**, 695–702 (1967).

PORTER, L. L.: Picrotoxinin and related substances. Chem. Rev. **67**, 441–464 (1967).

PRECHT, W., BAKER, R.: Synaptic organization of the vestibulo-trochlear pathway. Exp. Brain Res. **14**, 158–184 (1972).

PRECHT, W., BAKER, R., OKADA, Y.: Evidence for GABA as the synaptic transmitter of the inhibitory vestibulo-ocular pathway. Exp. Brain Res. in press (1973).

PRECHT, W., BAKER, R., YOSHIDA, M.: Blockage of caudate-evoked inhibition of neurons in the substantia nigra by picrotoxin. Brain Res. **32**, 229–233 (1971).

PURPURA, D. P., GIRADO, M., SMITH, T. G., CALLAN, D. A., GRUNDFEST, H.: Structure activity determinants of pharmacological effects of amino acids and related compounds on central synapses. J. Neurochem. **3**, 238–268 (1959).

RAMWELL, P. W., SHAW, J. E.: Some observations on the physical and pharmacological properties of picrotoxin solutions. J. Pharm. Pharmacol. **15**, 611–619 (1963).

RICHTER, J. J., WAINER, A.: Evidence for separate systems for the transport of neutral and basic amino acids across the blood brain barrier. J. Neurochem. **18**, 613–620 (1971).

ROBERTS, E.: Formation and liberation of γ-aminobutyric acid in brain. In: Progress in neurobiology; 1. Neurochemistry, eds. S. R. Korey and J. I. Nurnberger, p. 11–25. New York: Hoeber-Harper 1956.

ROBERTS, E.: Inhibition in the nervous system and gamma-aminobutyric acid. New York: Pergamon Press 1960.

ROBERTS, E.: Some biochemical-physiological correlations in studies of γ-aminobutyric acid. In: Structure and function of inhibitory neuronal mechanisms, eds. C. von Euler, S. Skoglund and U. Söderberg, p. 401–418. Oxford: Pergamon Press 1968.

ROBERTS, E.: The GABA system in brain development. In: Chemistry and brain development, advances in experimental medicine and biology, vol. 13, eds. R. Paoletti and A. N. Davison, p. 207–214. New York: Plenum Press 1971.

ROBERTS, E., EIDELBERG, E.: Metabolic and neurophysiological roles of γ-aminobutyric acid. Int. Rev. Neurobiol. **2**, 279–332 (1960).

ROBERTS, E., KURIYAMA, K.: Biochemical-physiological correlations in studies of the γ-aminobutyric acid system. Brain Res. **8**, 1–35 (1968).

ROBERTS, P. J., MITCHELL, J. F.: The release of amino acids from the hemisected spinal cord during stimulation. J. Neurochem. **19**, 2473–2481 (1972).

RUBIN, R. P.: The role of calcium in the release of neurotransmitter substances and hormones. Pharmacol. Rev. **22**, 389–428 (1970).

RYALL, R. W.: The subcellular distributions of acetylcholine, substance P, 5-hydroxytryptamine, γ-aminobutyric acid and glutamic acid in brain homogenates. J. Neurochem. **11**, 131–145 (1964).

RYALL, R. W.: Renshaw cell mediated inhibition of Renshaw cells: patterns of excitation and inhibition from impulses in motor axon collaterals. J. Neurophysiol. **32**, 256–270 (1970).

RYALL, R. W., DE GROAT, W. C.: The microelectrophoretic administration of noradrenaline, 5-hydroxytryptamine, acetylcholine and glycine to sacral parasympathetic preganglionic neurones. Brain Res. **37**, 345–347 (1972).

RYALL, R. W., PIERCEY, M. F., POLOSA, C.: Strychnine-resistant mutual inhibition of Renshaw cells. Brain Res. **41**, 119–129 (1972).

SALVADOR, R. A., ALBERS, R. W.: The distribution of glutamic-γ-aminobutyric transaminase in the nervous system of the Rhesus monkey. J. biol. Chem. **234**, 922–925 (1959).

SAUERLAND, E. K., MIZUNO, N.: Cortically induced presynaptic inhibition of trigeminal proprioceptive afferents. Brain Res. **13**, 556–568 (1969).

SCHENCK, J. R.: A note on the distribution of β-alanine. Proc. Soc. exp. Biol. (N. Y.) **54**, 6–7 (1943).

SCHMIDT, R. F.: Pharmacological studies on the primary afferent depolarization of the toad spinal cord. Pflügers Arch. ges. Physiol. **277**, 325–346 (1963).

SCHMIDT, R. F.: Presynaptic inhibition in the vertebrate central nervous system. Ergebn. Physiol. **63**, 20–101 (1971).

SCHOLES, N. W., ROBERTS, E.: Pharmacological studies of the optic system of the chick: effect of γ-aminobutyric acid and pentobarbital. Biochem. Pharmacol. **13**, 1319–1329 (1964).

SCHON, F., IVERSEN, L. L.: Selective accumulation of [^{3}H] GABA by stellate cells in rat cerebellar *in vivo*. Brain Res. **42**, 503–507 (1972).

SCHUBERT, P., KREUTZBERG, G. W., LUX, H. D.: Use of microelectrophoresis in the autoradiographic demonstration of fiber projections. Brain Res. **39**, 274–277 (1972).

SEILER, N., WIECHMANN, M.: Zum Vorkommen der γ-amino-buttersäure und der γ-amino-β-hydroxy-buttersäure in tierischem Gewebe. Hoppe-Seyler's Z. physiol. Chem. **350**, 1493–1500 (1969).

SEMBA, T., KANO, M.: Glycine in the spinal cord of cats with local tetanus rigidity. Science **164**, 571–572 (1969).

SETA, K., SERSHEN, H., LAJTHA, A.: Cerebral amino acid uptake *in vivo* in newborn mice. Brain Res. **47**, 415–425 (1972).

SHANK, R. P., APRISON, M. H.: The metabolism *in vivo* of glycine and serine in eight areas of the rat central nervous system. J. Neurochem. **17**, 1461–1475 (1970).

SHANK, R. P., APRISON, M. H.: Post mortem changes in the content and specific radioactivity of several amino acids in four areas of the rat brain. J. Neurobiol. **2**, 145–151 (1971).

SHANK, R. P., APRISON, M. H., BAXTER, C. F.: Precursors of glycine in the nervous system: comparison of specific activities in glycine and other amino acids after administration of [U-^{14}C]glucose, [3.4-^{14}C]glucose, [1-^{14}C]glucose, [U-^{14}C]serine or [1.5-^{14}C]citrate to the rat. Brain Res. **52**, 301–308 (1973).

SHAW, R. K., HEINE, J. D.: Ninhydrin positive substances present in different areas of normal rat brain. J. Neurochem. **12**, 151–155 (1965).

SHEPHERD, G. M.: Synaptic organization of the mammalian olfactory bulb. Physiol. Rev. **52**, 864–917 (1972).

SHERIDAN, J. J., SIMS, K. L., PITTS, F. N.: Brain γ-aminobutyrate-α-oxoglutarate transaminase—II. Activities in twenty-four regions of human brain. J. Neurochem. **14**, 571–578 (1967).

SHIMADA, M., KIHARA, T., KURIMOTO, K., MARUYAMA, Y., IJICHI, H.: Topographic separation of lyophilized mouse brain and gas liquid chromatographic analysis of free amino acids in brain divisions. Acta anat. nippon., **47**, 138–153 (1972)

SHIMADA, M., KURIMOTO, K., WADA, F., HINO, O., SUGINOSHITA, K., KIHARA, T.: Quantitative trial of free amino acid concentrations in the mouse brain loci by gas-liquid chromatography. Acta anat. nippon., **46**, 125–135 (1971).

SIGGINS, G. R., OLIVER, A. P., HOFFER, B. J., BLOOM, F. E.: Cyclic adenosine monophosphate and norepinephrine: effects on transmembrane properties of cerebellar Purkinje cells. Science **171**, 192–194 (1971).

SIMS, K. L., WEITSEN, H. A., BLOOM, F. E.: Histochemical localization of brain succinic semialdehyde dehydrogenase-A γ-aminobutyric acid degradative enzyme. J. Histochem. Cytochem. **19**, 405–415 (1971).

SIMMS, H. S.: The effect of salts on weak electrolytes. I. Dissociation of weak electrolytes in the presence of salts. J. Phys. Chem. **32**, 1121–1141 (1928).

SMALL, N. A., HOLTON, J. B., ANCILL, R. J.: *In vitro* inhibition of serotonin and γ-aminobutyric acid synthesis in rat brain by histidine metabolites. Brain Res. **21**, 55–62 (1970).

SMITH, S. E.: Kinetics of neutral amino acid transport in rat brain *in vitro*. J. Neurochem. **14**, 291–300 (1967).

SMITH, T. G., WUERKER, R. B., FRANK, K.: Membrane impedance changes during synaptic transmission in cat spinal motoneurons. J. Neurophysiol. **30**, 1072–1096 (1967).

SMYTHIES, J. R.: The chemical anatomy of synaptic mechanisms: receptors. Int. Rev. Neurobiol. **14**, 233–331 (1971).

SNODGRASS, S. R., CUTLER, R. W. P., KANG, E. S., LORENZO, A. V.: Transport of neutral amino acids from feline cerebrospinal fluid. Amer. J. Physiol. **217**, 974–980 (1969).

SNODGRASS, S. R., IVERSEN, L. L.: A sensitive double isotope derivative assay to measure release of amino acids from brain *in vivo*. Nature (Lond.) New Biol. **241**, 154–156 (1973a).

SNODGRASS, S. R., IVERSEN, L. L.: Effects of amino-oxyacetic acid on [^{3}H]GABA uptake by rat brain slices. J. Neurochem. **20**, 431–439 ((1973b).

SNYDER, S. H., LOGAN, W. J., BENNETT, J. P., ARREGUI, A.: Amino acids as central nervous transmitters: Biochemical studies. Neurosciences Research **5**, (1972).

SOMJEN, G. G.: Evoked sustained focal potentials and membrane potential of neurons and of unresponsive cells of the spinal cord. J. Neurophysiol. **33**, 562–582 (1970).

SOTELO, C., PRIVAT, A., DRIAN, M.-J.: Localization of [^{3}H] GABA in tissue culture of rat cerebellum using electron microscopy radioautography. Brain Res. **45**, 302–308 (1972).

SRIMAL, R.C., BHARGAVA, K.P.: Peripheral neural effects of gamma aminobutyric acid. Arch. int. Pharmacodyn. **164**, 444–450 (1966).

STARR, M.S., VOADEN, M.J.: The uptake of [^{14}C] γ-aminobutyric acid by the isolated retina of the rat. Vision Res. **12**, 549–557 (1972a).

STARR, M.S., VOADEN, M.J.: The uptake, metabolism and release of ^{14}C-taurine by rat retina *in vitro*. Vision Res. **12**, 1261–1269 (1972b).

STEFANIS, C.: Hippocampal neurons: their responsiveness to microelectrophoretically administered endogenous amines. The Pharmacologist, **6**, 171 (1964).

STEFANIS, C.: Discussion. In: The interneuron, ed. M.A. Brazier, p. 442–447. Los Angeles: Univ. Calif. Press 1969.

STEINER, F.A., MEYER, M.: Actions of L-glutamate, acetylcholine and dopamine on single neurones in the nuclei cuneatus and gracilis of the cat. Experientia (Basel) **22**, 58–59 (1966).

STEINER, F.A., RUF, K.: Interactions of L-glutamic acid, gamma-aminobutyric acid and pyridoxal-5'-phosphate at the neuronal level. Schweiz. Arch. Neurol. Psychiat. **100**, 310–320 (1967).

STERN, P., HADŽOVIĆ, S.: Effect of glycine on experimental hindlimb rigidity in rats. Life Sci. **9**, Part I, 955–959 (1970).

STEWARD, E.G., PLAYER, R., QUILLIAM, J.P., BROWN, D.A., PRINGLE, M.J.: Molecular conformation of GABA. Nature (Lond.) New Biol. **233**, 87–88 (1971).

STEWART, C.N., COURSIN, D.B., BHAGAVAN, H.N.: Electroencephalographic study of L-glutamate induced seizures in rats. Toxicol. appl. Pharmacol. **23**, 635–639 (1972).

STORM-MATHISEN, J.: Glutamate decarboxylase in the rat hippocampal region after lesions of the afferent fibre systems. Evidence that enzyme is localized in intrinsic neurones. Brain Res. **40**, 215–235 (1972).

STORM-MATHISEN, J., FONNUM, F.: Quantitative histochemistry of glutamate decarboxylase in the rat hippocampal region. J.Neurochem. **18**, 1105–1111 (1971).

STORM-MATHISEN, J., FONNUM, F.: Localization of transmitter candidates in the hippocampal region. Progr. Brain Res. **36**, 40–57 (1972).

STRASCHILL, M.: Action of drugs on single neurons in the cat's retina. Vision Res. **8**, 35–47 (1968).

STRASCHILL, M., PERWEIN, J.: The inhibition of retinal ganglion cells by catecholamines and γ-aminobutyric acid. Pflügers Arch. **312**, 45–54 (1969).

STRAUGHAN, D.W., LEGGE, K.F.: The pharmacology of amygdaloid neurones. J. Pharm. Pharmacol. **17**, 675–677 (1965).

STRAUGHAN, D.W., NEAL, M.J., SIMMONDS, M.A., COLLINS, G.G.S., HILL, R.G.: Evaluation of bicuculline as a GABA antagonist. Nature (Lond.) **233**, 352–354 (1971).

SUGAR, O., GERARD, R.W.: Spinal cord regeneration in the rat. J.Neurophysiol. **3**, 1–19 (1940).

SVERDLOV, YU. S., ALEKSEEVA, V.I.: Effect of tetanus toxin on presynaptic inhibition in the spinal cord. Fed. Proc. **25**, 931–935 (1966).

SVERDLOV, YU.S., KOZHECHKIN, S.N.: On presynaptic inhibitory mechanisms of monosynaptic reflexes. In: Mechanisms of the nervous system, eds. D.A. Birjukova, D.G. Kvasova and N.V. Zimkina, p. 136–143. Leningrad: Nauka 1969.

SYTINKSY, I.A.: The gamma-aminobutyric acid (GABA) system of the cerebral white matter and of brain tumours. Neuropath. pol. **7**, 371–375 (1969).

SYTINKSY, I.A.: γ-Aminobutyric acid on the activity of the nervous system. Leningrad: Nauka 1972.

SYTINKSY, I.A., VASILIJEV, V.Y.: Some catalytic properties of purified γ-aminobutyrate-α-oxoglutarate transminase from rat brain. Enzymologia **39**, 1–11 (1970).

SZE, P.Y.: Possible repression of L-glutamic acid decarboxylase by gamma-aminobutyric acid in developing mouse brain. Brain Res. **19**, 322–325 (1970).

SZE, P.Y., LOVELL, R.A.: Reduction of level of L-glutamate decarboxylase by γ-aminobutyric acid in mouse brain. J. Neurochem. **17**, 1657–1664 (1970).

SZENTÁGOTHAI, J.: Propriospinal pathways and their synapses. In: Organization of the spinal cord, eds. J.C. Eccles and J.P. Schade, p. 155–174. Amsterdam: Elsevier 1964.

SZENTÁGOTHAI, J.: Structure-function relationships in inhibitory synapses. In: Advances in cytopharmacology, vol. 1, eds. F. Clementi and B. Ceccarelli, p. 401–417. New York: Raven Press 1971.

TACHIKI, K.H., DEFEUDIS, F.V., APRISON, M.H.: Studies on the subcellular distribution of γ-aminobutyric acid in slices of rat cerebral cortex. Brain Res. **36**, 215–217 (1972).

TAKANO, K., NEUMANN, K.: Effect of glycine upon stretch reflex tension. Brain Res. **36**, 474–475 (1972).

Tallan, H. H.: A survey of amino acids and related compounds in nervous tissue. In: Amino acid pools, ed. J. T. Holden, p. 471–485. Amsterdam: Elsevier 1962.
Tallan, H. H., Moore, S., Stein, W. H.: Studies on the free amino acids and related compounds in the tissues of the cat. J. biol. Chem. **211**, 927–939 (1954).
Tebēcis, A. K.: Effects of monoamines and amino acids on medial geniculate neurones of the cat. Neuropharmacol. **9**, 381–390 (1970).
Tebēcis, A. K.: Studies on optic nerve transmitters. Proc. aust. physiol. pharmacol. Soc. **4**, 86–87 (1973a).
Tebēcis, A. K.: Transmitters and reticulospinal neurones. Exp. Neurol. **40**, 297–308 (1973b).
Tebēcis, A. K., Di Maria, A.: A re-evaluation of the mode of action of 5-hydroxytryptamine on lateral geniculate neurones: comparison with catecholamines and LSD. Exp. Brain Res. **14**, 480–493 (1972a).
Tebēcis, A. K.: Di Maria, A.: Strychnine-sensitive inhibition in the medullary reticular formation: evidence for glycine as an inhibitory transmitter. Brain Res. **40**, 373–383 (1972b).
Tebēcis, A. K., Hösli, L., Haas, H. L.: Bicuculline and the depression of medullary reticular neurones by GABA and glycine. Experientia (Basel) **27**, 548 (1971).
Tebēcis, A. K., Ishikawa, T.: Glycine and GABA as inhibitory transmitters in the medullary reticular formation: studies involving intra- and extra-cellular recording. Pflügers Arch. **338**, 273–278 (1973).
Tebēcis, A. K., Phillis, J. W.: The use of convulsants in studying possible functions of amino acids in the toad spinal cord. Comp. Biochem. Physiol. **28**, 1303–1315 (1969).
Tews, J. K., Carter, S. H., Roa, P. D., Stone, W. E.: Free amino acids and related compounds in dog brain. Postmortem and anoxic changes, effects of ammonium chloride infusion, and levels during seizures induced by picrotoxin and by pentylenetetrazol. J. Neurochem. **10**, 641–653 (1963).
Tomita, K.: Crystal data and some structural feature of γ-aminobutyric acid, 3-aminopropane sulfonic acid and their derivatives. Tetrahedron Letters No. **27**, 2587–2588 (1971).
Tunnicliff, G., Wein, J., Roberts, E.: Effects of imidazole-acetic acid on brain amino acids and body temperature in mice. J. Neurochem. **19**, 2017–2023 (1972).
Uhr, M. L.: Glycine decarboxylase in the central nervous system. J. Neurochem. **20**, 1005–1010 (1973).
Uhr, M. L., Sneddon, M. K.: Glycine and serine inhibition of D-glycerate dehydrogenase and 3-phosphoglycerate dehydrogenase. FEBS Letters, **17**, 137–140 (1971).
Uhr, M. L., Sneddon, M. K.: The regional distribution of D-glycerate dehydrogenase and 3-phosphoglycerate dehydrogenase in the cat central nervous system: correlation with glycine levels. J. Neurochem. **19**, 1495–1500 (1972).
Utley, J.: Gamma-aminobutyric acid and 5-hydroxytryptamine concentration in neurons and glial cells in the medial geniculate body of the cat. Biochem. Pharmacol. **12**, 1228–1230 (1963).
Van den Berg, C. J.: Glutamate and glutamine. In: Handbook of neurochemistry, vol. 3, Metabolic reactions in the nervous system, ed. A. Lajtha, p. 355–379. New York: Plenum Press 1970.
Van den Berg, C. J.: A model of compartmentation in mouse brain based on glucose and acetate metabolism. In: Metabolic compartmentation in the brain, eds. R. Balázs and J. E. Cremer, p. 129–136. London: Macmillan 1973.
Van Gelder, N. M.: The histochemical demonstration of γ-aminobutyric acid metabolism by reduction of a tetrazolium salt. J. Neurochem. **12**, 231–237 (1965a).
Van Gelder, N. M.: A comparison of γ-aminobutyric acid metabolism in rabbit and mouse nervous tissue. J. Neurochem. **12**, 239–244 (1965b).
Van Gelder, N. M: The effect of aminooxyacetic acid on the metabolism of γ-aminobutyric acid in brain. Biochem. Pharmacol. **15**, 533–539 (1966).
Van Gelder, N. M.: Antagonism by taurine of cobalt induced epilepsy in cat and mouse. Brain Res. **47**, 157–165 (1972).
Van Harreveld, A.: Compounds in brain extracts causing spreading depression of cerebral cortical activity and contraction of crustacean muscle. J. Neurochem. **3**, 300–315 (1959).
Van Harreveld, A.: A mechanism for fluid shifts specific for the central nervous system. In: Current research in neurosciences, topical problems in psychiatric neurology, vol. 10, ed. H. T. Wycis, p. 62–70. Basel-New York: Karger 1970.
Van Harreveld, A., Fifkova, E.: Glutamate release from the retina during spreading depression. J. Neurobiol. **2**, 13–29 (1970).

VAN HARREVELD, A., FIFKOVA, E.: Effects of glutamate and other amino acids on the retina. J. Neurochem. **18**, 2145–2154 (1971).

VAN HARREVELD, A., FIFKOVA, E.: Effects of metabolic inhibitors on the release of glutamate from the retina. J. Neurochem. **19**, 1439–1450 (1972).

VAN HARREVELD, A., KOOIMAN, M.: Amino acid release from the cerebral cortex during spreading depression and asphyxiation. J. Neurochem. **12**, 431–439. (1965).

VOADEN, M.J., STARR, M.S.: The efflux of radioactive GABA from rat retina *in vitro*. Vision Res. **12**, 559–566 (1972).

VON BAUMGARTEN, R., BLOOM, F.E., OLIVER, A.P., SALMOIRAGHI, G.C.: Response of individual olfactory nerve cells to microelectrophoretically administered chemical substances. Pflügers Arch. ges. Physiol. **277**, 125–140 (1963).

VON BAUMGARTEN, R., GREEN, J.D., MANCIA, M.: Recurrent inhibition in the olfactory bulb. II. Effects of antidromic stimulation of commisural fibres. J. Neurophysiol. **25**, 489–500 (1962).

VORKEL, W., HANITZSCH, R.: Effect of strychnine on the electroretinogram of the isolated rabbit retina. Experientia (Basel) **27**, 296–297 (1971).

VYKLICKÝ, L., SYKOVA, E., KŘIŽ, N., UJEC, E.: Post-stimulation changes of extracellular potassium concentration in the spinal cord of the rat. Brain Res. **45**, 608–611 (1972).

WAKSMAN, A., BLOCH, M.: Identification of multiple forms of aminobutyrate transaminase in mouse and rat brain: subcellular localization. J. Neurochem. **15**, 99–105 (1968).

WAKSMAN, A., ROBERTS, E.: Purification and some properties of mouse brain γ-aminobutyric-α-ketoglutaric acid transminase. Biochemistry, **4**, 2132–2139 (1965).

WALBERG, F., JANSEN, J.: Cerebellar cortico-vestibular fibers in the cat. Exp. Neurol. **3**, 32–52 (1961).

WALSH, D.A., SALLACH, H.J.: Comparative studies on the pathways for serine biosynthesis in animal tissues. J. biol. Chem. **241**, 4068–4076 (1966).

WATANABE, T., SIMADA, Z.: Picrotoxin: effect on collicular auditory neurons. Brain Res. **28**, 586–590 (1971).

WATKINS, J.C.: Metabolic regulation in the release and action of excitatory and inhibitory amino acids in the central nervous system. Biochem. Soc. Symp. **36**, 33–37 (1972).

WATKINS, J.C., CURTIS, D.R., BISCOE, T.J.: Central effects of β-N-oxalyl-α,β-diaminopropionic acid and other Lathyrus factors. Nature (Lond.) **211**, 637 (1966).

WEIGHT, F.F.: Mechanisms of synaptic transmission. In: Neurosciences research, vol. 4, p. 1–27. New York: Academic Press 1971.

WEINSTEIN, H., VARON, S., ROBERTS, E.: Effects of imipramine on the Na^+-dependent exchange and retention of γ-aminobutyric acid by mouse brain subcellular particles. Biochem. Pharmacol. **20**, 103–117 (1971).

WELCH, A. D., HENDERSON, V. E.: A comparative study of hydrastine, bicuculline and adlumine. J. Pharmacol. exp. Ther. **51**, 482–491 (1934).

WERMAN, R.: Criteria for identification of a central nervous system transmitter. Comp. Biochem. Physiol. **18**, 745–766 (1966).

WERMAN, R.: Amino acids as central neurotransmitters. Res. Publ. Ass. nerv. ment. Dis. **50**, 147–180 (1972).

WERMAN, R., DAVIDOFF, R. A., APRISON, M. H.: The inhibitory action of the cystathionine. Life Sci. **5**, 1431-1440 (1966).

WERMAN, R., DAVIDOFF, R. A., APRISON, M. H.: Evidence for glycine as the principal transmitter mediating postsynaptic inhibition in the spinal cord of the cat. J. gen. Physiol. **50**, 1093–1094 (1967).

WERMAN, R., DAVIDOFF, R. A., APRISON, M. H.: Inhibitory action of glycine on spinal neurons in the cat. J. Neurophysiol. **31**, 81–95 (1968).

WEST, R.: A pharmacological study of derivatives of two specimens of tubo-curare, and an examination of four members of genus *Strychnos* and one rubiaceous plant associated with the curares of British Guiana. Arch. int. Pharmac. Ther. **56**, 81–116 (1937).

WESTECKER, M. E.: Reciprocal activation of two evoked potential components in the olfactory bulb. Pflügers Arch. **324**, 297–310 (1971).

WHEELER, D. D., BOYARSKY, L. L.: Influx of glutamic acid in peripheral nerve. Energy, ionic, and pH dependence. J. Neurobiol. **2**, 181–190 (1971).

WHEELER, D. D., BOYARSKY, L. L., BROOKS, W. H.: Release of amino acids from nerve during stimulation. J. cell. Physiol. **67**, 141–148 (1966).

WHITTAKER, V. P.: The application of subcellular fractionation techniques to the study of brain function. Prog. Biophys. molec. Biol. **15**, 39–96 (1965).

WHITTAKER, V. P.: The subcellular distribution of amino acids in brain and its relation to a possible transmitter function for these compounds. In: Structure and function of inhibitory neuronal mechanisms, eds. C. von Euler, S. Skoglund, and U. Söderberg, p. 487–504. Oxford: Pergamon Press 1968.

WHTTAKER, V. P., BARKER, L. A.: The subcellular fractionation of brain tissue with special reference to the preparation of synaptosomes and their component organelles. In: Methods of neurochemistry, vol. 2, ed. R. Fried, p. 1–52. New York: Marcel Dekker 1972.

WILSON, V. J.: The vestibular nuclei and spinal motor activity. J. psychiat. Res. **8**, 259–272 (1971).

WILSON, V. J.: Physiological pathways through the vestibular nuclei. Int. Rev. Neurobiol. **15**, 27–81 (1972).

WILSON, V. J., TALBOT, W. H.: Integration at an inhibitory interneurone: inhibition of Renshaw cells. Nature (Lond.) **200**, 1325–1327 (1963).

WOFSEY, A. R., KUHAR, M. J., SNYDER, S. H.: A unique synaptosomal fraction, which accumulates glutamic and aspartic acids, in brain tissue. Proc. nat. Acad. Sci. (Wash.) **68**, 1102–1106 (1971).

WOLMAN, M.: A fluorescent histochemical procedure for gamma-aminobutyric acid. Histochemie, **28**, 118–130 (1971).

WOODWARD, D. J., HOFFER, B. J., SIGGINS, G. R., BLOOM, F. E.: The ontogenetic development of synaptic junctions, synaptic activation and responsiveness to neurotransmitter substances in rat cerebellar Purkinje cells. Brain Res. **34**, 73–97 (1971).

WOODWARD, D. J., HOFFER, B. J., SIGGINS, G. R., OLIVER, A. P.: Inhibition of Purkinje cells in the frog cerebellum. II. Evidence for GABA as the inhibitory transmitter. Brain Res. **33**, 91–100 (1971).

WYSSBROD, H. R., SCOTT, W. N., BRODSKY, W. A., SCHWARTZ, I. L.: Carrier-mediated transport processes. In: Handbook of neurochemistry, vol. 5, Part B, Metabolic turnover in the nervous system, ed. A. Lajtha, p. 683–819. New York-London: Plenum Press 1971.

YORK, D. H.: A microiontophoretic study of neurones in the globus pallidus-putamen complex. Aust. J. Exp. Biol. med. Sci. **46**, P3–4 (1968).

YORK, D. H.: Dopamine receptor blockade—a central action of chlorpromazine on striatal neurones. Brain Res. **37**, 91–99 (1972).

YOSHIDA, M., PRECHT, W.: Monosynaptic inhibition of neurons of the substantia nigra by caudato-nigral fibers. Brain Res. **32**, 225–228 (1971).

YOSHIDA, M., RABIN, A., ANDERSON, M.: Monosynaptic inhibition of pallidal neurons by axon collaterals of caudato-nigral fibers. Exp. Brain Res. **15**, 333–347 (1972).

YOSHINO, Y., DEFEUDIS, F. V., ELLIOTT, K. A. C.: Omega-amino acids in rat brain. Canad. J. Biochem. **48**, 147–148 (1970).

YUDILEVICH, D. L., DE ROSE, N., SEPÚLVEDA, F. V.: Facilitated transport of amino acids through the blood-brain barrier of the dog studied in a single capillary circulation. Brain Res. **44**, 569–578 (1972).

ZIEGLGÄNSBERGER, W., PUIL, E. A.: Tetrodotoxin interference of CNS excitation by glutamic acid. Nature (Lond.) New Biol. **239**, 204–205 (1972).

ZIEGLGÄNSBERGER, W., PUIL, E. A.: Actions of glutamic acid on spinal neurones. Exp. Brain Res. **17**, 35–49 (1973).

Namenverzeichnis/Author Index

Die *kursiven* Seitenzahlen beziehen sich auf das Literaturverzeichnis
Page numbers in *italics* refer to bibliography

Åberg, G., Eränkö, O. 9, *40*
Abraham, S., s. Bartley, J. 56, *86*
Abraham, S., s. Bartley, J.C. 65, *86*
Abraham, S., s. Bortz, W. 60, 65, 84, *86*
Abraham, S., s. Madsen, J. 56, *91*
Abraham, S., s. Matthes, K.J. 79, *92*
Abraham, S., s. Sabine, J.R. 64, *94*
Abraham, S., s. Smith, S. 59, 60, 64, 80, 81, 82, 84, *95*
Abrahamsson, H. 30, *40*
Adachi, K., s. Inoue, H. 59, *90*
Adams, P.R., Brown, D.A. 163, *165*
Adolph, A.R. 132, *165*
Aghajanian, G.K., Haigler, H.J., Bloom, F.E. 144, *165*
Agrawal, H.C., Davis, J.M., Himwich, W.A. 113, *165*
Agrawal, H.C., Davison, A.N., Kaczmarek, L.K. 114, 120, *165*
Agrawal, H.C., s. Kaczmarek, L.K. 122, *177*
Ahlquist, R.P. 18, *40*
Ahmad, F., Jacobson, B., Wood, H.G. 68, *85*
Ahmad, F., s. Jacobson, B. 68, *90*
Ahmed, A.I., s. Parry Jones, G. 110, *182*
Ailhaud, G.P., s. Vagelos, P.R. 76, 82, *96*
Ajmone-Marsan, C. 126, *165*
Akagami, H., s. Ohtsu, E. 71, 75, *93*
Akhtar, M., Bloxham, D.P. 83, *85*
Albers, R.W., Brady, R.O. 124, 127, 130, 135, 137, 152, *165*
Albers, R.W., s. Salvador, R.A. 124, 127, 130, 135, 137, 139, 152, *183*
Albert, A. 113, *165*
Alberts, A.W., Gordon, S.G., Vagelos, P.R. 76, *85*
Alberts, A.W., Nervi, A.M., Vagelos, P.R. 68, 76, *85*
Alberts, A.W., Vagelos, P.R. 68, 76, *86*
Alberts, A.W., s. Nervi, A.M. 68, 76, *93*
Alberts, A.W., s. Vagelos, P.R. 69, 72, 75, 76, 82, *96*
Alekseeva, V.I., s. Sverdlov, Yu.S. 161, *185*
Alexander, H.M. s. Johnston, G.A.R. 108, 149, 151, *177*
Allmann, D.W., Gibson, D.M. 60, 81, 82, 84, *86*
Allmann, D.W., Hubbard, D.D., Gibson, D.M. 60, 81, 84, *86*
Allmann, D.W., s. Gibson, D.M. 58, 82, *88*
Allmann, D.W., s. Hicks, S.E. 64, 82, *90*
Allred, J.B., Roehrig, K.L. 83, *86*
Altmann, H., Bruggencate, G.Ten, Sonnhof, U. 140, 144, *165*
Altmann, H., Steinberg, R., Bruggencate, G.Ten, Sonnhof, U. 140, 144, *165*
Alvarez, W.C. 2, 33, *40*
Amaldi, P., s. Levi, G. 110, *179*
Ambache, N., Freeman, M.A. 17, *40*
Ames, A., III, Pollen, D.A. 132, *165*
Ancill, R.J., s. Small, N.A. 115, *184*
Andersen, P., Curtis, D.R. 135, 136, *165*, *166*
Andersen, P., Eccles, J.C., Løyning, Y. 131, *166*
Andersen, P., Eccles, J.C., Løyning, Y., Voorhoeve, P.E. 129, 131, 136, *166*
Andersen, P., Eccles, J.C., Oshima, T., Schmidt, R.E. 146, *166*
Andersen, P., Etholm, B., Gordon, G. 146, *166*
Anderson, E.G., s. Bell, J.A. 160, *167*
Anderson, E.G., s. Haas, H.L. 115, 145, *175*
Anderson, E.G., s. Levy, R.A. 160, *179*
Anderson, H.K., s. Langley, J.N. 2, 3, 5, 22, 23, 28, 30, *47*
Anderson, M., s. Yoshida, M. 139, *188*
Andersson, R., Mohme-Lundholm, E. 19, 20, *40*
Andersson, S. 38, *40*
Anerbach 6
Angielski, S., Szutowicz, A. 77, *86*
Anker, H.S., s. Swanson, R.F. 76, *95*
Annison, E.F., Hill, K.J., Lewis, D. 58, *86*
Aprison, M.H. 109, 152, *166*
Aprison, M.H., Davidoff, R.A., Werman, R. 100, *166*
Aprison, M.H., McBride, W.J. 151, *166*
Aprison, M.H., Shank, R.P., Davidoff, R.A. 122, 128, 134, 142, 147, 151, *166*
Aprison, M.H., Werman, R. 99, *166*
Aprison, M.H., s. Davidoff, R.A. 116, 148, 149, 150, 151, 153, 157, 158, *171*

Aprison, M.H., s. Graham, L.T., Jr. 106, 117, 148, 149, *174*
Aprison, M.H., s. Johnson, J.L. 117, 121, 127, 135, 145, 148, 163, *177*
Aprison, M.H., s. Shank, R.P. 100, 107, 108, 114, 115, 122, 127, 128, 134, 135, 142, 147, 151, 153, *184*
Aprison, M.H., s. Tachiki, K.H. 101, 124, *185*
Aprison, M.H., s. Werman, R. 109, 115, 156, *187*
Arai, K. 36, *40*
Araki, T., s. Endo, K. 156, *173*
Argui, A., s. Snyder, S.H. 99, 101, 105, *184*
Arias, I.M., Doyle, D., Schimke, R.T. 64, *86*
Arinze, J.C., Mistry, S.P. 60, 84, *86*
Armstrong, D.G., Blaxter, K.L. 58, *86*
Arnfred, T., Herz, L. 118, *166*
Arregui, A., Logan, W.J., Bennett, J.P., Snyder, S.H. 109, 142, 150, 151, 153, *166*
Asanuma, H., s. Brooks, V.B. 126, *168*
Åström, A. 2, *40*
Atkinson, D.E., Walton, G.M. 59, *86*
Aumann, K.W., s. Youmans, W.B. 32, 34, *51*
Awapura, J. 113, *166*
Awapara, J., s. Peck, E.J. 114, *182*
Axelrod, J., s. Wolfe, D.E. 12, *51*
Azuma, H., s. Mori, J. 21, 22, *48*

Babkin, B.P. 26, *40*
Bachrach, W.H., s. Ivy, A.C. 32, *46*
Bădescu, A., s. Hauliča, I. 137, *175*
Badiani, G., s. Guidotti, A. 113, 123, 134, 135, *174*
Bailey, E., s. Lockwood, E.A. 59, 60, 80, 81, 84, *91*
Bailey, E., s. Taylor, C.B. 59, 84, *95*
Bak, I.J., s. Okada, Y. 123, 128, 130, 134, 137, 138, 142, 143, *182*
Bak, L.J., s. Kim, J.S. 138, 139, *178*
Baker, R., Mendel, D. 25, *40*
Baker, R., s. Llinas, B. 143, *179*
Baker, R., s. Precht, W. 142, 143, *183*
Baker, W.W., Kratky, M., Benedict, F. 131, *166*
Balázs, R., Machiyama, Y., Hammond, B.J., Julian, T., Richter, D. 111, *166*
Balázs, R., Patel, A.J., Richter, D. 117, *166*
Balcar, V.J., Johnston, G.A.R. 105, 106, 109, 115, 116, 118, 119, 120, 121, 150, 151, 153, 154, *166*
Baldes, E.J., s. Herrick, J.H. 35, 36, *45*
Ball, E.G., s. Flatt, J.P. 79, *88*
Ball, E.G., s. Kornacker, M.S. 58, 84, *90*
Ball, E.G., s. Wise, E.M. 80, 84, *96*
Ballard, F.J., Hanson, R.W. 58, 59, 84, *86*
Ballard, F.J., Hanson, R.W., Kronfeld, D.S., Raggi, F. 77, 86
Ballard, F.J., s. Hanson, R.W. 57, 58, *89*
Banerhee, G., s. Weber, G. 81, *96*
Banna, N.R., Jabbur, S.J. 146, 147, *166*
Banna, N.R., Naccache, A., Jabbur, S.J. 147, *166*
Baños, G., Daniel, P.M., Moorhouse, S.R., Pratt, O.E. 107, 114, 115, *166*
Barcroft, J., Florey, H. 36, *40*
Barden, R.E., Cleland, W.W. 71, *86*
Barden, R.E.; Fung, C.H., Utter, M.F., Scrutton, M.C. 68, *86*
Barden, R.E., s. Zahler, W.L. 71, *96*
Bardhanabaedya, S., s. Green, H.D. 25, *45*
Barker, J.L., Crayton, J.W., Nicoll, R.A. 137, *166*
Barker, J.L., Nicoll, R.A. 153, 160, *166*
Barker, J.L., s. Nicoll, R.A. 137, *181*
Barker, L.A., s. Whittaker, V.P. 101, *188*
Barrnett, R.J., s. Van Orden, L.S. 12, *51*
Barth, C.C., Sladek, M., Decker, K. 57, *86*
Bartley, J., Abraham, S., Chaikoff, I.L. 56, *86*
Bartley, J.C., Abraham, S. 65, *86*
Bartley, W., s. Taylor, C.B. 59, 84, *95*
Bartsch, G.E., s. Katz, J. 79, *90*
Basch, S., von 2, *40*
Battistin, L., Grynbaum, A., Lajtha, A. 106, 114, 121, 123, 127, 128, 134, *167*
Baum, H., s. Butterworth, P.H.W. 82, 84, *87*
Baumann, J.A. 5, *41*
Baumgarten, H.G. 11, 22, *41*
Baumgarten, H.G., Holstein, A.-F., Owman, Ch. 2, 12, 13, *41*
Baumgarten, H.G., Lange, W. 11, 21, 22, *41*
Baxter, C.F. 100, 110, 111, 115, *167*
Baxter, C.F., Tewari, S. 117, *167*
Baxter, C.F., Tewari, S., Raeburn, S. 111, *167*
Baxter, C.F., s. Graham, L.T., Jr. 131, *174*
Baxter, C.F., s. Shank, R.P. 108, 151, *184*
Bayliss, W.M. 33, *41*
Bayliss, W.M., Starling, E.H. 2, 27, *41*
Bean, J.W., s. Sidky, M. 27, *50*
Beani, L., Bianchi, C., Crema, A. 14, *41*
Beart, P.M., Curtis, D.R., Johnston, G.A.R. 110, *167*
Beart, P.M., Johnston, G.A.R. 105, 111, 112, 115, *167*

Beart, P.M., Johnston, G.A.R., Uhr, M.L. 110, *167*
Beart, P.M., s. Johnston, G.A.R. 109, 113, 159, *177*
Becker, H., s. Ratzenhofer, M. 11, *49*
Beenakkers, A.M.Th., Klingenberg, M. 57, *86*
Bell, J.A., Anderson, E.G. 160, *167*
Benedict, F., s. Baker, W.W. 131, *166*
Benjamin, A.M., Quastel, J.H. 102, *167*
Benjamin, W., s. Gellhorn, A. 64, *88*
Bennett, G.W., s. Edwardson, J.A. 137, *173*
Bennett, J.P., Jr., Logan, W.J., Snyder, S.H. 109, 118, 121, 122, *167*
Bennett, J.P., s. Arregui, A. 109, 142, 150, 151, 153, *166*
Bennett, J.P., s. Snyder, S.H. 99, 101, 105, *184*
Bennett, M.R., Burnstock, G., Holman, M.E. 14, *41*
Bennett, M.R., Rogers, D.C. 9, *41*
Benuck, M., Stern, F., Lajtha, A. 108, 151, *167*
Benzi, G., Berte, F., Crema, A., Frigo, G.M. 22, *41*
Bergeret, B., Chatagne, F. 115, 120, *167*
Berl, S., Clarke, D.D. 102, *167*
Berl, S., McMurtry, J.G. 120, 121, *167*
Berl, S., Nicklas, W.J., Clarke, D.D. 117, *167*
Berlin, C.M., Schimke, R.T. 66, *86*
Berry, K., s. Perry, T.L. 100, 115, 116, 121, 123, 127, 128, 134, 138, 139, 140, *182*
Berte, F., s. Benzi, G. 22, *41*
Bhaduri, A., Srere, P.A. 56, *86*
Bhaduri, A., s. Srere, P.A. 56, *95*
Bhagavan, H.N., Coursin, D.B., Stewart, C.N. 117, *167*
Bhagavan, H.N., s. Stewart, C.N. 117, *185*
Bhargava, K.P., Srivastava, R.K. 159, *167*
Bhargava, K.P., s. Srimal, R.C. 163, *185*
Bianchi, C., s. Beani, L. 14, *41*
Biscoe, T.J., Curtis, D.R. 126, 157, 158, *167*
Biscoe, T.J., Duggan, A.W., Lodge, D. 125, 126, 156, 157, 158, 159, *167*
Biscoe, T.J., Straughan, D.W. 130, 131, *167*
Biscoe, T.J., s. Duggan, A.W. 143, 144, *172*
Biscoe, T.J., s. Watkins, J.C. 118, *187*
Bisti, S., Iosif, G., Marchesi, G.F., Strata, P. 129, *167*
Blaho, G., s. Varro, V. 27, *51*
Blakemore, A.H., s. Smythe, S.McC. 36, *50*
Blasberg, R.G. 109, *167*
Blaxter, K.L., s. Armstrong, D.G. 58, *86*
Bloch, K., s. Vance, D. 83, *96*
Bloch, M., s. Waksman, A. 111, *187*
Block, M., s. De Groat, W.C. 163, *172*
Bloom, B., s. Foster, D.W. 79, *88*
Bloom, F.E. 99, *167*
Bloom, F.E., Costa, E., Salmoiraghi, G.C. 139, *168*
Bloom, F.E., s. Aghajanian, G.K. 144, *165*
Bloom, F.E., s. Iversen, L.L. 112, 124, 151, 153, *176*
Bloom, F.E., s. Siggins, G.R. 112, 129, *184*
Bloom, F.E., s. Sims, K.L. 111, *184*
Bloom, F.E., s. von Baumgarten, R. 133, *187*
Bloom, F.E., s. Van Orden, L.S. 12, *51*
Bloom, F.E., s. Woodward, D.J. 127, 129, *188*
Bloxham, D.P., s. Akhtar, M. 83, *85*
Boakes, R.J., Bradley, P.B., Briggs, I., Dray, A. 119, 144, *168*
Boehme, D.H., Fordice, M.W., Marks, N., Vogel, W. 123, 128, 134, 151, *168*
Bogdanski, D.F., s. Brodie, B.B. 2, *41*
Bonnycastle, D.D., s. Mirkin, B.L. 2, *48*
Bonomi, L., s. Brodie, B.B. 2, *41*
Bordeianu, A., s. Hauliča, I. 137, *175*
Bornschein, H., Heiss, W.D. 132, *168*
Borre, C., Geneser-Jensen, F.A. 130, *168*
Bortoff, A. 21, *41*
Bortz, W., Abraham, S., Chaikoff, I.L. 60, 65, 84, *86*
Bortz, W.M. 77, *86*
Bortz, W.M., Lynen, F. 65, 71, 77, *86*
Bortz, W.M., s. Numa, S. 65, 66, 71, 72, 75, *93*
Botha, G.S.M. 22, *41*
Boullin, D.J., Costa, E., Brodie, B.B. 16, *41*
Bovet, D., s. Longo, V.G. 158, *179*
Bowery, N.G., Brown, D.A. 162, 163, *168*
Bowman, R.H., s. Parmeggiani, A. 77, *93*
Bowman, W.C., Hall, M.T. 19, 20, *41*
Boyarsky, L.L., s. Wheeler, D.D. 150, *187*
Boyd, E.S., Meritt, D.A., Gardner, C. 147, *168*
Boyd, T.E., s. Carlson, A.J. 22, 34, *42*
Boyer, P.D., s. Kaziro, Y. 67, *90*
Braam-Houckgeest, J.P., van 2, *41*
Bradford, H.F. 102, 116, 118, *168*
Bradford, H.F., s. Edwardson, J.A. 137, *173*
Bradley, K., Easton, D.M., Eccles, J.C. 157, *168*
Bradley, P.B., s. Boakes, R.J. 119, 144, *168*
Bradley, R.M., s. Robinson, J.D. 83, *94*
Bradley, S.E. 23, *41*

Brady, R.O., s. Albers, R.W. 124, 127, 130, 135, 137, 152, *165*
Brady, R.O., s. Robinson, J.D. 83, *94*
Brand, K., s. Kather, H. 81, *90*
Bremer, J. 56, *86, 87*
Brendel, K., s. Bressler, R. 57, *87*
Bressler, R., Brendel, K. 57, *87*
Bricker, L.A., Levey, G.S. 83, *87*
Bricks, P.K., s. Reid, R.L. 58, *94*
Bridgers, W.F. 108, *168*
Briel, G., Neuhoff, V. 100, 153, *168*
Briggs, I. 127, *168*
Briggs, I., s. Boakes, R.J. 119, 144, *168*
Bright, E.M., s. Cannon, W.B. 28, 32, *42*
Brodal, A., Pompeiano, O., Walberg, A. 140, *168*
Broderick, D.S., Candland, K.L., North, J.A., Mangum, J.H. 108, *168*
Brodie, B.B., Bogdanski, D.F., Bonomi, L. 2, *41*
Brodie, B.B., s. Boullin, D.J. 16, *41*
Brodie, B.B., s. Schanker, L.S. 27, *50*
Brodsky, W.A., s. Wyssbrod, H.R. 104, *188*
Brody, J.F., DeFeudis, P.A., DeFeudis, F.V. 137, *168*
Brody, M.J., Diamond, J. 19, *41*
Brody, T.M., s. Diamond, J. 19, *43*
Bronstein, S.B., s. Weber, G. 81, *96*
Brooks, V.B., Asanuma, H. 126, *168*
Brooks, V.B., Curtis, D.R., Eccles, J.C. 159, *168*
Brooks, W.H., s. Wheeler, D.D. 150, *187*
Brown, D.A., s. Adams, P.R. 163, *165*
Brown, D.A., s. Bowery, N.G. 162, 163, *168*
Brown, D.A., s. Steward, E.G. 110, *185*
Brown, F.C., Gordon, P.H. 115, *168*
Brown, J., McLean, P. 59, 84, *87*
Brücke, F.T. van, Stern, F. 22, *41*
Bruggencate, G.Ten, Engberg, J. 109, 112, 141, 142, 156. 157, *168*
Bruggencate, G.Ten, Sonnhoff, U. 143, 144, *168*
Bruggencate, G.Ten, s. Altmann, H. 140, 144, *165*
Bruun, A., Ehinger, B. 131, 132, *168*
Buckler, R.T., s. Jarboe, C.H. 113, *177*
Bueding, E., Bülbring, E., Gercken, G., Hawkins, J.T., Kuriyama, H. 20, *41*
Bueding, E., Bülbring, E., Kuriyama, H., Gercken, G. 20, *41*
Bueding, E., Butcher, R.W., Hawkins, J., Timms, A.R., Sutherland, E.W. 19, 20, *41*
Bülbring, E. 20, *41*
Bülbring, E., Burn, J.H. 25, *41*
Bülbring, E., Casteels, R., Kuriyama, H. 20, *42*
Bülbring, E., Tomita, T. 14, 19, 20, *42*
Bülbring, E., s. Bueding, E. 20, *41*
Bunch, J.L. 1, 2, 28, *42*
Burke, J.P., s. Van Orden, L.S. 12, *51*
Burke, R.E., Fedina, L., Lundberg, A. 157, *169*
Burkhardt, D.A. 132, *169*
Burn, J.H., s. Bülbring, E. 25, *41*
Burns, G.P., Schenk, W.G. 35, 36, *42*
Burnstock, G. 12, 21, 40, *42*
Burnstock, G., Campbell, G., Rand, M.J. 16, *42*
Burnstock, G., Robinson, P.M. 12, *42*
Burnstock, G., s. Bennett, M.R. 14, *41*
Burnstock, G., s. Furness, J.B. 14, 18, 34, *44*
Burnstock, G., s. Read, J.B. 9, *49*
Burton, D.N., Collins, J.M., Kennan, A.L., Porter, J.W. 82, 84 *87*
Burton, D.N., Collins, J.M., Porter, J.W. 82, *87*
Buskirk, C., van, s. Kuntz, A. 30, *47*
Butcher, R.W., s. Bueding, E. 19, 20, *41*
Butcher, R.W., s. Robinson, G.A. 83, *94*
Butcher, R.W., s. Robison, G.A. 19, 20, *49*
Butterworth, P.H.W., Guchhait, R.B., Baum, H., Olson, E.B., Margolis, S.A., Porter, J.W. 82, 84, *87*

Caciappo, F., Pandolfo, L., Di Chiara, G. 111, *169*
Cajal, S., Ramon Y. 5, 11, *42*
Callan, D.A., s. Purpura, D.P. 121, *183*
Campbell, G., 16, *42*
Campbell, G., s. Burnstock, G. 16, 40, *42*
Campbell, W.R., s. Markowitz, J. 36, *48*
Candland, K.L., s. Broderick, D.S. 108, *168*
Cangiano, A., s. Cook, W.A., Jr. 160, *169*
Cannon, W.B. 28, *42*
Cannon, W.B., Murphy, F.T. 28, 36, *42*
Cannon, W.B., Newton, H.F., Bright, E.M., Menkin, V., Moore, R.M. 28, 32, *42*
Cannon, W.B., Rosenblueth, A. 2, *42*
Căpâlnă, S., s. Hauliča, I., 137, *175*
Capurso, L., Friedmann, C.A., Parks, A.G., 11, *42*
Carlson, A.J., Boyd, T.E., Pearcy, J.F. 22, 34, *42*
Carpenter, F.W. 5, *42*
Carter, S.H., s. Tews, J.K. 100, *186*
Casola, L., Di Matteo, G. 135, *169*
Casteels, R. 19, *42*
Casteels, R., Kuriyama, H. 19, *42*
Casteels, R., s. Bülbring, E. 20, *42*

Catchpole, B., s. Neely, J. 37, *48*
Cavanaugh, J.R. 107, *169*
Celander, O. 16, *42*
Celander, O., Folkow, B. 29, *42*
Chaikoff, I.L., s. Bartley, J. 56, *86*
Chaikoff, I.L., s. Bortz, W. 60, 65, 84, *86*
Chaikoff, I.L., s. Fitch, W.M. 80, 84, *88*
Chaikoff, I.L., s. Madsen, J. 56, *91*
Chaikoff, I.L., s. Matthes, K.J. 79, *92*
Chambers, G., Eldred, E., Eggett, C. 149, *169*
Chang, A.Y., Schneider, D.I. 80, 84, *87*
Chang, H.-C., Seidman, I., Teebor, G., Lane, M.D. 60, 78, 81, 84, *87*
Chang, H.C., s. Gregolin, C. 70, 73, 75, 76, *89*
Chang, H.C., s. Ryder, E. 70, 71, *94*
Chang, P.-Y., Hsu, F.-Y. 32, *42*
Chapman, J.B., McCance, I. 127, *169*
Chappell, J.B. 58, *87*
Chappell, J.B., Haarhoff, K.N. 58, *87*
Chappell, J.B., Robinson, B.H. 58, *87*
Chappell, J.B., s. Robinson, B.H. 57, *94*
Chatagne, F., s. Bergeret, B. 115, 120, *167*
Chen, G., Hauck, F.P. 158, *169*
Chisholm, E.M., s. Rudack, D. 81, 84, *94*
Christensen, H.N. 105, *169*
Chu, S. 132, *169*
Citters, R.L. von, s. Rushmer, R.F. 35, *49*
Clark, G.A. 25, *42*
Clark, W.G., s. Nakajima, T. 115, *181*
Clarke, D.D., s. Berl. S. 102, 117, *167*
Clarke, G., Hill, R.G. 125, *169*
Cleland, W.W., s. McClure, W.R. 68, *92*
Cleland, W.W., s. Zahler, W.L. 71, *96*
Clelang, W.W., s. Barden, R.E. 71, *86*
Cohen, M.I., Gootman, P.M. 29, *42*
Cohen, M.I., s. Gootman, P.M. 29, *45*
Cohen, S.R., Lajtha, A. 105, 111, 114, 116, 118, *169*
Coles, H., s. Williamson, J.R. 77, *96*
Collins, D.A., s. Hamilton, A.S. 34, *45*
Collins, G.G.S. 110, *169*
Collins, G.G.S., s. Straughan, D.W. 126, *185*
Collins, J.M., s. Burton, D.N. 82, 84, *87*
Connar, J.D. 139, *169*
Constantinescu, E., Florescu, D., Hateganu, D. 120, *169*
Cook, W.A., Jr., Cangiano, A. 160, *169*
Coombs, J.S., Eccles, J.C., Fatt, P. 157, *169*
Cooper, J.R., s. Nadler, J.V. 116, *181*
Cornick, L.R., s. Cutler, R.W.P. 152, *171*
Corrigan, J.J. 120, *169*
Costa, E., s. Bloom, F.E. 139, *168*
Costa, E., s. Boullin, D.J. 16, *41*
Costa,, M., Furness, J.B., 5, 6, 7, 14, 16, 22, *43*
Costa, M., Furness, J.B., Gabella, G. 5, *43*
Costa, M., Gabella, G. 5, 9, 11, 16, 21, 22, *43*
Costa, M., s. Furness, J.B. 4, 5, 9, 11, 14, 16, 21, 22, 30, *44*
Costa, M., s. Gabella, G. 9, 11, *44*
Côté, L.J., s. Fahn, S. 123, 124, 128, 134, 137, 138, 143, *173*
Cotman, C.W. 101, *169*
Cotman, C.W., Herschman, H., Taylor, D. 101, *169*
Coursin, D.B., s. Bhagavan, H.N. 117, *167*
Coursin, D.B., s. Stewart, C.N. 117, *185*
Courtade, D., Guyon, J.F. 2, *43*
Craig, M.C., Nepokroeff, C.M., Lakshmane, M.R., Porter, J.W. 82, *87*
Crawford, J.M. 116, 117, 122, *169*
Crawford, J.M., Curtis, D.R. 121, 125, *169*
Crawford, J.M., Curtis, D.R., Voorhoeve, P.E., Wilson, V.J. 126, 127, 129, 169, *170*
Crawford, J.M., s. Curtis, D.R. 99, *170*
Crayton, J.W., s. Barker, J.L. 137, *166*
Crema, A., Frigo, G.M., Lecchini, S. 17, *43*
Crema, A., s. Beani, L. 14, *41*
Crema, A., s. Benzi, G. 22, *41*
Cross, B.A., s. Moss, R.L. 137, *180*
Crout, J.R., s. Heimbach, D.M. 37, *45*
Crowcroft, P.J., Holman, M.E., Szurszewski, J.H. 31, *43*
Crowshaw, K., Jessup, S.J., Ramwell, P.W. 106, *170*
Csernay, L., s. Varro, V. 27, *51*
Csillik, B., Gerebtzoff, A.M., Kiss, J., Knyihár, E. 111, *170*
Cumming, J.F., s. Petras, J.M. 32, *49*
Curry, W.M., s. Swanson, R.F. 76, *95*
Curtis, D.R. 99, 103, 104, 118, 119, 154, 157, 158, *170*
Curtis, D.R., Crawford, J.M. 99, *170*
Curtis, D.R., Davis, R. 135, 136, *170*
Curtis, D.R., De Groat, W.C. 159, 161, *170*
Curtis, D.R., Duggan, A.W. 109, 119, 158, *170*
Curtis, D.R., Duggan, A.W., Felix, D. 141, *170*
Curtis, D.R., Duggan, A.W., Felix, D., Johnston, G.A.R. 109, 113, 114, 115, 158, 159, 160, *170*

Curtis, D.R., Duggan, A.W., Felix, D., Johnston, G.A.R., McLennan, H. 114, 115, 125, 129, 131, 136, 141, *170*
Curtis, D.R., Duggan, A.W., Felix, D., Johnston, G.A.R., Tebēcis, A.K., Watkins, J.C. 119, 122, 136, 154, 155, *170*
Curtis, D.R., Duggan, A.W., Johnston, G.A.R. 104, 105, 106, 109, 110, 118, 127, 153, 154, 155, 157, 158, *170*
Curtis, D.R., Felix, D. 126, 129, *170*
Curtis, D.R., Felix, D., Game, C.J.A., McCulloch, R.M. 112, 126, 129, 155, 161, *170*
Curtis, D.R., Felix, D., McLennan, H. 131, 159, *170*
Curtis, D.R., Game, C.J.A., Johnston, G.A.R., McCulloch, R.M., Maclachlan, R.M. 113, 126, 158, 159, *170*
Curtis, D.R., Game, C.J.A., McCulloch, R.M. 158
Curtis, D.R., Hösli, L., Johnston, G.A.R. 109, 110, 114, 115, 126, 156, 159, *170*
Curtis, D.R., Hösli, L., Johnston, G.A.R., Johnston, I.H. 109, 112, 115, 156, 157, 159, *171*
Curtis, D.R., Johnston, G.A.R. 97, 99, 103, 109, 113, 129, *171*
Curtis, D.R., Johnston, G.A.R., Game, C.J.A., McCulloch, R.M. 155, *171*
Curtis, D.R., Phillis, J.W., Watkins, J.C. 116, 118, 120, 153, 154, 156, 159, *171*
Curtis, D.R., Ryall, R.W. 153, 158, 160, *171*
Curtis, D.R., Tebēcis, A.K. 114, 136, *171*
Curtis, D.R., Watkins, J.C. 99, 105, 115, 118, 120, 121, 122, 125, 135, 153, 156, 162, *171*
Curtis, D.R., s. Andersen, P. 135, 136, *165*, *166*
Curtis, D.R., s. Beart, P.M. 110, *167*
Curtis, D.R., s. Biscoe, T.J. 126, 157, 158, *167*
Curtis, D.R., s. Brooks, V.B. 159, *168*
Curtis, D.R., s. Crawford, J.M. 121, 125, 126, 127, 129, *169*, *170*
Curtis, D.R., s. Johnston, G.A.R. 109, 113, 115, 159, 161, *177*
Curtis, D.R., s. Watkins, J.C. 118, *187*
Cutler, R.W.P., Hammerstad, J.P., Cornick, L.R., Murray, J.E. 152, *171*
Cutler, R.W.P., s. Hammerstad, J.P. 109, 152, *175*
Cutler, R.W.P., s. Murray, J.E. 107, *181*
Cutler, R.W.P., s. Snodgrass, S.R. 106, *184*
Cutting, R.A., Ochsner, A. 36, *48*

D'adamo, A.F., Haft, D.E. 56, *87*
Dahlen, J.V., Kennan, A.L., Porter, J.W. 81, *87*
Daikuhara, Y., Tsunemi, T., Takeda, Y. 57, *87*
Daikuhara, Y., s. Suzuki, F. 58, *95*
Dakshinamurti, K., Desjardins, P.R. 78, 84, *87*
Dale, H.H. 2, 25, *43*, 98, 99, 147, *171*
Daniel, P.M., s. Baños, G. 107, 114, 115, *166*
Danish, R., s. Williamson, J.R. 77, *96*
Datta, R.K., s. Marks, N. 147, *180*
David, V.C., Loring, M. 36, *43*
Davidoff, R.A. 113, 153, 158, 159, 160, *171*
Davidoff, R.A., Aprison, M.H. 158, *171*
Davidoff, R.A., Aprison, M.H., Werman, R. 157, *171*
Davidoff, R.A., Graham, L.T., Jr., Shank, R.P., Werman, R., Aprison, M.H. 116, 148, 149, 150, 151, 153, *171*
Davidoff, R.A., Silvey, G.E., Kobetz, S.A., Spira, H.M. 160, *171*
Davidoff, R.A., s. Aprison, M.H. 100, 122, 128, 134, 142, 147, 151, *166*
Davidoff, R.A., s. Werman, R. 109, 115, 156, *187*
Davidson, N., Reisine, H. 147, *171*
Davidson, N., Southwick, C.A.P. 146, *171*
Davies, J., Watkins, J.C. 119, 122, 146, *171*
Davies, L.P., Johnston, A.R. 108, 149, 151, *171*
Davis, J.M., s. Agrawal, H.C. 113, *165*
Davis, R., Huffman, R.D. 140, *171*
Davis, R., Vaughan, P.C. 140, *172*
Davis, R., s. Curtis, D.R. 135, 136, *170*
Davison, A.N., s. Agrawal, H.C. 114, 120, *165*
Davison, A.N., s. Kaczmarek, L.K. 114, 115, 122, 124, *177*, *178*
Davson, H. 107, *172*
Deal, C.P., Green, H.D. 25, *43*
Deal, C.P., s. Green, H.D. 25, *45*
Debakey, M., s. Ochsner, A. 28, *48*
Decker, K., s. Barth, C.C. 57, *86*
DeFeudis, F.V. 150, *172*
DeFeudis, F.V., Delgado, J.M.R., Roth, R.H. 124, 139, *172*
DeFeudis, F.V., s. Brody, J.F. 137, *168*
DeFeudis, F.V., s. Tachiki, K.H. 101, 124, *185*
DeFeudis, F.V., s. Yoshino, Y. 115, *188*
DeFeudis, P.A., s. Brody, J.F. 137, *168*
De Groat, W.C. 157, 158, 159, 162, 163, *172*
De Groat, W.C., Lalley, P.M., Block, M. 163, *172*
De Groat, W.C., Lalley, P.M., Saum, W.R. 162, *172*

De Groat, W.C., Ryall, R.W. 153, *172*
De Groat, W.C., s. Curtis, D.R. 159, 161, *170*
De Groat, W.C., s. Johnston, G.A.R. 159, 161, *177*
De Groat, W.C., s. Ryall, R.W. 156, *183*
Dehlinger, P., s. Tweto, J. 83, *95*
Dehlinger, P.J., Schimke, R.T. 83, *87*
Delgado, J.M.R., s. De Feudis, F.V. 124, 139, *172*
De Marchi, W.J., Johnston, G.A.R. 108, 149, 151, *172*
Denison, A.B., s. Green, H.D. 25, *45*
Dennison, M.E., s. Matus, A.I. 101, 109, 151, *180*
Derissen, J.L., Endeman, H.J., Peerdeman, A.F. 116, *172*
Dermietzel, R. 13, *43*
De Robertis, E. 101, 124, *172*
De Rose, N., s. Yudilevich, D.L. 107, *188*
Desjardins, P.R., s. Dakshinamurti, K. 78, 84, *87*
De Villiers, D.C., s. Dixit, P.K. 77, *87*
Devine, C.E., Simpson, F.O. 13, *43*
Dhawan, B.N., Sharma, J.N., Srimal, R.C. 156, 159, *172*
Diamant, S., Gorin, E., Shafrir, E. 59, 60, 80, 81, 82, 84, *87*
Diamond, J. 103, *172*
Diamond, J., Brody, T.M. 19, *43*
Diamond, J., s. Brody, M.J. 19, *41*
Diamond, S., s. Perry, T.L. 115, 116, 121, 123, 127, 128, 134, 138, 139, 140, *182*
Dice, J.F., Schimke, R.T. 83, *87*
Di Chiara, G., s. Caciappo, F. 111, *169*
Dickie, M.M., s. Ingalls, A.M. 59, *90*
Dickinson, W.L., s. Langley, J.N. 2, 3, *47*
Di Maria, A., s. Tebēcis, A.K. 136, 145, *186*
Di Matteo, G., s. Casola, L. 135, *169*
Dimroth, P., Guchhait, R.B., Stoll, E., Lane, M.D. 68, 76, *87*
Dixit, P.K., De Villiers, D.C., Lazarow, A. 77, *87*
Do Carmo, R.J., Leão, A.A.P. 121, 172
Dogiel, A.S. 5, *43*
Dolének, A., s. Kubiček, R. 135, *179*
Donaldson, W.E., Mueller, N.S., Mason, J.V. 60, *88*
Dorsey, J.A., Porter, J.W. 71, 83, *88*
Douglas, D.M., Mann, F.C. 36, *43*
Dowling, J.E. 131, *172*
Downman, C.B.B. 6, *43*
Doyle, D., s. Arias, I.M. 64, *86*
Doyle, D., s. Schimke, R.T. 62, 81, *94*
Doyon, M. 22, *43*
Drahota, Z., s. Hahn, P. 59, 84, *89*
Dray, A., s. Boakes, R.J. 119, 144, *168*
Dreifuss, J.J., Kelly, J.S. 137, *172*
Dreifuss, J.J., Kelly, J.S., Krnjevič, K. 125, *172*
Dreifuss, J.J., Matthews, E.K. 137, *172*
Dresel, P., Folkow, B., Wallentin, I. 24, 25, *43*
Dresel, P., Wallentin, I. 23, 24, 25, *43*
Drian, M.J., s. Sotelo, C. 129, *184*
Droz, B., s. Taxi, J. 7, 9, 12, *50*
Dudar, J.D. 130, *172*
Dugan, R.E., Porter, J.W. 83, *88*
Dugan, R.E., s. Porter, J.W. 82, *94*
Duggan, A.W. 153, *172*
Duggan, A.W., Johnston, G.A.R. 147, 148, 150, *172*
Duggan, A.W., Lodge, D., Biscoe, T.J. 143, 144, *172*
Duggan, A.W., McLennan, H. 136, *172*
Duggan, A.W., s. Biscoe, T.J. 125, 126, 156, 157, 158, 159, *167*
Duggan, A.W., s. Curtis, D.R. 104, 105, 106, 109, 110, 113, 114, 115, 118, 119, 122, 125, 127, 129, 131, 136, 141, 153, 154, 155, 157, 158, 159, 160, *170*
Dyball, R.E.J., s. Moss, R.L. 137, *180*

Easton, D.M., s. Bradley, K. 157, *168*
Eccles, J.C. 18, *43*, 103, 154, 156, 157, *172*
Eccles, J.C., Fatt, P., Koketsu, K. 157, 158, *173*
Eccles, J.C., Ito, M., Szentagothal, J. 129, 141, *173*
Eccles, J.C., Schmidt, R.F., Willis, W.D. 158, 160, *173*
Eccles, J.C., s. Andersen, P. 129, 131, 136, 146, *166*
Eccles, J.C., s. Bradley, K. 157, *168*
Eccles, J.C., s. Brooks, V.B. 159, *168*
Eccles, J.C., s. Coombs, J.S. 157, *169*
Edwards, D.A.W., Rowlands, E.N. 21, *43*
Edwards, J., s. Lane, M.D. 71, *91*
Edwards, J.B., s. Stoll, E. 69, *95*
Edwardson, J.A., Bennett, G.W., Bradford, H.F. 137, *173*
Eggett, C., s. Chambers, G. 149, *169*
Ehinger, B. 132, *173*
Ehinger, B., Falck, B. 132, *173*
Ehinger, B., s. Bruun, A. 131, 132, *168*
Eidelberg, E., s. Roberts, E. 99, *183*
Eldred, E., s. Chambers, G. 149, *169*
Ellaway, P.H. 157, *173*
Elliott, K.A.C., Jasper, H.H. 99, *173*
Elliott, K.A.C., Page, I.H., Quastel, J.H. 98, *173*
Elliott, K.A.C., Van Gelder, N.M. 111, *173*

Elliott, K.A.C., s. Gottesfeld, Z. 112, *174*
Elliott, K.A.C., s. Jasper, H.H. 121, 124, *177*
Elliott, K.A.C., s. Lovell, R.A. 134, 138, *179*
Elliott, K.A.C., s. Yoshino, Y. 115, *188*
Elliott, T.R. 2, 3, 22, 28, 35, *43*
Elwood, J.C., Morris, H.P. 64, *88*
Endeman, H.J., s. Derissen, J.L. 116, *172*
Endo, K., Araki, T. 156, *173*
Engberg, I., Thaller, A. 158, *173*
Engberg, J., s. Bruggencate, G.Ten 109, 112, 141, 142, 156, 157, *168*
Eränkö, O., s. Åberg, G. 9, *40*
Essex, H.E., s. Herrick, J.H. 35, 36, *45*
Etholm, B., s. Andersen, P. 146, *166*
Euler, U.S.von 2, *43*
Evans, M.H., s. Franz, D.N. 32, *44*
Eyerman, G.S., s. Lowe, I.P. 124, 127, 135, 139, *179*

Fahn, S., Côté, L.J. 123, 124, 128, 134, 137, 138, 143, *173*
Falck, B., s. Ehinger, B. 132, *173*
Fang, M., Lowenstein, J.M. 77, *88*
Fang, M., s. Watson, J.A. 57, *96*
Fara, J.W. 25, *43*
Fara, J.W., Ross, G. 24, *43*
Farley, I.J., s. Hockman, C.H. 139, *176*
Fastier, F.N. 113, *173*
Fatt, P. 158, *173*
Fatt, P., s. Coombs, J.S. 157, *169*
Fatt, P., s. Eccles, J.C. 157, 158, *173*
Faupel, R.P., Seitz, H.J., Tarnowski, W. 77, *88*
Fedina, L., s. Burke, R.E. 157, *169*
Fedinec, A.A., Shank, R.P. 159, 161, *173*
Feigin, R.D., s. Olney, J.W. 117, *182*
Felicioli, R.A., Gabrielli, F. 59, 84, *88*
Felix, D., McLennan, H. 132, 133, *173*
Felix, D., s. Curtis, D.R. 109, 112, 113, 114, 115, 119, 122, 125, 126, 129, 131, 136, 141, 154, 155, 158, 159, 160, 161, *170*
Felpel, L.P. 142, *173*
Feltz, P. 139, *173*
Fifkova, E., s. VanHarreveld, A. 121, 132, *186, 187*
Fine, J. 35, *44*
Finkleman, B. 2, *44*
Fitch, W.M., Chaikoff, I.L. 80, 84, *88*
Fitzpatrick, H.F., s. Smythe, S.McC. 36, *50*
Flatt, J.P., Ball, E.G. 79 *88*
Fletcher, J.E., s. Spector, A.A. 71, *95*
Florescu, D., s. Constantinescu, E. 120, *169*
Florey, E. 99, *173*
Florey, H., s. Barcroft, J. 36, *40*
Foerster, O., Gagel, O., Sheehan, D. 32, *44*
Folkow, B. 29, *44*
Folkow, B., Frost, J., Uvnäs, B. 25, *44*
Folkow, B., Lewis, D.H., Lundgren, O., Mellander, S., Wallentin, I. 23, 24, 25, *44*
Folkow, B., s. Celander, O. 29, *42*
Folkow, B., s. Dresel, P. 24, 25, *43*
Fonnum, F. 100, 116, 128, 130, 138, 140, 143, *173*
Fonnum, F., Storm-Mathisen, J., Walberg, F. 140, 141, *173*
Fonnum, F., Walberg, F. 100, 128, 141, *173*
Fonnum, F., s. Storm-Mathisen, J. 130, *185*
Fordice, M.W., s. Boehme, D.H. 123, 128, 134, 151, *168*
Formica, J.V. 56, *88*
Foster, D.O., s. Shrago, E. 80, 84, *94*
Foster, D.W., Bloom, B. 79, *88*
Foster, D.W., McWhorter, W.P. 72, *88*
Foster, D.W., Srere, P.A. 59, *88*
Foster, D.W., s. Srere, P.A. 59, *95*
Frank, K., s. Smith, T.G. 157, *184*
Franklin, D., s. Vatner, S.F. 35, *51*
Franklin, D.L., s. Rushmer, R.F. 35, *49*
Franz, D.N., Evans, M.H., Perl, E.R. 32, *44*
Freedland, R.A., s. Murad, S. 59, 80, 84, *92*
Freeman, M.A., s. Ambache, N. 17, *40*
Freinkel, N., s. Herrera, E. 77, *90*
Friedmann, C.A. 23, *44*
Friedmann, C.A., s. Capurso, L. 11, *42*
Frigo, G.M., s. Benzi, G. 22, *41*
Frigo, G.M., s. Crema, A. 17, *43*
Fritz, I.B. 56, *88*
Fritz, I.B., Hsu, M.P. 71, 72, *88*
Fritz, I.B., Yue, K.T.N. 56, *88*
Fritz, I.B., s. Halperin, M.L. 58, *89*
Frost, J., s. Folkow, B. 25, *44*
Fujii, Y., s. Semba, T. 27, *50*
Fujiwara, M., s. Mori, J. 21, 22, *48*
Fujiwara, M., s. Muryobayashi, T. 11, *48*
Fujimori, H., s. Kajimoto, Y. 113, *178*
Fukuda, J., Highstein, S.M., Ito, M. 142, *173*
Fukunishi, K., s. Inoue, H. 59, *90*
Fukunishi, K., s. Suzuki, F. 58, *95*
Fung, C.H., s. Barden, R.E. 68, *86*
Furchgott, R.F. 18, *44*
Furness, J.B. 5, 6, 9, 11, 14, 23, *44*
Furness, J.B., Burnstock, G. 14, 18, 34, *44*

Furness, J.B., Costa, M. 4, 5, 9, 11, 14, 16, 21, 22, 30, *44*
Furness, J.B., Marshall, J.M. 25, *44*
Furness, J.B., s. Costa, M. 5, 6, 7, 14, 16, 22, *43*

Gabella, G. 9, 12, 13, *44*
Gabella, G., Costa, M. 9, 11, *44*
Gabella, G., s. Costa, M. 5, 9, 11, 16, 21, 22, *43*
Gabrielli, F., s. Felicioli, R.A. 59, 84, *88*
Gage, I.M., s. Ochsner, A. 36, *48*
Gagel, O., s. Foerster, O. 32, *44*
Gaitonde, M.K. 113, 114, 115, 120, 122, *173*
Galindo, A. 146, *173*
Galindo, A., Krnjević, K., Schwartz, S. 145, 146, *174*
Game, C.J.A., s. Curtis, D.R. 112, 113, 126, 129, 155, 158, 159, 161, *170*, *171*
Game, C.J.A., s. Johnston, G.A.R. 109, 113, 159, *177*
Ganguly, J. 78, *88*
Gardner, C., s. Boyd, E.S. 147, *168*
Garland, P.B., s. Tubbs, P.K. 77, 83, *95*
Garrett, J.R., s. Howard, E.R. 9, 22, *46*
Garry, R.C. 1, 28, *44*
Gehuchten, V., v. Kölliker 5,
Gellhorn, A., Benjamin, W. 64, *88*
Geneser-Jensen, F.A., s. Borre, C. 130, *168*
Gerard, R.W., s. Sugar, O. 149, *185*
Gercken, G., s. Bueding, E. 20, *41*
Gerebtzoff, A.M., s. Csillik, B. 111, *170*
Geroch, M.E., s. Miller, A.L. 71, *92*
Gershon, M.D. 14, 16, 21, *44*
Gershon, M.D., s. Ross, L.L. 9, *49*
Gerwin, B.I., Jacobson, B., Wood, H.G. 68, *88*
Gerwin, B.I., s. Jacobson, B. 68, *90*
Gessi, T., Rabini, C., Volta, F. 131, *174*
Gfeller, E., Kuhar, M.J., Snyder, S.H. 139, *174*
Giacobino, J.-P. 83, *88*
Gibson, D.M., Hubbard, D.D. 81, 84, *89*
Gibson, D.M., Hicks, S.E., Allmann, D.W. 58, 82, *88*
Gibson, D.M., Lyons, R.T., Scott, D.F., Muto, Y. 58, 59, 81, 82, *89*
Gibson, D.M., Titchener, E.B., Wakil, S.J. 79, *89*
Gibson, D.M., s. Allmann, D.W. 60, 81, 82, 84, *86*
Gillespie, J.S. 9, 14, 16, *44*, *45*
Gillespie, J.S., Mackenna, B.R. 3, *45*
Gillespie, J.S., Maxwell, J.D. 21, 22, *45*
Ginsborg, B.L. 103, *174*
Girado, M., s. Purpura, D.P. 121, *183*
Giarman, N.J., s. Van Orden, L.S. 12, *51*
Gibson, D.M., s. Hicks, S.E. 64, 82, *90*
Gibson, D.M., s. Muto, Y. 60, 84, *92*
Gibson, D.M., s. Wakil, S.J. 66, *96*
Gjesdahl, P., s. Gjessing, L.R. 106, *174*
Gjessing, L.R., Gjesdahl, P., Shaastad, O. 106, *174*
Gjessing, L.R., Torvick, A. 123, 128, 138, *174*
Glagoleva, A.A. 116, 117, *174*
Globus, A., Lux, H.D., Schubert, P. 101, 156, *174*
Gogolak, G., s. Herz, A. 130, 131, *175*
Goldberg, I., s. Vance, D. 83, *96*
Goldberg, L.J. 143, *174*
Goldman, J.K., s. Wakil, S.J. 82, *96*
Goodchild, M., Neal, M.J. 131, *174*
Goodman, D.S. 71, *89*
Gootman, P.M., Cohen, M.I. 29, *45*
Goodridge, A.G. 59, 84, *89*
Goodridge, A.G., s. Silpananta, P. 81, 84, *94*
Gootman, P.M., s. Cohen, M.I. 29, *42*
Gordon, G., s. Andersen, P. 146, *166*
Gordon, P.H., s. Brown, F.C. 115, *168*
Gordon, S.G., s. Alberts, A.W. 76, *85*
Gorin, E., s. Diamant, S. 59, 60, 80, 81, 82, 84, *87*
Goto, T., Ringelmann, E., Riedel, B., Numa, S. 73, *89*
Goto, T., s. Numa, S. 69, 75, *93*
Gottesfeld, Z., Elliott, K.A.C. 112, *174*
Gottesfeld, Z., Kelly, J.S., Rayner, C.N. 153, *174*
Gottesfeld, Z., Kelly, J.S., Renaud, L.P. 125, *174*
Gottesfeld, Z., Kelly, J.S., Schon, F. 161, 162, *174*
Govier, W.C., Sugrue, M.F., Shore, P.A. 2, 3, *45*
Gozukara, E.M., s. Rudack, D. 81, 84, *94*
Graham, L.T. Jr. 131, *174*
Graham, L.T., Jr., Aprison, M.H. 149, *174*
Graham, L.T., Jr., Baxter, C.F., Lolley, R.N. 131, *174*
Graham, L.T., Jr., Pong, S.F. 132, *174*
Graham, L.T., Jr., Shank, R.P., Werman, R., Aprison, M.H. 106, 117, 148, *174*
Graham, L.T., Jr., s. Davidoff, R.A. 116, 148, 149, 150, 151, 153, *171*
Graham, L.T., Jr., s. Pong, S.F. 113, *182*
Granit, R. 160, *174*
Gravlin, L., s. Perry, T.L. 100, 115, 116, 121, 123, 127, 128, 138, 139, *182*
Grayson, J., Mendel, D. 2, 23, 29, *45*
Green, H.D., Deal, C.P., Bardhanabaedya, S., Denison, A.B. 25, *45*
Green, H.D., Kepchar, J.H. 23, *45*
Green, H.D., s. Deal, C.P. 25, *43*
Green, J.D., Mancia, M., von Baumgarten, R. 133, *174*

Green, J.D., s. von Baumgarten, R. 133, *187*
Green, N.M. 68, *89*
Greenbaum, A.L., s. McLean, P. 85, *92*
Greenbaum, A.L., s. Saggerson, E.D. 60, 82, 84, *94*
Greenberg, D.M., s. Hirsch, H. 107, *175*
Greenspan, M., Lowenstein, J.M. 69, *89*
Greenspan, M.D., Lowenstein, J.M. 69, 72, *89*
Gregolin, C., Ryder, E., Kleinschmidt, A.K., Warner, R.C., Lane, M.D. 69, 73, 74, 75, *89*
Gregolin, C., Ryder, E., Lane, M.D. 66, 69, 73, *89*
Gregolin, C., Ryder, E., Warner, R.S., Kleinschmidt, A.K., Chang, H.-C., Lane, M.D. 70, 73, 75, 76, *89*
Gregolin, C., Ryder, E., Warner, R.C., Kleinschmidt, A.K., Lane, M.D. 65, 73, 75, *89*
Gregolin, C., s. Ryder, E. 70, 71, *94*
Grillo, M.A., Palay, S.L. 13, *45*
Grim, E. 2, 23, *45*
Grim, E., Lindseth, E.O. 35, *45*
Grollman, A. 35, 36, *45*
Grossman, M.I., s. Ivy, A.C. 32, *46*
Grossman, M.I., s. Jacobson, E.D. 26, *46*
Grundfest, H. 99, *174*
Grundfest, H., s. Purpura, D.P. 121, *183*
Grynbaum, A., s. Battistin, L. 106, 114, 121, 123, 127, 128, 134, *167*
Guchhait, R.B., Moss, J., Sokolski, W., Lane, M.D. 68, 76, *89*
Guchhait, R.B., s. Butterworth, P.H.W. 82, 84, *87*
Guchhait, R.B., s. Dimroth, P. 68, 76, *87*
Guchhait, R.B., s. Polakis, S.E. 67, *94*
Guidotti, A., Badiani, G., Pepeu, G. 113, 123, 134, 135, *174*
Gumaa, K.A., s. McLean, P. 85, *92*
Gunn, M. 11, *45*
Gushchin, I.S., Kozhechkin, S.N., Sverdlov, Y.S. 159, *174*
Guynn, R.W., Veloso, D., Veech, R.I. 59, *89*
Guyon, J.F., s. Courtade, D. 2, *43*

Haarhoff, K.N., s. Chappell, J.B. 58, *87*
Haas, H.L., Anderson, E.G., Hösli, L. 115, 145, *175*
Haas, H.L., Hösli, L. 145, *175*
Haas, H.L., s. Tebecis, A.K. 145, *186*
Haber, B., Kuriyama, K., Roberts, E. 111, *175*
Haber, B., s. Kuriyama, K. 111, 127, 131, *179*
Hadžovič, S., s. Stern, P. 156, *185*
Haft, D.E., s. D'adamo, A.F. 56, *87*
Hagen, J. 113, *175*
Hager, H., Tafuri, W. 12, *45*
Hahn, P., Drahota, Z. 59, 84, *89*
Haigler, H.J., s. Aghajanian, G.K. 144, *165*
Håkanson, R., Lilja, B., Owman, C. 11, *45*
Haldeman, S., Huffman, R.D., Marshall, K.C., McLennan, H. 119 135, 136, *175*
Haldeman, S., McLennan, H. 116, 135, 136, 155, *175*
Hall, M.T., s. Bowman, W.C. 19, 20, *41*
Halperin, M.L., Robinson, B.H., Fritz, I.B. 58, *89*
Hamberger, A., s. Henn, F.A. 105, *175*
Hamberger, B., Norberg, K.-A. 4, *45*
Hamberger, B., s. Norberg, K.-A. 4, *48*
Hamilton, A.S., Collins, D.A., Oppenheimer, M.J. 34, *45*
Hamilton, J.G., s. Sullivan, A.C. 57, *95*
Hammerschlag, R., Weinreich, D. 99, *175*
Hammerstad, J.P., Murray, J.E., Cutler, R.W.P. 109, 152, *175*
Hammerstad, J.P., s. Cutler, R.W.P. 152, *171*
Hammond, B.J., s. Balázs, R. 111, *166*
Haney, H.F., s. Youmans, W.B. 34, *51*
Hanitzsch, R., s. Vorkel, W. 132, *187*
Hankin, H., s. Maragoudakis, M.E. 78, *92*
Hansen, S., s. Perry, T.L. 100, 115, 116, 121, 123, 127, 128, 138, 139, *182*
Hanson, R.W., Ballard, F.J. 57, 58, *89*
Hanson, R.W., s. Ballard, F.J. 58, 59, 77, 84, *86*
Harris, M., Hopkin, J.M., Neal, M.J. 112, *175*
Harvey, J.A., McIlwain, H. 119, *175*
Hashimoto, T., Iritani, N., Nakanishi, S., Numa, S. 66, 69, 76, *89*
Hashimoto, T., Isano, H., Iritani, N., Numa, S. 65, 66, 68, 69, 70, 71, 73, 76, 78, *89*
Hashimoto, T., Numa, S. 65, 66, 68, 69, 70, 76, 78, *89*
Hashimoto, T., s. Numa, S. 60, 61, 62, 66, 68, 69, 70, 73, 76, 78, *93*
Hass, H.L., s. Hösli, L. 109, 145, 153, *176*
Hass, L.F., s. Kaziro, Y. 67, *90*
Hassler, R., s. Kim, J.S. 138, 139, *178*
Hassler, R., s. Okada, Y. 123, 128, 130, 134, 137, 138, 139, 142, 143, *182*
Hateganu, D., s. Constantinescu, E. 120, *169*
Hauck, F.P., s. Chen, G. 158, *169*
Hauliča, I., Căpâlnă, S., Nestianu,V., Bordeianu, A., Bădescu, A. 137, *175*
Hawkins, J., s. Bueding, E. 19, 20, *41*
Hawkins, J.T., s. Bueding, E. 20, *41*

Hayashi, T. 99, 115, 121, *175*
Heimbach, D.M., Crout, J.R. 37, *45*
Heine, J.D., s. Shaw, R.K. 100, 114, 115, 119, 122, 127, 128, *184*
Heiss, W.D., s. Bornschein, H. 132, *168*
Heller, A., s. Margolis, R.K. 135, *180*
Helm, D. van der, s. Patterson, A.L. 68, *93*
Henderson, V.E., s. Welch, A.D. 113, *187*
Hendrick, J.L., Sallach, H.J. 108, *175*
Henle, J. 5, *45*
Henn, F.A., Hamberger, A. 105, *175*
Hennecke, H., Wiechert, P. 117, *175*
Henniger, G. 73, 74, *89*
Henniger, G., Numa, S. 73, 74, *89*
Henning, U., Lynen, F., Weiland, O., Neufeldt, I. 81, 84, *89*
Henrich, H., Lutz, J. 24, *45*
Henschen, A., Söderberg, U. 107, *175*
Herczeg, B., s. Williamson, J.R. 77, *96*
Herrera, E., Freinkel, N. 77, *90*
Herrick, J.H., Essex, H.E., Mann, F.C., Baldes, E.J. 35, 36, *45*
Herschman, H., s. Cotman, C.W. 101, *169*
Hertz, A.F. 35, 37, *45*
Herz, A., Gogolak, G. 130, 131, *175*
Herz, A., Nacimiento, A. 130, *175*
Herz, A., von Freytag-Loringhoven, Hj. 139, *175*
Herz, L., s. Arnfred, T. 118, *166*
Heymans, C., Neil, E. 29, *45*
Hicks, S.E., Allmann, D.W., Gibson, D.M. 64, 82, *90*
Hicks, S.E., s. Gibson, D.M. 58, 82, *88*
Higgins, C.B., s. Vatner, S.F. 35, *51*
Highstein, S.M., Ito, M., Tsuchiya, T. 142, 143, *175*
Highstein, S.M., s. Fukuda, J. 142, *173*
Highstein, S.M., s. Obata, K. 143, *181*
Hill, K.J., s. Annison, E.F. 58, *86*
Hill, R.G., Simmonds, M.A., Straughan, D.W. 113, 126, *175*
Hill, R.G. s. Clarke, G. 125, *169*
Hill, R.G., s. Straughan, D.W. 126, *185*
Himwich, W.A., s. Agrawal, H.C. 113, *165*
Hino, O., s. Shimada, M. 100, 119, 120, *184*
Hinrichsen, J., Ivy, A.C. 28, *45*
Hirota, K., s. Kajimoto, Y. 113, *178*
Hirsch, H., Greenberg, D.M. 107, *175*
Hirsch, H.E., Robins, E. 124, 127, *176*
Ho, O.L., s. Olney, J.W. 116, 117, 120, *182*
Hochheuser, W., Weiss, H., Wieland, O. 56, *90*
Hockman, C.H., Lloyd, K.G., Farley, I.J., Hornykiewicz, O. 139, *176*
Hökfelt, T. 12, *45*
Hökfelt, T., Jonsson, G., Ljungdahl, A. 129, *176*
Hökfelt, T., Ljungdahl, A. 100, 101, 105, 109, 124, 129, 131, 151, *176*
Hökfelt, T., s. Hösli, E. 152, 153, *176*
Hökfelt, T., s. Ljungdahl, A. 152, *179*
Hösli, E., Ljungdahl, A., Hökfelt, T., Hösli, L. 152, 153, *176*
Hösli, L., Hass, H.L. 109, 145, 153, *176*
Hösli, L., Tebēcis, A.K. 144, 145, *176*
Hösli, L., Tebēcis, A.K., Haas, H.L. 145, *176*
Hösli, L., s. Curtis, D.R. 109, 110, 112, 114, 115, 126, 156, 157, 159, *170*, *171*
Hösli, L., s. Haas, H.L. 115, 145, *175*
Hösli, L., s. Hösli, E. 152, 153, *176*
Hösli, L., s. Tebēcis, A.K. 145, *186*
Hoffer, B.J., s. Siggins, G.R. 112, 129, *184*
Hoffer, B.J., s. Woodward, D.J. 127, 129, *188*
Hoffmann, K.-P., Stone, J., Sherman, S.M. 135, *176*
Hogan, J.P., s. Reid, R.L. 58, *94*
Hogben, C.A.M., s. Schanker, L.S. 27, *50*
Holman, M.E., s. Bennett, M.R. 14, *41*
Holman, M.E., s. Crowcroft, P.J. 31, *43*
Holstein, A.-F., s. Baumgarten, H.G. 2, 12, 13, *41*
Holten, D., s. Rudack, D. 81, 84, *94*
Holton, J.B., s. Small, N.A. 115, *184*
Hongo, T., Ryall, R.W. 153, *176*
Honjin, R., Takahashi, A., Shimasaki, S. Maruyama, H. 7, 13, *45*
Hopkin, J., Neal, M.J. 109, 114, 118, 152, *176*
Hopkin, J.M., s. Harris, M. 112, *175*
Hornykiewicz, O., s. Hockman, C.H. 139, *176*.
Horton, R.W., s. Meldrum, B.S. 113, *180*
Hotz, G. 36, *46*
Howanitz, P.J., Levy, H.R. 59, 60, 84, *90*
Howard, E.R., Garrett, J.R. 9, 22, *46*
Hsu, F.-Y., s. Chang, P.-Y. 32, *42*
Hsu, M.P., s. Fritz, I.B. 71, 72, *88*
Hubbard, D.D., s. Allmann, D.W. 60, 81, 84, *86*
Hubbard, D.D., s. Gibson, D.M. 81, 84, *89*
Huffman, R.D., McFadin, L.S. 160, *176*
Huffman, R.D., Yim, G.K.W. 142, *176*
Huffman, R.D., s. Davis, R. 140, *171*
Huffman, R.D., s. Haldeman, S. 119, 135, 136, *175*
Huffman, R.D., s. McLennan, H. 135, *180*

Hukuhara, T., Nakayama, S., Nanba, R. 32, *46*
Hukuhara, T., Nakayama, S., Yamagami, M., Miyake, T. 32, *46*
Hultborn, H., Jankowska, E., Lindström, S. 158, *176*
Hultén, L. 29, 30, 34, *46*
Hultén, L., Jodal, M. 15, 25, *46*
Hultén, L., Jodal, M., Lundgren, O. 23, 25, *46*

Iggo, A., Vogt, M. 29, 30, *46*
Ijichi, H., s. Shimada, M. 123, 128, 134, 138, 139, 142, *184*
Ingalls, A.M., Dickie, M.M., Snell, G.D. 59, *90*
Ingelfinger, F.J. 28, *46*
Inoue, H., Lowenstein, J.M. 73, 76, *90*
Inoue, H., Suzuki, F., Fukunishi, K., Adachi, K., Takeda, Y. 59, *90*
Iosif, G., s. Bisti, S. 129, *167*
Iritani, N., Nakanishi, S., Numa, S. 75, 76, *90*
Iritani, N., s. Hashimoto, T. 65, 66, 68, 69, 70, 71, 73, 76, 78, *89*
Iritani, N., s. Numa, S. 60, 61, 62, 69, 76, 78, *93*
Iriki, M., s. Walther, O.-E. 35, *51*
Isano, H., s. Hashimoto, T. 65, 66, 68, 69, 70, 71, 73, 76, 78, *89*
Ishikawa, T., s. Tebēcis, A.K. 145, *186*
Ito, M., Udo, M., Mano, N. 145, *176*
Ito, M., Yoshida, M. 141, *176*
Ito, M., Yoshida, M., Obata, K., Kawai, N., Udo, M. 128, *176*
Ito, M., s. Eccles, J.C. 129, 141, *173*
Ito, M., s. Fukuda, J. 142, *173*
Ito, M., s. Highstein, S.M. 142, 143, *175*
Ito, M., s. Obata, K. 112, 141, *181*
Ito, S., Tani, M., Sato, S., Oda, M. 121, *176*
Iversen, L.L. 99, 110, 111, *176*
Iversen, L.L., Bloom, F.E. 112, 124, 151, 153, *176*
Iversen, L.L., Johnston, G.A.R. 109, 112, 115, 129, 151, 153, *176*
Iversen, L.L., Mitchell, J.F., Neal, M.J., Srinivasan, V. 124, *177*
Iversen, L.L., Mitchell, J.F., Srinivasan, V. 102, 112, 124, *177*
Iversen, L.L., Neal, M.J. 105, 111, *177*
Iversen, L.L., s. Johnston, G.A.R. 109, 114, 115, 122, 129, 142, 151, *177*
Iversen, L.L., s. Neal, M.J. 101, 112, 124, 132, *181*
Iversen, L.L., s. Schon, F. 129, *184*
Iversen, L.L., s. Snodgrass, S.R. 100, 112, 125, *184*
Ivy, A.C., Grossman, M.I., Bachrach, W.H. 32, *46*
Ivy, A.C., s. Hinrichsen, J. 28, *45*
Izquierdo, J.J., Koch, E. 28, *46*

Jabbur, S.J., s. Banna, N.R. 146, 147, *166*
Jacobowitz, C. 4, 5, 9, *46*
Jacobs, R., Kilburn, E., Majerus, R.W. 78, 79, *90*
Jacobs, R., s. Majerus, P.W. 64, *91*
Jacobs, M.W., s. Kuntz, A. 4, *47*
Jacobson, B., Gerwin, B.I., Ahmad, F., Waegell, P., Wood, H.G. 68, *90*
Jacobson, B., s. Ahmad, F. 68, *85*
Jacobson, B., s. Gerwin, B.I. 68, *88*
Jacobson, E.D. 26, *46*
Jacobson, E.D., Linford, R.H., Grossman, M.I. 26, *46*
Jacobson, E.D., s. Shehadeh, Z. 24, 27, *50*
Jankowska, E., Lindström, S. 158, *177*
Jankowska, E., Roberts, W.J. 157, 158, *177*
Jankowska, E., s. Hultborn, H. 158, *176*
Jansen, J., s. Walberg, F. 140, *187*
Jansson, G. 15, 16, 34, *46*
Jansson, G., Lisander, B. 15, 16, 28, 29, 34, *46*
Jansson, G., Lisander, B., Martinson, J. 15, 32, 34, *46*
Jansson, G., Martinson, J. 15, 21, 29, 32, 33, *46*
Jarboe, C.H., Porter, L.A., Buckler, R.T. 113, *177*
Jasper, H.H., Khan, R.T., Elliott, K.A.C. 121, 124, *177*
Jasper, H.H., Koyama, I. 106, 112, 114, 118, 121, 124, *177*
Jasper, H.H., s. Elliott, K.A.C. 99, *173*
Jasper, H.H., s. Koyama, I. 124, *178*
Jenkinson, D.H., Morton, I.K.M. 18, 19, *46*
Jessup, S.J., s. Crowshaw, K. 106, *170*
Job, C., Lundberg, A. 31, *46*
Jodal, M., s. Hultén, L. 15, 23, 25, *46*
Johansson, B., Jonsson, O., Ljung, B. 32, *46*
Johansson, B., Langston, J.B. 32, *46*
John, K., s. Spector, A.A. 71, *95*
Johnson, B.C., Sassoon, H.F. 80, 81, 84, *90*
Johnson, C.K., s. Patterson, A.L. 68, *93*
Johnson, E.S., Roberts, M.H.T., Straughan, D.W. 110, 125, 126, *177*
Johnson, J.L. 99, 117, *177*
Johnson, J.L., Aprison, M.H. 117, 121, 127, 135, 145, 148, 163, *177*
Johnson, P.C., s. Richardson, D.R. 25, *49*
Johnston, G.A.R. 106, 116, 117,148, *177*
Johnston, G.A.R., Beart, P.M., Curtis, D.R., Game, C.J.A., McCulloch, R.M., MacLachlan, R.M. 109, 113, 159, *177*

Johnston, G.A.R., Curtis, D.R. 113, 115, 159, *177*
Johnston, G.A.R., De Groat, W.C., Curtis, D.R. 159, 161, *177*
Johnston, G.A.R., Iversen, L.L. 109, 114, 115, 122, 129, 142, 151, *177*
Johnston, G.A.R., Vitali, M.V. 108, 151, *177*
Johnston, G.A.R., Vitali, M.V., Alexander, H.M. 108, 149, 151, *177*
Johnston, G.A.R., s. Balcar, V.J. 105, 106, 109, 115, 116, 118, 119, 120, 121, 150, 151, 153, 154, *166*
Johnston, G.A.R., s. Beart, P.M. 105, 110, 111, 112, 115, *167*
Johnston, G.A.R., s. Curtis, D.R. 104, 105, 106, 109, 110, 112, 113, 114, 115, 118, 119, 122, 125, 126, 127, 129, 131, 136, 141, 153, 154, 155, 156, 157, 158, 159, 160, *170*, *171*
Johnston, A.R., s. Davies, L.P. 108, 149, 151, *171*
Johnston, G.A.R., s. De Marchi, W.J. 108, 149, 151, *172*
Johnston, G.A.R., s. Duggan, A.W. 147, 148, 150, *172*
Johnston, G.A.R., s. Iversen, L.L. 109, 112, 115, 129, 151, 153, *176*
Johnston, I.H., s. Curtis, D.R. 97, 99, 103, 109, 112, 113, 115, 156, 157, 159, *171*
Jonsson, G., s. Hökfelt, T. 129, *176*
Jonsson, O., s. Johansson, B. 32, *46*
Jordan, C.C., Webster, R.A. 102, 109, 152, *177*
Joshi, V.C., s. Plate, C.A. 82, *94*
Jütting, G., s. Lynen, F. 67, 68, *91*
Julian, T., s. Balázs, R. 111, *166*
Jung, E., s. McGeer, P.L. 137, 139, *180*
Jung, I., s. Varro, V. 27, *51*
Kaczmarek, L.K., Agrawal, H.C., Davison, A.N. 122, *177*
Kaczmarek, L.K., Davison, A.N. 114, 115, 124, *178*
Kaczmarek, L.K., s. Agrawal, H.C. 114, 120, *165*
Kahlson, G., Rosengren, E. 115, *178*
Kajimoto, Y., Fujimori, H., Hirota, K. 113, *178*
Kakefuda, T., s. Kuriyama, K. 112, *179*
Kallen, R.G., Lowenstein, J.M. 65, *90*
Kanazawa, I., Miyata, Y., Toyokura, Y., Otsuka, M. 137, 138, *178*
Kandera, J., Levi, G., Lajtha, A. 115, 127, 128, *178*
Kang, E.S., s. Snodgrass, S.R. 106, *184*
Kano, M., s. Semba, T. 159, 161, *184*
Karstens, A.I., s. Youmans, W.B. 32, *51*
Karush, F., Sonenberg, M. 71, *90*
Kather, H., Rivera, M., Brand, K. 81, *90*
Katz, B. 18, *46*
Katz, J., Landau, B.R., Bartsch, G.E. 79, *90*
Katz, J., Rognstad, R. 79, *90*
Katz, J., s. Rognstad, R. 56, *94*
Kauzmann, W. 69, *90*
Kawai, N., s. Ito, M. 128, *176*
Kawamura, H., Provini, L. 127, 129, *178*
Kawamura, Y., s. Morimoto, T. 144, *180*
Kaziro, Y., Hass, L.F., Boyer, P.D., Ochoa, S. 67, *90*
Kaziro, Y., Ochoa, S. 66, *90*
Kee, R.D., Wells, J.N., Yim, G.K.M. 113, 142, *178*
Kellerth, J.-O. 158, 160, *178*
Kellerth, J.-O., Szumski, A.J. 160, *178*
Kelly, J.S., Krnjević, K. 109, 125, *178*
Kelly, J.S., Krnjevic, K., Morris, M.E., Yim, G.K.W. 125, *178*
Kelly, J.S., Renaud, L.P. 146, 147, *178*
Kelly, J.S., s. Dreifuss, J.J. 125, 137, *172*
Kelly, J.S., s. Gottesfeld, Z. 125, 153, 161, 162, *174*
Kennan, A.L., s. Burton, D.N. 82, 84, *87*
Kennan, A.L., s. Dahlen, J.V. 81, *87*
Kepchar, J.H., s. Green, H.D. 23, *45*
Kerr, D.I.B. 113,
Kewenter, J. 15, 29, 34, *46*
Khan, R.P., s. Pande, S.V. 80, 84, *93*
Khan, R.T., s. Jasper, H.H. 121, 124, *177*
Kidokoso, J., Kubota, K., Shuto, S., Sumino, R. 143, *178*
Kihara, T., s. Shimada, M. 100, 119, 120, 123, 128, 134, 138, 139, 142, *184*
Kilburn, E., s. Jacobs, R. 78, 79, *90*
Kilburn, E., s. Majerus, P.W. 60, 61, 62, 63, *91*
Kim, J.S., Bak, L.J., Hassler, R., Okada, Y. 138, 139, *178*
Kim, J.S., s. Okada, Y. 123, 128, 130, 134, 137, 138, 142, 143, *182*
Kim, T.S., Shulman, J., Levine, R.A. 19, *46*
King, E.J. 107, 110, 113, *178*
Kisch, B. 33, *46*
Kishida, K., Naka, K.I. 132, *178*
Kiss, J., s. Csillik, B. 111, *170*
Kleinschmidt, A.K., Moss, J., Lane, M.D. 73, *90*
Kleinschmidt, A.K., s. Gregolin, C. 65, 69, 70, 73, 74, 75, 76, *89*
Kleinschmidt, A.K., s. Moss, J. 65, 69, 72, 73, 75, *92*
Klethi, J., s. Pasantes-Morales, H. 131, *182*
Klingenberg, M. 58, *90*
Klingenberg, M., s. Beenakkers, A.M.Th. 57, *86*
Klingman, J.D., s. McBride, W.J. 162, 163, *180*
Kloster, M., s. Perry, T.L. 100, 115, 116, 121, 123, 127, 128, 138, 139, *182*

Knappe, J. 66, *90*
Knoll, J., Vizi, E.S. 14, *46*
Knappe, J., s. Lynen, F. 67, 68, *91*
Kneifel, H.P., s. McClure, W.R. 68, *92*
Knoll, J., s. Vizi, E.S. 14, *51*
Knyihár, E., s. Csillik, B. 111, *170*
Kobayashi, N., s. Oomura, Y. 137, *182*
Kobetz, S.A., s. Davidoff, R.A. 160, *171*
Koch, A.R. 63, *90*
Koch, E., s. Izquierdo, J.J. 28, *46*
Kock, N.G. 27, 32, 34, *47*
Kölliker, V., s. Gehuchten, V. 5,
Koeppe, R.E., s. Mushahwar, I.K. 117, *181*
Koizumi, K., Suda, I. 29, *47*
Koketsu, K., s. Eccles, J.C. 157, 158, *173*
Konishi, S., Otsuka, M. 153, *178*
Konishi, S., s. Otsuka, M. 153, *182*
Kooiman, M., s. VanHarreveld, A. 121, *187*
Korchak, H.M., Masoro, E.J. 60, 65, 71, *90*
Kornacker, M.S., Ball, E.G. 58, 84, *90*
Kornacker, M.S., Lowenstein, J.M. 56, 58, 59, 78, 84, *90, 91*
Kosterlitz, H.W. 3, 38, *47*
Koyama, I. 121, 123, *178*
Koyama, I., Jasper, H.H. 124, *178*
Koyama, I., s. Jasper, H.H. 106, 112, 114, 118, 121, 124, *177*
Kozhechkin, S.N., s. Gushchin, I.S. 159, *174*
Kozhechkin, S.N., s. Sverdlov, Yu.S. 161, *185*
Kratky, M., s. Baker, W.W. 131, *166*
Kravitz, E.A., Potter, D.D. 141, *178*
Krebs, H.A., Lowenstein, J.M. 56, *91*
Kremer, M., Wright, S. 28, *47*
Kreutzberg, G.W., s. Schubert, P. 101, *184*
Kříž, N., s. Vyklicky, L. 160, *187*
Krnjević, K. 99, 103, 121, 122, 125, *178*
Krnjević, K., Morris, M.E. 160, *178*
Krnjević, K., Phillis, J.W. 121, 125, *178*
Krnjević, K., Randić, M., Straughan, D.W. 126, *178*
Krnjević, K., Schwartz, S. 112, 118, 122, 125, *178, 179*
Krnjević, K., Whittaker, V.P. 120, *179*
Krnjević, K., s. Dreifuss, J.J. 125, *172*
Krnjević, K., s. Galindo, A. 145, 146, *174*
Krnjević, K., s. Kelly, J.S. 109, 125, *178*
Kronfeld, D.S., s. Ballard, F.J. 77, *86*
Kržalić, L., Mandić, V., Mihailovic, L. 134, *179*
Kubiček, R., Dolének, A. 135, *179*
Kubota, K., s. Kidokoso, J. 143, *178*
Kuhar, M.J., s. Gfeller, E. 139, *174*
Kuhar, M.J., s. Wofsey, A.R. 116, 118, 120, *188*
Kullmann, R., Schönung, W., Simon, E. 35, *47*
Kumar, S., s. Lin, C.Y. 82, *91*
Kumar, S., s. Nandedkar, A.K.N. 82, *93*
Kumar, S., s. Porter, J.W. 82, *94*
Kuno, M., Muneoka, A. 159, *179*
Kuntz, A. 2, 4, 30, 31, *47*
Kuntz, A., Buskirk, C. van 30, *47*
Kuntz, A., Jacobs, M.W. 4, *47*
Kuntz, A., Saccomanno, G. 30, 31, *47*
Kurimoto, K., s. Shimada, M. 100, 119, 120, 123, 128, 134, 138, 139, 142, *184*
Kuriyama, H., Osa, T., Toida, N. 14, *47*
Kuriyama, H., s. Bueding, E. 20, *41*
Kuriyama, H., s. Bülbring, E. 20, *42*
Kuriyama, H., s. Casteels, R. 19, *42*
Kuriyama, K., Haber, B., Roberts, E. 111, *179*
Kuriyama, K., Haber, B., Sisken, B., Roberts, E. 127, *179*
Kuriyama, K., Roberts, E., Kakefuda, T. 112, *179*
Kuriyama, K., Sisken, B., Haber, B., Roberts, E. 131, *179*
Kuriyama, K., s. Haber, B. 111, *175*
Kuriyama, K., s. Roberts, E. 110, 111, 112, *183*

La Bella, F., Vivian, S., Queen, G. 115, *179*
Lachance, J.-P., s. Lynen, F. 67, 68, *91*
Lacroix, E. 23, *47*
Lähdesmäki, P., Oja, S.S. 114, 124, *179*
Lajtha, A., s. Battistin, L. 106, 114, 121, 123, 127, 128, 134, *167*
Lajtha, A., s. Benuck, M. 108, 151, *167*
Lajtha, A., s. Cohen, S.R. 105, 111, 114, 116, 118, *169*
Lajtha, A., s. Kandera, J. 115, 127, 128, *178*
Lajtha, A., s. Marks, N. 147, *180*
Lajtha, A., s. Seta, K. 107, *184*
Lakshmane, M.R., s. Craig, M.C. 82, *87*
Lalley, P.M., s. De Groat, W.C. 162, 163, *172*
Lam, D.M.K. 131, *179*
Lam, D.M.K., Steinmann, L. 132, *179*
Landau, B.R., s. Katz, J. 79, *90*
Lane, M.D., Edwards, J., Stoll, E., Moss, J. 71, *91*
Lane, M.D., Moss, J. 72, *91*
Lane, M.D., s. Chang, H.-C. 60, 78, 81, 84, *87*
Lane, M.D., s. Dimroth, P. 68, 76, *87*
Lane, M.D., s. Gregolin, C. 65, 66, 69, 70, 73, 74, 75, 76, *89*

Lane, M.D., s. Guchhait, R.B. 68, 76, *89*
Lane, M.D., s. Kleinschmidt, A.K. 73, *90*
Lane, M.D., s. Moss, J. 65, 66, 69, 70, 72, 73, 75, *92*
Lane, M.D., s. Polakis, S.E. 67, *94*
Lane, M.D., s. Ryder, E. 70, 71, *94*
Lane, M.D., s. Stoll, E. 69, *95*
Lange, W., s. Baumgarten, H.G. 11, 21, 22, *41*
Langemann, H., s. Müller, P.B. 124, 135, 137, 139, *181*
Langley, J.N. 5, 23, *47*
Langley, J.N., Anderson, H.K. 2, 3, 5, 22, 23, 28, 30, *47*
Langley, J.N., Dickinson, W.L. 2, 3, *47*
Langston, J.B., s. Johansson, B. 32, *46*
Lardy, H.A., s. McClure, W.R. 68, *92*
Lardy, H.A., s. Shrago, E. 59, 80, 84, *94*
Lardy, H.A., s. Young, J.W. 80, 84, *96*
Larrabee, A.R., s. Tweto, J. 82, 83, *95*, *96*
Larrabee, A.R., s. Vagelos, P.R. 76, 82, *96*
Larson, M.D. 157, 159, *179*
La Villa, I. 5, *47*
Lazarow, A., s. Dixit, P.K. 77, *87*
Leão, A.A.P., s. Do Carmo, R.J. 121, *172*
Learmonth, J.R., Markowitz, J. 22, *47*
Lecchini, S., s. Crema, A. 17, *43*
Lee, C.Y. 14, *47*
Legge, K.F., Randić, M., Straughan, D.W. 130, 131, *179*
Legge, K.F., s. Straughan, D.W. 130, 131, *185*
Lembeck, F., Zetler, G. 153, *179*
Léránth, C., s. Ungváry, G. 31, *51*
Lesk, D., s. Perry, T.L. 100, 115, 116, 121, 123, 127, 128, 138, 139, *182*
Levey, G.S., s. Bricker, L.A. 83, *87*
Levi, G. 114, *179*
Levi, G., Amaldi, P., Morisi, G. 110, *179*
Levi, G., Raiteri, M. 112, *179*
Levi, G., s. Kandera, J. 115, 127, 128, *178*
Levine, R.A., s. Kim, T.S. 19, *46*
Levy, B., s. Wilkenfeld, B.E. 19, 20, *51*
Levy, H.R. 71, *91*
Levy, H.R., s. Howanitz, P.J. 59, 60, 84, *90*
Levy, H.R., s. Miller, A.L. 65, 71, 72, *92*
Levy, R.A., Anderson, E.G. 160, *179*
Lewis, D., s. Annison, E.F. 58, *86*
Lewis, D.H., s. Folkow, B. 23, 24, 25, *44*
Li, P.L. 11, *47*
Liberati, M., s. Tweto, J. 83, *96*
Liere, E.J., van, s. Pearcy, J.F. 30, *49*
Lilja, B., s. Håkanson, R. 11, *45*
Lin, C.Y., Kumar, S. 82, *91*
Linford, R.H., s. Jacobson, E.D. 26, *46*
Lindseth, E.O., s. Grim, E. 35, *45*
Lindström, S., s. Hultborn, H. 158, *176*
Lindström, S., s. Jankowska, E. 158, *177*
Lipman, F., s. Orrego, F. 117, *182*
Lisander, B., s. Jansson, G. 15, 16, 28, 29, 32, 34, *46*
Lister, J. 2, *47*
Ljung, B., s. Johansson, B. 32, *46*
Ljungdahl, A., Hökfelt, T. 152, *179*
Ljungdahl, A., s. Hökfelt, T. 100, 101, 105, 109, 124, 129, 131, 151, *176*
Ljungdahl, A., s. Hösli, E. 152, 153, *176*
Llinas, B., Baker, R. 143, *179*
Llinas, R. 18, *47*
Lloyd, K.G., s. Hockman, C.H. 139, *176*
Lockwood, E.A., Bailey, E., Taylor, C.B. 59, 60, 80, 81, 84, *91*
Lodge, D., s. Biscoe, T.J. 125, 126, 156, 157, 158, 159, *167*
Lodge, D., s. Duggan, A.W. 143, 144, *172*
Lodoen, F.V., s. Van Orden, L.S. 12, *51*
Logan, W.J., Snyder, S.H. 109, 114, 115, 116, 118, 150, 151, *179*
Logan, W.J., s. Arregui, A. 109, 142, 150, 151, 153, *166*
Logan, W.J., s. Bennett, J.P., Jr. 109, 118, 121, 122, *167*
Logan, W.J., s. Snyder, S.H. 99, 101, 105, *184*
Lolley, R.N., s. Graham, L.T., Jr. 131, *174*
Lolloy, R.N. 131, *179*
Longo, V.G., Pinto Corrado, A. 158, *179*
Longo, V.G., Silvestrini, B., Bovet, D. 158, *179*
Longo, V.G., s. Pinto Corrado, A. 158, *182*
Lorch, E., s. Lynen, F. 67, 68, *91*
Lorenzo, A.V., s. Snodgrass, S.R. 106, *184*
Loring, M., s. David, V.C. 36, *43*
Loughridge, J.S., s. M'Fadden, G.D.F. 3, 28, *48*
Lovell, R.A., Elliott, K.A.C. 134, 138, *179*
Lovell, R.A., s. Sze, P.Y. 111, *185*
Lowe, I.P., Robins, E., Eyerman, G.S. 124, 127, 135, 139, *179*
Lowenstein, J.M. 56, 57, 59, 78, 79, 84, *91*
Lowenstein, J.M., s. Fang, M. 77, *88*
Lowenstein, J.M., s. Greenspan, M. 69, *89*
Lowenstein, J.M., s. Greenspan, M.D. 69, 72, *89*
Lowenstein, J.M., s. Inoue, H. 73, 76, *90*

Lowenstein, J.M., s. Kallen, R.G. 65, *90*
Lowenstein, J.M., s. Krebs, H.A. 56, *91*
Lowenstein, J.M., s. Kornacker, M.S. 56, 58, 59, 78, 84, *90*, *91*
Lowenstein, J.M., s. Spencer, A.F. 56, 59, 77, 78, 84, *95*
Lowenstein, J.M., s. Watson, J.A. 57, *96*
Løyning, Y., s. Andersen, P. 129, 131, 136, *166*
Luck, J.M., s. Teresi, J.D. 71, *95*
Ludwig, C. 2, *47*
Lundberg, A., s. Burke, R.E. 157, *169*
Lundberg, A., s. Job, C. 31, *46*
Lundgren, O., s. Folkow, B. 23, 24, 25, *44*
Lundgren, O., s. Hultén, L. 23, 25, *46*
Lutz, J., s. Henrich, H. 24, *45*
Lux, H.D., s. Globus, A. 101, 156, *174*
Lux, H.D., s. Schubert, P. 101, *184*
Lynen, F. 54, 66, 77, 82, *91*
Lynen, F., Knappe, J., Lorch, E., Jütting, G., Ringelmann, E., Lachance, J.-P. 67, 68, *91*
Lynen, F., Matsuhashi, M., Numa, S., Schweizer, E. 65, 66, 68, 69, 72, 75, 77, *91*
Lynen, F., s. Bortz, W.M. 65, 71, 77, *86*
Lynen, F., s. Henning, U. 81, 84, *89*
Lynen, F., s. Matsuhashi, M. 65, 66, 68, 69, 72, 75, *92*
Lynen, F., s. Numa, S. 60, 65, 66, 67, 71, 72, 75, 78, 81, 84, *93*
Lynen, F., s. Wieland, O. 60, 84, *96*
Lyons, R.T., s. Gibson, D.M. 58, 59, 81, 82, *89*

Machado, A.B.M. 12, *47*
Machiyama, Y., s. Balázs, R. 111, *166*
Mackenna, B.R., s. Gillespie, J.S. 3, *45*
MacLachlan, R., Searle, P., Stephanson, A., Walsh, H. 165
MacLachlan, R.M., s. Curtis, D.R. 113, 158, 159, *170*
MacLachlan, R.M., s. Johnston, G.A.R. 109, 113, 159, *177*
MacLean, A.B. 37, *47*
Madsen, J., Abraham, S., Chaikoff, I.L. 56, *91*
Majerus, P.W., Jacobs, R., Smith, M.B. 64, *91*
Majerus, P.W., Kilburn, E. 60, 61, 62, 63, *91*
Majerus, P.W., Vagelos, P.R. 76, 82, *91*
Majerus, P.W., s. Jacobs, R. 78, 79, *90*
Majerus, P.W., s. Vagelos, P.R. 76, 82, *96*
Malmfors, T. 6, *47*
Mancia, M., s. Green, J.D. 133, *174*
Mancia, M., s. von Baumgarten, R. 133, *187*
Mandel, P., Mark, J. 120, *180*
Mandel, P., s. Pasantes-Morales, H. 131, *182*
Mandić, V., s. Kržalić, L. 134, *179*
Mangan, J.L., Whittaker, V.P. 120, 124, *180*
Mangum, J.H., s. Broderick, D.S. 108, *168*
Mann, F.C., s. Douglas, D.M. 36, *43*
Mann, F.C., s. Herrick, J.H. 35, 36, *45*
Mann, M., West, G.B. 2, *48*
Mano, N., s. Ito, M. 145, *176*
Maragoudakis, M.E. 78, *91*, *92*
Maragoudakis, M.E., Hankin, H. 78, *92*
Maragoudakis, M.E., Hankin, H., Wasvary, J.M. 78, *92*
Marchesi, G.F., s. Bisti, S. 129, *167*
Marcucci, F., s. Mussini, E. 120, *181*
Margolis, R.K., Heller, A., Moore, R.Y. 135, *180*
Margolis, S.A., s. Butterworth, P.H.W. 82, 84, *87*
Mark, J., s. Mandel, P. 120, *180*
Markowitz, J., Campbell, W.R. 36, *48*
Markowitz, J., s. Learmonth, J.R. 22, *47*
Marks, N. 108, *180*
Marks, N., Datta, R.K., Lajtha, A. 147, *180*
Marks, N., s. Boehme, D.H. 123, 128, 134, 151, *168*
Marshall, J.M., s. Furness, J.B. 25, *44*
Marshall, K.C., s. Haldeman, S. 119, 135, 136, *175*
Marshall, K.C., s. McLennan, H. 135, *180*
Martin, D.B., Vagelos, P.R. 65, 69, *92*
Martin, D.B., s. Vagelos, P.R. 69, 72, 75, *96*
Martinson, J., s. Jansson, G. 15, 21, 29, 32, 33, 34, *46*
Maruyama, H., s. Honjin, R. 7, 13, *45*
Maruyama, Y., s. Shimada, M. 123, 128, 134, 138, 139, 142, *184*
Masi, I., Paggi, P., Pocchiar, R., Toschi, G. 162, 163, *180*
Mason, J.V., s. Donaldson, W.E. 60, *88*
Masoro, E.J., s. Korchak, H.M. 60, 65, 71, *90*
Matsuda, T., s. Yugari, Y. 71, *96*
Matsuhashi, M., Matsuhashi, S., Lynen, F. 65, 66, 68, 69, 72, 75, *92*
Matsuhashi, M., Matsuhashi, S., Numa, S., Lynen, F. 65, *92*
Matsuhashi, M., s. Lynen, F. 65, 66, 68, 69, 72, 75, 77, *91*
Matsuhashi, M., s. Numa, S. 60, 78, 81, 84, *93*
Matsuhashi, S., s. Matsuhashi, M. 65, 66, 68, 69, 72, 75, *92*
Matsushita, A., Smith, C.M. 149, *180*
Matthes, K.J., Abraham, S., Chaikoff, I.L. 79, *92*
Matthews, E.K., s. Dreifuss, J.J. 137, *172*
Matthies, H., s. Popov, N. 132, *183*

Matus, A.I., Dennison, M.E. 101, 109, 151, *180*
Max, S.R., Scorpio, R.M., Purvis, J.L. 58, *92*
Maxwell, J.D., s. Gillespie, J.S. 21, 22, *45*
McBride, W.J., Klingman, J.D. 162, 163, *180*
McBride, W.J., s. Aprison, M.H. 151, *166*
McCance, I., Phillis, J.W. 127, *180*
McCance, I., Phillis, J.W., Tebēcis, A.K., Westerman, R.A. 135, *180*
McCance, I., s. Chapman, J.B. 127, *169*
McClure, W.R., Lardy, H.A., Cleland, W.W. 68, *92*
McClure, W.R., Lardy, H.A., Kneifel, H.P. 68, *92*
McClure, W.R., Lardy, H.A., Wagner, M., Cleland, W.W. 68, *92*
McCrea, E. d'A. 28, *48*
McCulloch, R.M., s. Curtis, D.R. 112, 113, 114, 126, 129, 155, 158, 159, 161, *170*, *171*
McCulloch, R.M., s. Johnston, G.A.R. 109, 113, 159, *177*
McDowall, R.J.S. 2, *48*
McFadin, L.S., s. Huffman, R.D. 160, *176*
McGeer, E.G., s. McGeer, P.L. 137, 139, *180*
McGeer, P.L., McGeer, E.G., Wada, J.A., Jung, E. 137, 139, *180*
McIlwain, H., s. Harvey, J.A. 119, *175*
McLean, P., Greenbaum, A.L., Gumaa, K.A. 85, *92*
McLean, P., s. Brown, J. 59, 84, *87*
McLennan, H. 122, 133, *180*
McLennan, H., Huffman, R.D., Marshall, K.C. 135, *180*
McLennan, H., Pascoe, J.E. 31, *48*
McLennan, H., York, D.H. 139, *180*
McLennan, H., s. Curtis, D.R. 114, 115, 125, 129, 131, 136, 141, *170*
McLennan, H., s. Duggan, A.W. 136, *172*
McLennan, H., s. Felix, D. 132, 133, *173*
McLennan, H., s. Haldeman, S. 116, 119, 135, 136, 155, *175*
McMurtry, J.G., s. Berl, S. 120, 121, *167*
McSwiney, B.A. 1, *48*
McWhorter, W.P., s. Foster, D.W. 72, *88*
Meister, A. 104, *180*
Meldrum, B.S., Horton, R.W. 113, *180*
Mellander, S., s. Folkow, B. 23, 24, 25, *44*
Mendel, D., s. Baker, R. 25, *40*
Mendel, D., s. Grayson, J. 2, 23, 29, *45*
Menkin, V., s. Cannon, W.B. 28, 32, *42*
Meritt, D.A., s. Boyd, E.S. 147, *168*
Meyer, M. 145, *180*
Meyer, M., s. Steiner, F.A. 145, *185*
M'Fadden, G.D.F., Loughridge, J.S., Milroy, T.H. 3, 28, *48*
Mihailovic, L., s. Kržalić, L. 134, *179*
Mildvan, A.S., Scrutton, M.C., Utter, M.F. 68, *92*
Miller, A.L., Geroch, M.E., Levy, H.R. 71, *92*
Miller, A.L., Levy, H.R. 65, 72, *92*
Miller, O.N., s. Sullivan, A.C. 57, *95*
Milroy, T.H., s. M'Fadden, G.D.F. 3, 28, *48*
Minkin, J.A., s. Patterson, A.L. 68, *93*
Mirkin, B.L., Bonnycastle, D.D. 2, *48*
Mistry, S.P., s. Arinze, J.C. 60, 84, *86*
Mitchell, G.A.G. 4, 22, *48*
Mitchell, J.F., Neal, M.J., Srinivasan, V. 114, 118, *180*
Mitchell, J.F., s. Iversen, L.L. 102, 112, 124, *177*
Mitchell, J.F., s. Roberts, P.J. 109, 152, *183*
Mitchell, Nancy 40
Mito, S., s. Nagasawa, J. 12, 13, *48*
Mitsuhashi, O., s. Vance, D. 83, *96*
Miyake, T., s. Hukuhara, T. 32, *46*
Miyata, Y., Obata, K., Tanaka, Y., Otsuka, M. 140, 143, *180*
Miyata, Y., Otsuka, M. 152, *180*
Miyata, Y., s. Kanazawa, I. 137, 138, *178*
Miyata, Y., s. Otsuka, M. 100, 124, 128, 140, 141, 143, 148, 152, 161, *182*
Mizuno, N., s. Sauerland, E.K. 143, *183*
Mohme-Lundholm, E., s. Anderson, R. 19, 20, *40*
Mok, C., s. Perry, T.L. 100, 115, 116, 121, 123, 127, 128, 134, 138, 139, 140, *182*
Monnier, M. 4, *48*
Moore, R.M. 28, *48*
Moore, R.M., s. Cannon, W.B. 28, 32, *42*
Moore, R.Y., s. Margolis, R.K. 135, *180*
Moore, S., s. Tallan, H.H. 100, *186*
Moorhouse, S.R., s. Baños, G. 107, 114, 115, *166*
Morgan, R., Vrbova, G., Wolstencroft, J.H. 135, *180*
Mori, J., Azuma, H., Fujiwara, M. 21, 22, *48*
Morimoto, T., Kawamura, Y. 144, *180*
Morimoto, T., Takata, M., Kawamura, Y. 144, *180*
Morin, G., Vial, J. 32, *48*
Morisi, G., s. Levi, G. 110, *179*
Morris, H.P., s. Elwood, J.C. 64, *88*
Morris, H.P., s. Sabine, J.R. 64, *94*
Morris, M.E., s. Kelly, J.S. 125, *178*
Morris, M.E., s. Krnjevic, K. 160, *178*
Morton, I.K.M., s. Jenkinson, D.H. 18, 19, *46*
Moss, J., Lane, M.D. 66, 70, 73, 75, *92*

Moss, J., Yamagishi, M., Kleinschmidt, A.K., Lane, M.D. 65, 69, 72, 73, 75, *92*
Moss, J., s. Guchhait, R.B. 68, 76, *89*
Moss, J., s. Kleinschmidt, A.K. 73, *90*
Moss, J., s. Lane, M.D. 71, 72, *91*
Moss, R.L., Dyball, R.E.J., Cross, B.A. 137, *180*
Moss, R.L., Urban, I., Cross, B.A. 137, *180*
Müller, L.R. 1, 5, *48*
Mueller, N.S., s. Donaldson, W.E. 60, *88*
Müller, O., s. Ratzenhofer, M. 11, *49*
Müller, P.B., Langemann, H. 124, 135, 137, 139, *181*
Mugnaini, E., Walberg, F. 140, *181*
Muneoka, A. 156, *181*
Muneoka, A., s. Kuno, M. 159, *179*
Munro, A.F. 22, *48*
Murad, S., Freedland, R.A. 59, 80, 84, *92*
Murakami, M., Ohtsu, K., Ohtsuka, T. 132, *181*
Murayama, S., Smith, C.M. 149, *181*
Murphy, F.T., s. Cannon, W.B. 28, 36, *42*
Murray, J.E., Cutler, R.W.P. 107, *181*
Murray, J.E., s. Cutler, R.W.P. 152, *171*
Murray, J.E., s. Hammerstad, J.P. 109, 152, *175*
Muryobayashi, T., Fujiwara, M., Shimamoto, K. 11, *48*
Mushahwar, I.K., Koeppe, R.E. 117, *181*
Mussini, E., Marcucci, E. 120, *181*
Muto, Y., Gibson, D.M. 60, 84, *92*
Muto, Y., s. Gibson, D.M. 58, 59, 81, 82, *89*

Naccache, A., s. Banna, N.R. 147, *166*
Nacimiento, A., s. Herz, A. 130, *175*
Nadler, J.V., Cooper, J.R. 116, *181*
Nagasawa, J., Mito, S. 12, 13, *48*
Nagata, Y., Yokoi, Y., Tsukada, Y. 162, 163, *181*
Nakajima, T., Wolfgram, F., Clark, W.G. 115, *181*
Naka, K.I., s. Kishida, K. 132, *178*
Nakamura, Y., Wu, C.Y. 143, *181*
Nakanishi, S., Numa, S. 60, 61, 62, 63, 73, 84, *92*
Nakanishi, S., s. Hashimoto, T. 66, 69, 76, *89*
Nakanishi, S., s. Iritani, N. 75, 76, *90*
Nakanishi, S., s. Numa, S., 60, 61, 62, 66, 68, 69, 70, 73, 76, 78, *93*
Nakayama, S., s. Hukuhara, T. 32, *46*
Nanba, R., s. Hukuhara, T. 32, *46*
Nandedkar, A.K.N., Kumar, S. 82, *93*
Nandedkar, A.K.N., Schirmer, E.W., Pynadath, T.I., Kumar, S. 82, *93*
Nathan, P.W., Smith, M.C. 151, *181*
Neal, M.J. 109, 122, 151, *181*
Neal, M.J., Iversen, L.L. 101, 112, 124, 132, *181*
Neal, M.J., Starr, M.S. 131, *181*
Neal, M.J., White, R.D. 131, *181*
Neal, M.J., s. Goodchild, M. 131, *174*
Neal, M.J., s. Harris, M. 112, *175*
Neal, M.J., s. Hopkin, J. 109, 114, 118, 152, *176*
Neal, M.J., s. Iversen, L.L. 105, 111, 124, *177*
Neal, M.J., s. Mitchell, J.F. 114, 118, *180*
Neal, M.J., s. Straughan, D.W. 126, *185*
Neely, J., Catchpole, B. 37, *48*
Neil, E., s. Heymans, C. 29, *45*
Nelson, E.E., s. Woods, G. 25, *51*
Nelson, V.E., s. Woods, G. 25, *51*
Némethy, G., Scheraga, H.A. 69, *93*
Nepokroeff, C.M., s. Craig, M.C. 82, *87*
Nervi, A.M., Alberts, A.W., Vagelos, P.R. 68, 76, *93*
Nervi, A.M., s. Alberts, A.W. 68, 76, *85*
Nestianu, V., s. Hauliča, I. 137, *175*
Neuberger, A. 117, *181*
Neufeldt, I., s. Henning, U. 81, 84, *89*
Neufeldt, I., s. Wieland, O. 60, 84, *96*
Neufeldt, I.E., s. Wieland, O. 60, 65, *96*
Neuhoff, V., s. Briel, G. 100, 153, *168*
Neumann, K., s. Takano, K. 156, *185*
Newsholme, E.A., s. Start, C. 77, *95*
Newton, H.F., s. Cannon, W.B. 28, 32, *42*
Nicklas, W.J., s. Berl, S. 117, *167*
Nicoll, R.A. 133, 134, 161, *181*
Nicoll, R.A., Barker, J.L. 137, *181*
Nicoll, R.A., s. Barker, J.L. 137, 153, 160, *166*
Nisimaru, N. 29, *48*
Nitsch-Hassler, C., s. Okada, Y. 123, 128, 130, 134, 137, 138, 142, 143, *182*
Noell, W.K. 132, *181*
Nomura, S., s. Vance, D. 83, *96*
Norberg, K.-A. 4, 5, 9, *48*
Norberg, K.-A., Hamberger, B. 4, *48*
Norberg, K.-A., s. Hamberger, B. 4, *45*
Nordlie, R.C., s. Shrago, E. 80, 84, *94*
North, J.A., s. Broderick, D.S. 108, *168*
Northrop, D.B. 68, *93*
Northrop, D.B., Wood, H.G. 68, *93*
Numa, S. 53, 73, *93*
Numa, S., Bortz, W.M., Lynen, F. 65, 66, 71, 72, 75, *93*

Numa, S., Goto, T., Ringelmann, E., Riedel, B. 69, 75, *93*
Numa, S., Hashimoto, T., Iritani, N. 73, *93*
Numa, S., Hashimoto, T., Nakanishi, S., Okazaki, T. 66, 68, 69, 70, 73, 76, *93*
Numa, S., Matsuhashi, M., Lynen, F. 60, 78, 81, 84, *93*
Numa, S., Nakanishi, S., Hashimoto, T., Iritani, N., Okazaki, T. 60, 61, 62, 66, 69, 76, 78, *93*
Numa, S., Nakanishi, S., Iritani, N. 60, 61, *93*
Numa, S., Ringelmann, E. 69, 72, 75, *93*
Numa, S., Ringelmann, E., Lynen, F. 65, 66, 67, 71, 72, 75, 78, *93*
Numa, S., Ringelmann, E., Riedel, B, 69, 73, 75, *93*
Numa, S., s. Goto, T. 73, *89*
Numa, S., s. Hashimoto, T. 65, 66, 68, 69, 70, 71, 73, 76, 78, *89*
Numa, S., s. Henninger, G. 73, 74, *89*
Numa, S., s. Iritani, N. 75, 76, *90*
Numa, S., s. Lynen, F. 65, 66, 68, 69, 72, 75, 77, *91*
Numa, S., s. Matsuhashi, M. 65, *92*
Numa, S., s. Nakanishi, S. 60, 61, 62, 63, 73, 84, *92*
Numa, S., s. Ohtsu, E. 71, 75, *93*
Numa, S., s. Okazaki, T. 73, *93*
Numa, S., s. Wieland, O. 60, 84, *96*

Obata, K. 99, 128, 142, 162, 163, *181*
Obata, K., Highstein, S.M. 143, *181*
Obata, K., Ito, M., Ochi, R., Sato, N. 112, 141, *181*
Obata, K., Takeda, K. 102, 112, 128, 142, *182*
Obata, K., Takeda, K., Shinozaki, H. 141, 142, *182*
Obata, K., s. Ito, M. 128, *176*
Obata, K., s. Miyata, Y. 140, 143, *180*
Obata, K., s. Otsuka, M. 124, 128, 140, 141, 148, 152, 161, *182*
Ochi, R., s. Obata, K. 112, 141, *181*
Ochoa, S., s. Kaziro, Y. 66, 67, *90*
Ochs, S., s. Phillis, J.W. 121, *182*
Ochsner, A., Debakey, M. 28, *48*
Ochsner, A., Gage, I.M. 36, *48*
Ochsner, A., Gage, I.M., Cutting, R.A. 36, *48*
Oda, M., s. Ito, S. 121, *176*
Ohkawa, H., Prosser, C.L. 14, 17, 18, *48*
Ohtsu, E., Akagami, H., Okazaki, T., Numa, S. 71, 75, *93*
Ohtsu, K., s. Murakami, M. 132, *181*
Ohtsuka, T., s. Murakami, M. 132, *181*
Oja, S.S., s. Lähdesmäki, P. 114, 124, *179*
Okada, Y., Hassler, R. 139, *182*
Okada, Y., Nitsch-Hassler, C., Kim, J.S., Bak, I.J., Hassler, R. 123, 128, 130, 134, 137, 138, 142, 143, *182*
Okada, Y., s. Kim, J.S. 138, 139, *178*
Okada, Y., s. Precht, W. 142, 143, *183*
Okaya, Y. 113, *182*
Okazaki, T., Numa, S. 73, *93*
Okazaki, T., s. Numa, S. 60, 61, 62, 66, 68, 69, 70, 73, 76, 78, *93*
Okazaki, T., s. Ohtsu, E. 71, 75, *93*
Okomoto, S. 121, *182*
Oldendorf, W.H. 107, *182*
Olivecrona, H. 28, 36, *48*
Oliver, A.P., s. Siggins, G.R. 112, 129, *184*
Oliver, A.P., s. von Baumgarten, R. 133, *187*
Oliver, A.P., s. Woodward, D.J. 129, *188*
Olney, J.W., Ho, O.L., Rhee, V. 116, 117, 120, *182*
Olney, J.W., Sharpe, L.G., Feigin, R.D. 117, *182*
Olney, J.W., s. Perez, V.J. 117, *182*
Olson, E.B., s. Butterworth, P.H.W. 82, 84, *87*
Omura, S., s. Vance, D. 83, *96*
Ono, M. 12, *48*
Ono, T., s. Oomura, Y. 137, *182*
Ono, T., s. Potter, R. 81, 84, *94*
Oomura, Y., Ooyama, H., Yamamoto, T., Ono, T., Kobayashi, N. 137, *182*
Ooyama, H., s. Oomura, Y. 137, *182*
Openchowski, T. 5, *48*
Oppenheimer, M.J., s. Hamilton, A.S. 34, *45*
Orden, L.S., s. van Orden, L.S. 12, *51*
Orrego, F., Lipmann, F. 117, *182*
Osa, T., s. Kuriyama, H. 14, *47*
Osborne, L.W., s. Silva, D.G. 9, 11, *50*
Ohsima, T., s. Andersen, P. 146, *166*
Otsuka, M. 99, *182*
Otsuka, M., Konishi, S., Takahashi, T. 153, *182*
Otsuka, M., Miyata, Y. 100, 143, 152, *182*
Otsuka, M., Obata, K., Miyata, Y., Tanaka, Y. 124, 128, 140, 141, 148, 152, 161, *182*
Otsuka, M., s. Kahazawa, I. 137, 138, *178*
Otsuka, M., s. Konishi, S. 153, *178*
Otsuka, M., s. Miyata, Y. 140, 143, 152, *180*
Ouchterlony 62,
Owman, Ch., s. Baumgarten, H.G. 2, 12, 13, *41*
Owman, C., s. Håkanson, R. 11, *45*

Page, I.H., s. Elliott, K.A.C. 98, *173*
Paggi, P., s. Masi, I. 162, 163, *180*
Palay, S.L., s. Grillo, M.A. 13, *45*

Palmieri, F., Stipani, I., Quagliariello, E. 58, *93*
Pande, S.V., Khan, R.P., Venkitasubramanian, T.A. 80, 84, *93*
Pandolfo, L., s. Caciappo, F. 111, *169*
Parks, A.G., s. Capurso, L. 11, *42*
Permeggiani, A., Bowman, R.H. 77, *93*
Parry Jones, G., Roberts, R.T., Ahmed, A.I. 110, *182*
Pasantes-Morales, H., Klethi, J., Urban, P.F., Mandel, P. 131, *182*
Pascoe, J.E., s. McLennan, H. 31, *48*
Patel, A.J., s. Balázs, R. 117, *166*
Paton, W.D.M., Vizi, E.S. 18, *48*
Paton, W.D.M., Vizi, E.S., Zar, M.A. 14, *49*
Paton, W.D.M., Zar, M.A. 14, 17, *49*
Patrick, T., s. Vatner, S.F. 35, *51*
Patterson, A.L., Johnson, C.K., Helm, D. van der, Minkin, J.A. 68, *93*
Pearce, J. 60, 84, *93*
Pearcy, J.F., Liere, E.J. van 30, *49*
Pearcy, J.F., s. Carlson, A.J. 22, 34, *42*
Peck, E.J., Awapara, J. 114, *182*
Peerdeman, A.F., s. Derissen, J.L. 116, *172*
Pepeu, G., s. Guidotti, A. 113, 123, 134, 135, *174*
Perez, V.J., Olney, J.W. 117, *182*
Perl, E.R., s. Franz, D.N. 32, *44*
Perry, T.L., Berry, K., Diamond, S., Mok, C. 115, 116, 121, 123, 127, 128, 134, 138, 139, 140, *182*
Perry, T.L., Hansen, S., Berry, K., Mok, C., Lesk, D. 100, 116, 127, 128, 139, *182*
Perry, T.L., Sanders, H.D., Hansen, S., Lesk, D., Kloster, M., Gravlin, L. 100, 115, 116, 121, 123, 127, 128, 138, 139, *182*
Persson, C.G.A. 21, 22, *49*
Perwein, J., s. Straschill, M. 132, *185*
Petras, J.M., Cumming, J.F. 32, *49*
Petri, G., Szenohradszky, J., Pórszász-Gibiszer, K. 37, *49*
Pflüger, E. 2, *49*
Phillis, J.W., Ochs, S. 121, *182*
Phillis, J.W., Tebēcis, A.K., York, D.H. 115, *182*
Phillis, J.W., s. Curtis, D.R. 116, 118, 120, 153, 154, 156, *171*
Phillis, J.W., s. Krnjević, K. 121, 125, *178*
Phillis, J.W., s. McCance, I. 127, 135, *180*
Phillis, J.W., s. Tebēcis, A.K. 153, 160, *186*
Pick, J. 12, 13, *49*
Piercey, M.F., s. Ryall, R.W. 159, *183*
Pinto Corrado, A., Longo, V.G. 158, *182*
Pinto Corrado, A., s. Longo, V.G. 158, *179*
Pitts, F.N., s. Sheridan, J.J. 139, *184*
Plate, C.A., Joshi, V.C., Sedgwick, B., Wakil, S.J. 82, *94*
Player, R., s. Steward, E.G. 110, *185*
Pocchiar, F., s. Masi, I. 162, 163, *180*
Pogell, B.M., s. Taketa, K. 71, *95*
Pohle, W., s. Popov, N. 132, *183*
Polakis, S.E., Guchhait, R.B., Lane, M.D. 67, *94*
Pollen, D.A., s. Ames, A. III 132, *165*
Polosa, C., s. Ryall, R.W. 159, *183*
Pomeranz, B., Wall, P.D., Weber, W.V. 32, *49*
Pompeiano, O., s. Brodal, A. 140, *168*
Pong, S.F., Graham, L.T., Jr. 113, *182*
Pong, S.F., s. Graham, L.T., Jr. 132, *174*
Ponnuswamy, P.K., Sasisekharan, V. 107, *183*
Popov, N., Pohle, W., Rösler, V., Matthies, H. 132, *183*
Pórszász-Gibiszer, K., s. Petri, G. 37, *49*
Porter, J.W., Kumar, S., Dugan, R.E. 82, *94*
Porter, J.W., s. Burton, D.N. 82, 84, *87*
Porter, J.W., s. Butterworth, P.H.W. 82, 84, *87*
Porter, J.W., s. Craig, M.C. 82, *87*
Porter, J.W., s. Dahlen, J.V. 81, *87*
Porter, J.W., s. Dorsey, J.A. 71, 83, *88*
Porter, J.W., s. Dugan, R.E. 83, *88*
Porter, L.A., s. Jarboe, C.H. 113, *177*
Porter, L. 113, *183*
Potter, D.D., s. Kravitz, E.A. 141, *178*
Potter, L.T., s. Wolfe, D.E. 12, *51*
Potter, R., Ono, T. 81, 84, *94*
Pratt, O.E., s. Banos, G. 107, 114, 115, *166*
Precht, W., Baker, R. 143, *183*
Precht, W., Baker, R., Okada, Y. 142, 143, *183*
Precht, W., Baker, R., Yoshida, M. 139, *183*
Precht, W., s. Yoshida, M. 139, *188*
Price, W.E., s. Shehadeh, Z. 24, 27, *50*
Pringle, M.J., s. Steward, E.G. 110, *185*
Privat, A., s. Sotelo, C. 129, *184*
Prosser, C.L., s. Ohkawa, H. 14, 17, 18, *48*
Provini, L., s. Kawamura, H. 127, 129, *178*
Pull, E.A., s. Zieglgänsberger, W. 118, 119, 154, *188*
Purpura, D.P., Girado, M., Smith, T.G., Callan, D.A., Grundfest, H. 121, *183*
Purvis, J.L., s. Max, S.R. 58, *92*
Pynadath, T.I., s. Nandedkar, A.K.N. 82, *93*

Quagliariello, E., s. Palmieri, F. 58, *93*
Quastel, J.H., s. Benjamin, A.M. 102, *167*
Quastel, J.H., s. Elliott, K.A.C. 98, *173*
Queen, G., s. LaBella, F. 115, *179*
Quilliam, J.P., s. Steward, E.G. 110, *185*

Rabin, A., s. Yoshida, M. 139, *188*
Rabini, C., s. Gessi, T. 131, *174*
Raeburn, S., s. Baxter, C.F. 111, *167*
Raggi, F., s. Ballard, F.J. 77, *86*
Raiteri, M., s. Levi, G. 112, *179*
Ramwell, P.W., Shaw, J.E. 113, *183*
Ramwell, P.W., s. Crowshaw, K. 106, *170*
Rand, M.J., s. Burnstock, G. 16, *42*
Randić, M., s. Krnjević, K. 126, *178*
Randić, M., s. Legge, K.F. 130, 131, *179*
Ratzenhofer, M., Müller, O., Becker, H. 11, *49*
Rayner, C.N., s. Gottesfeld, Z. 153, *174*
Read, J.B., Burnstock, G. 9. *49*
Reed, J.D., Sanders, D.J. 16, *49*
Reed, J.D., Sanders, D.J., Thorpe, V. 26, *49*
Reid, R.L., Hogan, J.P., Bricks, P.K. 58, *94*
Reisine, H., s. Davidson, N. 147, *171*
Renaud, L.P., s. Gottesfeld, Z. 125, *174*
Renaud, L.P., s. Kelly, J.S. 146, 147, *178*
Renold, A.E., s. Winegrad, A.I. 79, *96*
Rhee, V., s. Olney, J.W. 116, 117, 120, *182*
Richardson, D.R., Johnson, P.C. 25, *49*
Richardson, D.R., Zweifach, B.W. 25, *49*
Richardson, K.C. 12, *49*
Richardson, K.C., s. Wolfe, D.E. 12, *51*
Richter, D., s. Balázs, R. 111, 117, *166*
Richter, J.J., Wainer, A. 107, 114, 115, *183*
Riedel, B., s. Goto, T. 73, *89*
Riedel, B., s. Numa, S. 69, 73, 75, *93*
Ringelmann, E., s. Goto, T. 73, *89*
Ringelmann, E., s. Lynen, F. 67, 68, *91*
Ringelmann, E., s. Numa, S. 65, 66, 67, 69, 71, 72, 73, 75, 78, *93*
Rintoul, J.R. 11, *49*
Rivera, M., s. Kather, H. 81, *90*
Roa, P.D., s. Tews, J.K. 100, *186*
Roberts, E. 98, 99, 122, *183*
Roberts, E., s. Scholes 132
Roberts, E., Eidelberg, E. 99, *183*
Roberts, E., Kuriyama, K. 110, 111, 112, *183*
Roberts, E., s. Haber, B. 111, *175*
Roberts, E., s. Kuriyama, K. 111, 112, 127, 131, *179*
Roberts, E., s. Tunnicliff, G. 115, *186*
Roberts, E., s. Waksman, A. 111, *187*
Roberts, E., s. Weinstein, H. 112, *187*
Roberts, M.H.T., s. Johnson, E.S. 110, 125, 126, *177*
Roberts, P.J., Mitchell, J.F. 109, 152, *183*
Roberts, R.T., s. Parry Jones, G. 110, *182*
Roberts, W.J., s. Jankowska, E. 157, 158, *177*
Robins, E., s. Hirsch, H.E. 124, 127, *176*
Robins, E., s. Lowe, I.P. 124, 127, 135, 139, *179*
Robinson, B.H., Cappell, J.B. 57, *94*
Robinson, B.H., Williams, G.R. 57, *94*
Robinson, B.H., s. Chappell, J.B. 58, *87*
Robinson, B.H., s. Halpern, M.L. 58, *89*
Robinson, G.A., Butcher, R.W., Sutherland, E.W. 83, *94*
Robinson, J.D., Brady, R.O., Bradley, R.M. 83, *94*
Robinson, P.M., s. Burnstock, G. 12, *42*
Robison, G.A., Butcher, R.W., Sutherland, E.W. 19, 20, *49*
Roehrig, K.L., s. Allred, J.B. 83, *86*
Rösler, V., s. Popov, N. 132, *183*
Rogers, D.C., s. Bennett, M.R. 9, *41*
Rognstad, R., Katz, J. 56, *94*
Rognstad, R., s. Katz, J. 79, *90*
Rosenblueth, A., s. Cannon, W.B. 2, *42*
Rosengren, E., s. Kahlson, G. 115, *178*
Ross, G. 24, 25, *49*
Ross, G., s. Fara, J.W. 24, *43*
Ross, G., s. Silva, D.G. 9, 11, *50*
Ross, J.G. 31, *49*
Ross, L.L., Gershon, M.D. 9, *49*
Roth, R.H., s. DeFeudis, F.V. 124, 139, *172*
Rous, S. 57, *94*
Rowlands, E.N., s. Edwards, D.A.W. 21, *43*
Rubin, R.P. 102, *183*
Rudack, D., Chisholm, E.M., Holten, D. 81, 84, *94*
Rudack, D., Gozukara, E.M. Chisholm, E.M., Holten, D. 81, 84, *94*
Ruf, K., s. Steiner, F.A. 130, 131, *185*
Rushmer, R.F., Franklin, D.L., Citters, R.L. von, Smith, O.A. 35, *49*
Ryall, R.W. 120, 124, 158, *183*
Ryall, R.W., DeGroat, W.C. 156, *183*
Ryall, R.W., Piercey, M.F., Polosa, C. 159, *183*
Ryall, R.W., s. Curtis, D.R. 153, 158, 160, *171*
Ryall, R.W., s. DeGroat, W.C. 153, *172*

Ryall, R.W., s. Hongo, T. 153, *176*
Ryder, E. 60, 79, 84, *94*
Ryder, E., Gregolin, C., Chang, H.C., Lane, M.D. 70, 71, *94*
Ryder, E., s. Gregolin, C. 65, 66, 69, 70, 73, 74, 75, 76, *89*
Ryder, E., s. Stoll, E. 69, *95*

Sabine, J.R., Abraham, S., Morris, H.P. 64, *94*
Saccomanno, G., s. Kuntz, A. 30, 31, *47*
Saggerson, E.D., Greenbaum, A.L. 60, 82, 84, *94*
Sallach, H.J., s. Hendrick, J.L. 108, *175*
Sallach, H.J., s. Walsh, D.A. 107, *187*
Salmoiraghi, G.C., s. Bloom, F.E. 139, *168*
Salmoiraghi, G.C., s. von Baumgarten, R. 133, *187*
Salvador, R.A., Albers, R.W. 124, 127, 130, 135, 137, 139, 152, *183*
Sanders, D.J., s. Reed, J.D. 16, 26, *49*
Sanders, H.D., s. Perry, T.L. 100, 115, 116, 121, 123, 127, 128, 138, 139, *182*
Sasaki, H., s. Semba, T. 27, *50*
Sasisekḷaran, V., s. Ponnuswamy, P.K. 107, *183*
Sassoon, H.F., s. Johnson, B.C. 80, 81, 84, *90*
Sato, A. 29, 30, *49*
Sato, N., s. Obata, K. 112, 141, *181*
Sato, S., s. Ito, S. 121, *176*
Sauerland, E.K., Mizuno, N. 143, *183*
Saum, W.R., s. DeGroat, W.C. 162, *172*
Schabadasch, A. 11, *49*
Schaefer, J.-M., s. Van Orden, L.S. 12, *51*
Schanker, L.S., Shore, P.A., Brodie, B.B., Hogben, C.A.M. 27, *50*
Schenck, J.R. 115, *183*
Schenk, W.G., s. Burns, G.P. 35, 36, *42*
Scheraga, H.A., s. Némethy, G. 69, *93*
Schimke, R.T. 64, 83, *94*
Schimke, R.T., Doyle, D. 62, 81, *94*
Schimke, R.T., s. Arias, I.M. 64, *86*
Schimke, R.T., s. Berlin, C.M. 66, *86*
Schimke, R.T., s. Dehlinger, P.J. 83, *87*
Schimke, R.T., s. Dice, J.F. 83, *87*
Schirmer, E.W., s. Nandedkar, A.K.N. 82, *93*
Schmidt, R.E., s. Andersen, P. 146, *166*
Schmidt, R.F. 103, 153, 159, 160, 161, *183*
Schmidt, R.F., s. Eccles, J.C. 158, 160, *173*
Schneider, D.I., s. Chang, A.Y. 80, 84, *87*
Schönung, W., s. Kullmann, R. 35, *47*
Schofield, G.C. 11, 31, *50*
Scholes, s. Roberts, E. 132
Schon, F., Iversen, L.L. 129, *184*
Schon, F., s. Gottesfeld, Z. 161, 162, *174*
Schubert, P., Kreutzberg, G.W., Lux, H.D. 101, *184*
Schubert, P., s. Globus, A. 101, 156, *174*
Schwartz, I.L., s. Wyssbrod, H.R. 104, *188*
Schwartz, S., s. Galindo, A. 145, 146, *174*
Schwartz, S., s. Krnjević, K. 112, 118, 122, 125, *178*, *179*
Schweizer, E., s. Lynen, F. 65, 66, 68, 69, 72, 75, 77, *91*
Scorpio, R.M., s. Max, S.R. 58, *92*
Scott, D.F., s. Gibson, D.M. 58, 59, 81, 82, *89*
Scott, W.N., s. Wyssbrod, H.R. 104, *188*
Scrutton, M.C., s. Barden, R.E. 68, *86*
Scrutton, M.C., s. Mildvan, A.S. 68, *92*
Searle, T., s. MacLachlan, R. 165,
Sedgwick, B., s. Plate, C.A. 82, *94*
Seeman, M., s. Tarnowski, W. 59, 77, *95*
Seidman, I., s. Chang, H.-C. 60, 78, 81, 84, *87*
Seiler, N., Weichmann, M. 115, *184*
Seitz, H.J., s. Faupel, R.P. 77, *88*
Semba, T. 31, *50*
Semba, T., Fujii, Y. 27, *50*
Semba, T., Kano, M. 159, 161, *184*
Semba, T., Sasaki, H. 27, *50*
Sepůlveda, F.V., s. Yudilevich, D.L. 107, *188*
Sershen, H., s. Seta, K. 107, *184*
Seta, K., Sershen, H., Lajtha, A. 107, *184*
Shaastad, O., s. Gjessing, L.R. 106, *174*
Shafrir, E., s. Diamant, S. 59, 60, 80, 81, 82, 84, *87*
Shank, R.P., Aprison, M.H. 100, 107, 108, 114, 115, 122, 127, 128, 134, 135, 142, 147, 151, 153, *184*
Shank, R.P., Aprison, M.H., Baxter, C.F. 108, 151, *184*
Shank, R.P., s. Aprison, M.H. 122, 128, 134, 142, 147, 151, *166*
Shank, R.P., s. Davidoff, R.A. 116, 148, 149, 150, 151, 153, *171*
Shank, R.P., s. Fedinec, A.A. 159, 161, *173*
Shank, R.P., s. Graham, L.T., Jr. 106, 117, 148, *174*
Shapiro, H., Woodward, E.R. 31, *50*
Sharma, J.N., s. Dhawan, B.N. 156, 159, *172*
Sharpe, L.G., s. Olney, J.W. 117, *182*
Shaw, J.E., s. Ramwell, P.W. 113, *183*
Shaw, R.K., Heine, J.D. 100, 114, 115, 119, 122, 127, 128, *184*
Sheehan, D., s. Foerster, O. 32, *44*
Shehadeh, Z., Price, W.E., Jacobson, E.D. 24, 27, *50*

Shepherd, G.M. 133, *184*
Shepherd, J.J., Wright, P.G. 23, *50*
Sheridan, J.J., Sims, K.L., Pitts, F.N. 139, *184*
Sherman, S.M., s. Hoffmann, K.-P. 135, *176*
Shimada, M., Kihara, T., Kurimoto, K., Maruyama, Y., Ijichi, H. 123, 128, 134, 138, 139, 142, *184*
Shimada, M., Kurimoto, K., Wada, F., Hino, O., Suginoshita, K., Kihara, T. 100, 119, 120, *184*
Shimamoto, K., s. Muryobayashi, T. 11, *48*
Shimasaki, S., s. Honjin, R. 7, 13, *45*
Shinozaki, H., s. Obata, K. 141, 142, *182*
Shore, P.A. 2, *50*
Shore, P.A., s. Govier, W.C. 2, 3, *45*
Shore, P.A., s. Schanker, L.S. 27, *50*
Shrago, E., Lardy, H.A. 59, 84, *94*
Shrago, E., Lardy, H.A., Nordlie, R.C., Foster, D.O. 80, 84, *94*
Shrago, E., s. Young, J.W. 80, 84, *96*
Shulman, J., s. Kim, T.S. 19, *46*
Shuto, S., s. Kidokoso, J. 143, *178*
Sidky, M., Bean, J.W. 27, *50*
Siggins, G.R., Oliver, A.P., Hoffer, B.J., Bloom, F.E. 112, 129, *184*
Siggins, G.R., s. Woodward, D.J. 127, 129, *188*
Silpananta, P., Goodridge, A.G. 81, 84, *94*
Silva, D.G., Ross, G., Osborne, L.W. 9, 11, *50*
Silvestrini, B., s. Longo, V.G. 158, *179*
Silvey, G.E., s. Davidoff, R.A. 160, *171*
Simada, Z., s. Watanabe, T. 143, *187*
Simmonds, M.A., s. Hill, R.G. 113, 126, *175*
Simmonds, M.A., s. Straughan, D.W. 126, *185*
Simms, H.S. 116, *184*
Simon, E., s. Kullmann, R. 35, *47*
Simon, E., s. Walther, O.-E. 35, *51*
Simpson, F.O., s. Devine, C.E. 13, *43*
Sims, K.L., Weitsen, H.A., Bloom, F.E. 111, *184*
Sims, K.L., s. Sheridan, J.J. 139, *184*
Sisken, B., s. Kuriyama, K. 127, 131, *179*
Skosey, J.L. 83, *94*
Sladek, M., s. Barth, C.C. 57, *86*
Small, N.A., Holton, J.B., Ancill, R.J. 115, *184*
Smets, W. 22, 35, *50*
Smith, C.M., s. Matsushita, A. 149, *180*
Smith, C.M., s. Murayama, S. 149, *181*
Smith, M.B., s. Majerus, P.W. 64, *91*
Smith, M.C., s. Nathan, P.W. 151, *181*
Smith, O.A., s. Rushmer, R.F. 35, *49*
Smith, S., Abraham, S. 59, 60, 64, 80, 81, 82, 84, *95*
Smith, S.E. 109, 114, *184*
Smith, T.G., Wuerker, R.B., Frank, K. 157, *184*
Smith, T.G., s. Purpura, D.P. 121, *183*
Smythe, S. McC., Fitzpatrick, H.F., Blakemore, A.H. 36, *50*
Smythies, J.R. 104, *184*
Sneddon, M.K., s. Uhr, M.L. 107, 151, *186*
Snell, G.D., s. Ingalls, A.M. 59, *90*
Snodgrass, S.R., Cutler, R.W.P., Kang, E.S., Lorenzo, A.V. 106, *184*
Snodgrass, S.R., Iversen, L.L. 100, 112, 125, *184*
Snyder, S.H., Logan, W.J., Bennett, J.P., Argui, A. 99, 101, 105, *184*
Snyder, S.H., s. Arregui, A. 109, 142, 150, 151, 153, *166*
Snyder, S.H., s. Bennett, J.P., Jr. 109, 118, 121, 122, *167*
Snyder, S.H., s. Gfeller, E. 139, *174*
Snyder, S.H., s. Logan, W.J. 109, 114, 115, 116, 118, 150, 151, *179*
Snyder, S.H., s. Wofsey, A.R. 116, 118, 120, *188*
Söderberg, U., s. Henschen, A. 107, *175*
Sokolski, W., s. Guchait, R.B. 68, 76, *89*
Somjen, G.G. 160, *184*
Sonenberg, M., s. Karush, F. 71, *90*
Sonnhof, U., s. Altmann, H. 140, 144, *165*
Sonnhof, U., s. Bruggencate, G. Ten 143, 144, *168*
Sotelo, C., Privat, A., Drian, M.J. 129, *184*
Southwick, C.A.P., s. Davidson, N. 146, *171*
Spector, A.A., John, K., Fletcher, J.E. 71, *95*
Spencer, A.F., Lowenstein, J.M. 56, 59, 77, 78, 84, *95*
Spira, H.M., s. Davidoff, R.A. 160, *171*
Srere, P.A. 56, 59, 71, *95*
Srere, P.A., Bhaduri, A. 56, *95*
Srere, P.A., Foster, D.W. 59, *95*
Srere, P.A., s. Bhaduri, A. 56, *86*
Srere, P.A., s. Foster, D.W. 59, *88*
Srimal, R.C., Bhargava, K.P. 163, *185*
Srimal, R.C., s. Dhawan, B.N. 156, 159, *172*
Srinivasan, V., s. Iversen, L.L. 102, 112, 124, *177*
Srinivasan, V., s. Mitchell, J.F. 114, 118, *180*
Srivastava, R.K., s. Bhargava, K.P. 159, *167*
Starling, E.H. 1, 2, *50*
Starling, E.H., s. Bayliss, W.M. 2, 27, *41*
Starr, M.S., Voaden, M.J. 131, *185*
Starr, M.S., s. Neal, M.J. 131, *181*
Starr, M.S., s. Voaden, M.J. 132, *187*

Start, C., Newsholme, E.A. 77, *95*
Stefanis, C. 130, 131, *185*
Stein, W.H., s. Tallan, H.H. 100, *186*
Steinberg, R., s. Altmann, H. 140, 144, *165*
Steiner, F.A., Meyer, M. 145, *185*
Steiner, F.A., Ruf, K. 130, 131, *185*
Steinmann, L., s. Lam, D.M.K. 132, *179*
Stephanson, A., s. MacLachlan, R. 165,
Stern, F., s. Benuck, M. 108, 151, *167*
Stern, F., s. van Brücke, F.T. 22, *41*
Stern, P., Hadžovič, S. 156, *185*
Steward, E.G., Player, R., Quilliam, J.P., Brown, D.A., Pringle, M.J. 110, *185*
Stewart, C.N., Coursin, D.B., Bhagavan, H.N. 117, *185*
Stewart, C.N., s. Bhagavan, H.N. 117, *167*
Stipani, I., s. Palmieri, F. 58, *93*
Stöhr, P. 11, *50*
Stoll, E., Ryder, E., Edwards, J.B., Lane, M.D. 69, *95*
Stoll, E., s. Dimroth, P. 68, 76, *87*
Stoll, E., s. Lane, M.D. 71, *91*
Stone, J., s. Hoffmann, K.-P. 135, *176*
Stone, W.E., s. Tews, J.K. 100, *186*
Storm-Mathisen, J. 130, *185*
Storm-Mathisen, J., Fonnum, F. 130, *185*
Storm-Mathisen, J., s. Fonnum, F. 140, 141, *173*
Straschill, M. 132, *185*
Straschill, M., Perwein, J. 132, *185*
Strata, P., s. Bisti, S. 129, *167*
Straughan, D.W., Legge, K.F. 130, 131, *185*
Straughan, D.W., Neal, M.J., Simmonds, M.A., Collins, G.G.S., Hill, R.G. 126, *185*
Straughan, D.W., s. Biscoe, T.J. 130, 131, *167*
Straughan, D.W., s. Hill, R.G. 113, 126, *175*
Straughan, D.W., s. Johnson, E.S. 110, 125, 126, *177*
Straughan, D.W., s. Krnjević, K. 126, *178*
Straughan, D.W., s. Legge, K.F. 130, 131, *179*
Stumpf, P.K. 82, *95*
Suda, I., s. Koizumi, K. 29, *47*
Suda, M., s. Yugari, Y. 71, *96*
Sugar, O., Gerard, R.W. 149, *185*
Suginoshita, K., s. Shimada, M. 100, 119, 120, *184*
Sugrue, M.F., s. Govier, W.C. 2, 3, *45*
Sullivan, A.C., Hamilton, J.G., Miller, O.N., Wheatley, V.R. 57, *95*
Sumino, R., s. Kidokoso, J. 143, *178*
Sutherland, E.W., s. Bueding, E. 19, 20, *41*
Sutherland, E.W., s. Robinson, G.A. 83, *94*
Sutherland, E.W., s. Robison, G.A. 19, 20, *49*
Suzuki, F., Fukunishi, K., Daikuhara, Y. 58, *95*
Suzuki, F., s. Inoue, H. 59, *90*
Sverdlov, Yu.S., Alekseeva, V.I. 161, *185*
Sverdlov, Yu.S., Kozhechkin, S.N. 161, *185*
Sverdlov, Y.S., s. Gushchin, I.S. 159, *174*
Swanson, R.F., Curry, W.M., Anker, H.S. 76, *95*
Sykova, E., s. Vyklický, L. 160, *187*
Sytinsky, I.A. 99, 111, 122, 124, *185*
Sytinsky, I.A., Vasilijev, V.Y. 111, 115, *185*
Szarvas, F., s. Varro, V. 27, *51*
Sze, P.Y. 111, *185*
Sze, P.Y., Lovell, R.A. 111, *185*
Szenohradszky, J., s. Petri, G. 37, *49*
Szentágothai, J. 103, 151, *185*
Szentagothal, N., s. Eccles, J.C. 129, 141, *173*
Szumski, A.J., s. Kellerth, J.-O. 160, *178*
Szurszewski, J.H., s. Crowcroft, P.J. 31, *43*
Szutowicz, A., s. Angielski, S. 77, *86*

Tachiki, K.H., DeFeudis, F.V., Aprison, M.H. 101, 124, *185*
Tafuri, W., s. Hager, H. 12, *45*
Tafuri, W.L. 13, *50*
Takahasi, A., s. Honjin, R. 7, 13, *45*
Takahasi, T., s. Otsuka, M. 153, *182*
Takano, K., Neumann, K. 156, *185*
Takata, M., s. Morimoto, T. 144, *180*
Takeda, K., s. Obata, K. 102, 112, 128, 141, 142, *182*
Takeda, Y., s. Daikuhara, Y. 57, *87*
Takeda, Y., s. Inoue, H. 59, *90*
Taketa, K., Pogell, B.M. 71, *95*
Talbot, W.H., s. Wilson, V.J. 158, *188*
Tallan, H.H. 114, 115, *186*
Tallan, H.H., Moore, S., Stein, W.H. 100, *186*
Tanaka, Y., s. Miyata, Y. 140, 143, *180*
Tanaka, Y., s. Otsuka, M. 124, 128, 140, 141, 148, 152, 161, *182*
Tani, M., s. Ito, S. 121, *176*
Tarnowski, W., Seemann, M. 59, 77, *95*
Tarnowski, W., s. Faupel, R.P. 77, *88*
Taylor, C.B., Bailey, E., Bartley,W. 59, 84, *95*
Taylor, C.B., s. Lockwood, E.A. 59, 60, 80, 81, 84, *91*
Taylor, D., s. Cotman, C.W. 101, *169*
Taxi, J. 12, *50*
Taxi, J., Droz, B. 7, 9, 12, *50*

Tebēcis, A.K. 135, 136, 144, 145, *186*
Tebēcis, A.K., Di Maria, A. 136, 145, *186*
Tebēcis, A.K., Hösli, L., Haas, H.L. 145, *186*
Tebēcis, A.K., Ishikawa, T. 145, *186*
Tebēcis, A.K., Phillis, J.W. 153, 160, *186*
Tebēcis, A.K., s. Curtis, D.R. 114, 119, 122, 136, 137, 154, 155, *170*, *171*
Tebēcis, A.K., s. Hösli, L. 144, 145, *176*
Tebēcis, A.K., s. McCance, I. 135, *180*
Tebēcis, A.K., s. Phillis, J.W. 115, *182*
Teebor, G., s. Chang, H.-C. 60, 78, 81, 84, *87*
Tepperman, H.M., Tepperman, J. 80, 81, 84, *95*
Tepperman, H.M., s. Tepperman, J. 80, 84, *95*
Tepperman, J., Tepperman, H.M. 80, 84, *95*
Tepperman, J., s. Tepperman, H.M. 80, 81, 84, *95*
Teresi, J.D., Luck, J.M. 71, *95*
Tewari, S., s. Baxter, C.F. 111, 117, *167*
Tews, J.K., Carter, S.H., Roa, P.D., Stone, W.E. 100, *186*
Thaller, A., s. Engberg, I. 158, *173*
Thoenen, H., s. Tranzer, J.P. 12, *50*
Thompson, J.E., Vane, J.R. 26, *50*
Thorpe, V., s. Reed, J.D. 26, *49*
Timms, A.R., s. Bueding, E. 19, 20, *41*
Titchener, E.B., s. Gibson, D.M. 79, *89*
Titchener, E.B., s. Wakil, S.J. 66, *96*
Toida, N., s. Kuriyama, H. 14, *47*
Tomita, K. 110, *186*
Tomita, T., s. Bülbring, E. 14, 19, 20, *42*
Toomey, R.E., s. Wakil, S.J. 82, *96*
Torvick, A., s. Gjessing, L.R. 123, 128, 138, *174*
Toschi G., s. Masi, I. 162, 163, *180*
Toyokura, Y., s. Kanazawa, I. 137, 138, *178*
Tranzer, J.P., Thoenen, H. 12, *50*
Trendelenburg, U. 18, *50*
Tsuchiya, T., s. Highstein, S.M. 142, 143, *175*
Tsukada, Y., s. Nagata, Y. 162, 163, *181*
Tsunemi, T., s. Daikuhara, Y. 57, *87*
Tubbs, P.K., Garland, P.B. 77, 83, *95*
Tuffery, A.R. 40
Tumen, H.J. 36, *50*
Tunnicliff, G., Wein, J., Roberts, E. 115, *186*
Tweto, J., Dehlinger, P., Larrabee, A.R. 83, *95*
Tweto, J., Larrabee, A.R. 82, 83, *96*
Tweto, J., Liberati, M., Larrabee, A.R. 83, *96*

Udo, M., s. Ito, M. 145, *176*
Uhr, M.L. 108, 151, *186*
Uhr, M.L., Sneddon, M.K. 107, 151, *186*
Uhr, M.L., s. Beart, P.M. 110, *167*
Ujec, E., s. Vyklický, L. 160, *187*
Ungváry, G., Léránth, C. 31, *51*
Urban, I., s. Moss, R.L. 137, *180*
Urban, P.F., s. Pasantes-Morales, H. 131, *182*
Utley, J. 134, 135, *186*
Utter, M.F., s. Barden, R.E. 68, *86*
Utter, M.F., s. Mildvan, A.S. 68, *92*
Uvnäs, B., s. Folkow, B. 25, *44*

Vagelos, P.R. 76, 82, *96*
Vagelos, P.R., Alberts, A.W., Martin, D.B. 69, 72, 75, *96*
Vagelos, P.R., Majerus, P.W., Alberts, A.W., Larrabee, A.R., Ailhaud, G.P. 76, 82, *96*
Vagelos, P.R., s. Alberts, A.W. 68, 76, *85*, *86*
Vagelos, P.R., s. Majerus, P. W. 76, 82, *91*
Vagelos, P.R., s. Martin, D.B. 65, 69, *92*
Vagelos, P.R., s. Nervi, A.M. 68, 76, *93*
Vance, D., Goldberg, I., Mitsuhashi, O., Bloch, K., Omura, S., Nomura, S. 83, *96*
Van den Berg, C.J. 102, 117, 121, *186*
Vane, J.R., s. Thompson, J.E. 26, *50*
Van Gelder, N.M. 111, 114, 122, 123, 130, *186*
Van Gelder, N.M., s. Elliott, K.A.C. 111, *173*
Van Harreveld, A. 121, *186*
Van Harreveld, A., Fifkova, E. 121, 132, *186*, *187*
Van Harreveld, A., Kooiman, M. 121, *187*
Van Orden, L.S., Orden, L.S., Bloom, F.E., Barrnett, R.J., Giarman, N.J. 12, *51*
Van Orden, L.S., Schaefer, J.-M., Burke, J.P., Lodoen, F.V. 12, *51*
Varon, S., s. Weinstein, H. 112, *187*
Varro, V., Blaho, G., Csernay, L., Jung, I., Szarvas, F. 27, *51*
Vasilijev, V.Y., s. Sytinsky, I.A. 111, 115, *185*
Vatner, S.F., Higgins, C.B., White, S., Patrick, T., Franklin, D. 35, *51*
Vaughan, P.C., s. Davis, R. 140, *172*
Veech, R.I., s. Guynn, R.W. 59, *89*
Veloso, D., s. Guynn, R.W. 59, *89*
Venkitasubramanian, T.A., s. Pande, S.V. 80, 84, *93*
Vial, J., s. Morin, G. 32, *48*
Vitali, M.V., s. Johnston, G.A.R. 108, 149, 151, *177*
Vivian, S., s. La Bella, F. 115, *179*
Vizi, E.S., Knoll, J. 14, *51*
Vizi, E.S., s. Knoll, J. 14, *46*

Vizi, E.S., s. Paton, W.D.M. 14, 18, *48*
Voaden, M.J., Starr, M.S. 132, *187*
Voaden, M.J., s. Starr, M.S. 131, *185*
Vogel, W., s. Boehme, D.H. 123, 128, 134, 151, *168*
Vogt, M., s. Iggo, A. 29, 30, *46*
Volta, F., s. Gessi, T. 131, *174*
von Baumgarten, R., Bloom, F.E., Oliver, A.P., Salmoiraghi, G.C. 133, *187*
von Baumgarten, R., Green, J.D., Mancia, M. 133, *187*
von Baumgarten, R., s. Green, J.D. 133, *174*
von Freytag-Loringhoven, Hj., s. Herz, A. 139, *175*
Voorhoeve, P.E., s. Andersen, P. 129, 131, 136, *166*
Voorhoeve, P.E., s. Crawford, J.M. 126, 127, 129, *169*, *170*
Vorkel, W., Hanitzsch, R. 132, *187*
Vrbova, G., s. Morgan, R. 135, *180*
Vyklický, L., Sykova, E., Křiž, N., Ujec, E. 160, *187*

Wada, F., s. Shimada, M. 100, 119, 120, *184*
Wada, J.A., s. McGeer, P.L. 137, 139, *180*
Waddell, M.C. 5, *51*
Waegell, P., s. Jacobson, B. 68, *90*
Wagner, G.A. 36, *51*
Wagner, M., s. McClure, W.R. 68, *92*
Wainer, A., s. Richter, J.J. 107, 114, 115, *183*
Waite, M. 65, *96*
Waite, M., Wakil, S.J. 65, *96*
Wakil, S.J. 54, 82, *96*
Wakil, S.J., Goldman, J.K., Williamson, I.P., Toomey, R.E. 82, *96*
Wakil, S.J., Titchener, E.B., Gibson, D.M. 66, *96*
Wakil, S.J., s. Gibson, D.M. 79, *89*
Wakil, S.J., s. Plate, C.A. 82, *94*
Wakil, S.J., s. Waite, M. 65, *96*
Waksman, A., Bloch, M. 111, *187*
Waksman, A., Roberts, E. 111, *187*
Walberg, A., s. Brodal, A. 140, *168*
Walberg, F., Jansen, J. 140, *187*
Walberg, F., s. Fonnum, F. 100, 128, 140, 141, *173*
Walberg, F., s. Mugnaini E. 140, *181*
Wall, P.D., s. Pomeranz, B. 32, *49*
Wallentin, I., s. Dresel, P. 23, 24, 25, *43*
Wallentin, I., s. Folkow, B. 23, 24, 25. *44*
Walsh, D.A., Sallach, H.J. 107, *187*
Walsh, H., s. MacLachlan, R. 165
Walther, O.-E., Iriki, M., Simon, E. 35, *51*
Walton, G.M., s. Atkinson, D.E. 59, *86*
Wangensteen, O.H. 36, *51*
Warner, R.C., s. Gregolin, C. 65, 69, 70, 73, 74, 76, *89*
Warner, R.S., s. Gregolin, C. 70, 73, 75, 76, *89*
Wasvary, J.M., Maragoudakis, M.E. 78, *92*
Watanabe, T., Simada, Z. 143, *187*
Watkins, J.C. 99, 117, *187*
Watkins, J.C., Curtis, D.R., Biscoe, T.J. 118, *187*
Watkins, J.C., s. Curtis, D.R. 99, 105, 115, 116, 118, 119, 120, 121, 122, 125, 135, 136, 153, 154, 155, 156, 162, *170*, *171*
Watkins, J.C., s. Davies, J. 119, 122, *171*
Watkins, J.L., s. Davies, J. 122, 146, *171*
Watson, J.A., Fang, M., Lowenstein, J.M. 57, *96*
Watson, J.A., Lowenstein, J.M. 57, *96*
Watt, A.J. 16, *51*
Weber, G. 80, 81, 84, *96*
Weber, G., Banerjee, G., Bronstein, S.B. 81, *96*
Weber, W.V., s. Pomeranz, B. 32, *49*
Webster, R.A., s. Jordan, C.C. 102, 109, 152, *177*
Weichmann, M., s. Seiler, N. 115, *184*
Weight, F.E. 103, *187*
Weiland, O., s. Henning, U. 81, 84, *89*
Wein, J., s. Tunnicliff, G. 115, *186*
Weinreich, D., s. Hammerschlag, R. 99, *175*
Weinstein, H., Varon, S., Roberts, E. 112, *187*
Weiss, H., s. Hochheuser, W. 56, *90*
Weiss, H., s. Wieland, O. 56, *96*
Weitsen, H.A., s. Sims, K.L. 111, *184*
Welch, A.D., Henderson, V.E. 113, *187*
Wells, J.N., s. Kee, R.D. 113, 142, *178*
Werman, R. 99, 103, 115, *187*
Werman, R., Davidoff, R.A., Aprison, M.H. 109, 115, 156, *187*
Werman, R., s. Aprison, M.H. 99, 100, *166*
Werman, R., s. Davidoff, R.A. 116, 148, 149, 150, 151, 153, 157, *171*
Werman, R., s. Graham, L.T., Jr. 106, 117, 148, *174*
West, G.B., s. Mann, M. 2, *48*
West, R. 158, *187*
Westecker, M.E. 133, *187*
Westerman, R.A., s. McCane, I. 135, *180*
Westphal, K. 22, *51*
Wheatley, V.R., s. Sullivan, A.C. 57, *95*
Wheeler, D.D., Boyarsky, L.L. 150, *187*
Wheeler, D.D., Boyarsky, L.L., Brooks, W.H. 150, *187*
White, R.D., s. Neal, M.J. 131, *181*
White, S., s. Vatner, S.F. 35, *51*

Whittaker, V.P. 101, 120, 124, *188*
Whittaker, V.P., Barker, L.A. 101, *188*
Whittaker, V.P., s. Krnjević, K. 120, *179*
Whittaker, V.P., s. Mangan, J.L. 120, 124, *180*
Wiechert, P., s. Hennecke, H. 117, *175*
Wieland, O., Neufeldt, I.E. 60, 65, *96*
Wieland, O., Neufeldt, I., Numa, S., Lynen, F. 60, 84, *96*
Wieland, O., Weiss, H. 56, *96*
Wieland, O., s. Hochheuser, W. 56, *90*
Wilkenfeld, B.E., Levy, B. 19, 20, *51*
Williams, G.R., s. Robinson, B.H. 57, *94*
Williamson, I.P., s. Wakil, S.J. 82, *96*
Williamson, J.R., Herczeg, B., Coles, H., Danish, R. 77, *96*
Willis, W.D., s. Eccles, J.C. 158, 160, *173*
Wilson, V.J. 142, *188*
Wilson, V.J., Talbot, W.H. 158, *188*
Wilson, V.J., s. Crawford, J.M. 126, 127, 129, *169*, *170*
Winegrad, A.I., Renold, A.E. 79, *96*
Wise, E.M., Ball, E.G. 80, 84, *96*
Wofsey, A.R., Kuhar, M.J., Snyder, S.H. 116, 118, 120, *188*
Wolfe, D.E., Potter, L.T., Richardson, K.C., Axelrod, J. 12, *51*
Wolfgram, F., s. Nakajima, T. 115, *181*
Wolman, M. 111, *188*
Wolstencroft, J.H., s. Morgan, R. 135, *180*
Wood, H.G., s. Ahmad, F. 68, *85*
Wood, H.G., s. Gerwin, B.I. 68, *88*
Wood, H.G., s. Jacobson, B. 68, *90*
Wood, H.G., s. Northrop, D.B. 68, *93*
Woods, G., Nelson, V.E., Nelson, E.E. 25, *51*
Woodward, D.J., Hoffer, B.J., Siggins, G.R., Bloom, F.E. 127, 129, *188*
Woodward, D.J., Hoffer, B.J., Siggins, G.R., Oliver, A.P. 129, *188*
Woodward, E.R., s. Shapiro, H. 31, *50*
Wright, P.G., s. Shepherd, J.J. 23, *50*
Wright, S., s. Kremer, M. 28, *47*
Wu, C.Y., s. Nakamura, Y. 143, *181*
Wuerker, R.B., s. Smith, T.G. 157, *184*
Wyatt, A.P. 22, *51*
Wynia, F., s. Youmans, W.B. 34, *51*
Wyssbrod, H.R., Scott, W.N., Brodsky, W.A., Schwartz, I.L. 104, *188*

Yamagami, M., s. Hukuhara, T. 32, *46*
Yamagishi, M., s. Moss, J 65, 69, 72, 73, 75, *92*
Yamamoto, T., s. Oomura, Y. 137, *182*
Yim, G.K.W., s. Huffman, R.D. 142, *176*
Yim, G.K.M., s. Kee, R.D. 113, 142, *178*
Yim, G.K.W., s. Kelly, J.S. 125, *178*
Yokoi, Y., s. Nagata, Y. 162, 163, *181*
York, D.H. 139, *188*
York, D.H., s. McLennan, H. 139, *180*
York, D.H., s. Phillis, J.W. 115, *182*
Yoshida, M., Precht, W. 139, *188*
Yoshida, M., Rabin, A., Anderson, M. 139, *188*
Yoshida, M., s. Ito, M. 128, 141, *176*
Yoshida, M., s. Precht, W. 139, *183*
Yoshino, Y., DeFeudis, F.V., Elliott, K.A.C. 115, *188*
Youmans, W.B. 2, 30, 33, *51*
Youmans, W.B., Aumann, K.W., Haney, H.F., Wynia, F. 34, *51*
Youmans, W.B., Karstens, A.I., Aumann, K.W. 32, *51*
Young, J.W. 59, 80, 81, 84, *96*
Young, J., Shargo, E., Lardy, H.A. 80, 84, *96*
Yudilevich, D.L., De Rose, N., Sepúlveda, F.V. 107, *188*
Yue, K.T.N., s. Fritz, I.B. 56, *88*
Yugari, Y., Matsuda, T., Suda, M. 71, *96*

Zahler, W.L., Barden, R.E., Cleland, W.W. 71, *96*
Zar, M.A., s. Patson, W.D.M. 14, 17, *49*
Zetler, G., s. Lembeck, F. 153, *179*
Zieglgänsberger, W., Pull, E.A. 118, 119, 154, *188*
Zweifach, B.W., s. Richardson, D.R. 25, *49*

Sachverzeichnis/Subject Index

acetyl carnitine 56
acetylcholine 121, 122
— release, inhibition of 14, 15, 18
— — and longitudinal muscle 17
— — and terminal polarization 18
N-acetyl-L-aspartate 116
acetyl-CoA 54, 55
— carboxylase 54
— —, activation of 69
— —, activity-structure relation 75
— —, allosteric nature 76
— —, apoenzyme 78, 79
— —, carboxylated active site 67
— — conformation of 75
— —, efficiency of 62, 65ff
— —, electron microscopy 73, 74
— —, hepatic 60, 62, 64, 68
— —, —, catalytic capacity 78
— —, —, specific activity 63
— —, —, specificity of activators 68
— —, holoenzyme 78, 79
— —, immunological determination 61
— —, molecular properties 73f
— —, sedimentation pattern 72, 73, 75
— —, subunit structure 76
— —, synthesis and degradation 62–64
— —, tissue level 60–62, 66, 84
— citrate 56
—, cytoplasmic, generation of 56f
—, mitochondrial 56
— synthetase 56
—, translocation of 56
acetyl pantetheine 69
actinomycin D 59, 64, 81, 82
activator specificity for acetyl-CoA carboxylase 68, 69
acyl carnitine 72
acyl-CoA, long chain 77, 83
— thioesters, long chain 65, 71, 76
adenosine-3′-5′-monophosphate see cyclic AMP
adenyl cyclase and β-receptors 19, 20
adipose tissue 60, 69, 75
adrenalectomy and intestinal reflex 31
adrenaline and cyclic AMP 19, 20
—, effect on sphincter 22, 23
—, effect on nonadrenergic neurones 21
— and gastric secretion 26
— release 2
—, reversal potential 19
adrenergic action, direct on smooth gut muscle 16, 17, 18
— axons in the gut 12, 13
— —, ultrastructure 7, 12, 13
— cell bodies, location of in gut 3f
— fibres in gut, ascending and of descending 6
— — to gut smooth muscle 5
— innervation of splanchnic vessels 28, 29
— nerves and absorption 27
— —, effects on cholinergic contractions 15, 17
— — to gut vessels 23f
— — and motility 27
— —, inhibiton of myogenic tone 15, 17
— —, stimulation of 23–26
— —, tonic action of 27, 28
— terminals 16
— —, ramifications of 5f
— vesicles 12, 13
alanine 109, 114, 153, 157
β-alanine 111, 115, 126, 129
β-alanine actions 132, 136, 141, 144
alanine levels 106, 123, 128, 134, 148
allosteric effectors 66
— regulators, 73
— —, kinetic effects 68–72
— —, tissue levels 77, 78
alloxan diabetes 53, 59, 60, 62, 63, 64, 65, 77, 84
amacrine cells and GABA 131, 132
amino acid, see also under transmitter
— — antagonists 103, 104, 105
— —, antagonist complex 104
— — depolarization. ionic mechanism 154
— — level 99, 102
— — —, determination of 100
— — —, regional variations 100, 106, 117
— — —, postmortem changes 100
— — precursors 107
— — on primary afferent terminals 103
— — receptors 104
— — synthesizing enzymes 100, 101
— — transmission, central 99

amino acid transport systems 101, 104
— — turnover rates 100
— — uptake 105
— — —, interference with 105
amino acids, antagonists of metabolising enzymes 102
— —, and autonomic ganglia 162, 163
— —, dendritic actions 103
— —, and dorsal root ganglia 161, 162
— —, enzyme levels 149
— —, GABA-like 110, 157
— —, glycine-like 109, 157
— —, inactivation of 154
— —, intraspinal distribution 147, 148
— —, subcellular studies 101
— —, synaptic release 102
γ-aminobutyric acid see GABA
γ-aminobutyrylcholine 115
ε-aminocaproic acid 141
γ-amino-β-hydroxybutyric acid 115
amino-oxyacetic acid 111, 112, 124, 125, 128, 141, 162
3-aminopropane sulphonate 105
aminopyrine 26
δ-amino valeric acid 110, 111, 141, 144
ammonia metabolism 117
amphetamine 14
amphibian spinal cord 153, 160
anal sphincter 21, 22
angiotensin 27
aortic occlusion 149, 150, 152
arginine, guanido-labelled 64
arteriolar pressure 25
L-aspartate antagonists 116, 122
— levels 106, 121, 127, 138, 148, 150, 161
— metabolism 116
—, neurochemistry 116
—, postsynaptic actions 133, 135, 139, 140, 144, 153
— release 102, 116, 121
—, storage of 120
— transport 116
— uptake 101, 105, 120, 150
ATP increase and adrenaline 20
atropine 15, 16
auerbach plexus, noradrenaline release 16
autonomic afferents and intestinal motility 33, 34
— —, stimulation of
— ganglia and amino acids 162, 163
autoradiography 101
autoregulation 24
autoregulatory escape 24, 25, 27
autotransplants, hepatic 65
avidin 70
axon reflex 30, 147
azapetine 25

baroreceptor reflexes 29, 33
basket cell inhibition 129
benzyl penicillin 112, 113
bethanidine 37
bicucine methyl ester 113
bicuculline 104, 105, 112, 113, 114, 125, 126, 129, 131, 133, 136, 137, 141, 142, 143, 145, 146, 147, 155, 159, 160, 162
— methiodide 113
— methochloride 113, 126, 129
biocytin 68
biotin 66, 67, 69, 76, 79
— carboxyl carrier protein 76
— carboxylase 67, 76
— enzymes 66
biotinyl prosthetic group 70, 76
blood brain barrier 104, 106, 107, 114, 117
— flow, intestinal, and gastric secretion 26
— pressure, systemic 29
body temperature and intestinal blood flow 35
brain stem and amino acids 142–145
bretylium 16
brucine 109, 158
n-butylmalonate 57
butyryl-CoA 82

Ca^{++} and catecholamine action 19
— contractile proteins 19
— transmitter release 102, 124
carboxyl transferase 76
cardiac output 35
carnitine acetyl transferase 57
catecholamine delpletion 6
— release in intestine 2
catecholamines, excitatory action 20
—, intestinal motility 34, 37
caudate nucleus and amino acids 139
caudato-nigral path 139
cerebellum, amino acid level 106
—, — acids in 127–130
—, basket cell inhibition
—, Golgi cell inhibition 129
cerebral cortex, amino acids 106, 120–127
— —, stellate cell inhibition 126
cerebrospinal fluid, amino acids 106, 107
cerulenin 83
chlorpromazine 109, 112, 118
cholinergic contraction, antagonism to 34
— nerves, inhibition of 38
— neurones, inhibition of 14, 15, 17, 18
— systems 115
— tone 33, 34
p-chloromercuriphenylsulphonate 106, 112, 118, 154
circular muscle 9, 11
citrate, acetyl-CoA carboxylase regulation 65, 68, 70, 74, 76

citrate cleavage enzyme 56, 57, 78, 80, 84
— — —, activity of 58, 59
— — —, control of fatty acid synthesis 58, 59
—, content in liver 77
— intermediate 56, 57, 58
— pathway in fatty acid synthesis 56, 57
— translocation, control of 58, 59
— transport, mitochondrial 58
citrate-malate cyclus 80
Cl^- conductance and adrenaline 19
— — and pregnancy 20
clearance, gastric 26
coeliac ganglion 3, 4
— plexus 22, 31, 32
colon, adrenergic innervation 11
—, vesicles in 13
convulsants 158
corlumine 113
cuneate terminals, excitability 146
cyclic AMP 83
— — in gut smooth muscle 19
cystathionine 109, 115, 153
— levels 123, 128, 134, 148
L-cysteate 120
L-cysteine sulphinate 120
cysteine sulphinic acid decarboxylase 122

deafferentation 152, 153
decentralization and intestinal reflex 30, 31
defense reaction 38
dehydrogenase counterstain 6
Deiters' nucleus 140
dendrobine 109, 158
depolarization block in ganglionic cells 18
2,4-diaminobutyric acid 115
diaboline 109, 158
diazoxide 19
dibenamine 25
dibutyryl cyclic AMP 111
digestion and intestinal blood flow 35
—, oxygen consumption 36
diphenylamino-ethanol 113
DLH see DL-homocysteate
dorsal column, glycine in 151
— — nuclei and amino acids 145ff
— root ganglia and amino acids 161, 162
— — section 149
dumping syndrome 38

end-product inhibition 65, 71
enteric ganglia, cell-types in 18
enterogastric reflex 31, 38
enzyme quantity and efficiency 60, 61
—, recycling 100
— synthesis, adaptive 64
enzymes in amino acid metabolism 149, 150, 151
ephedrine 2
ergotamine 15
ergothioneine 127
ergotoxin 25
ethionine 81
2-ethylcitrate 57
exercise and intestinal blood flow 35
extraocular motoneurones and GABA 143
extravascular space, fluid transfer

fasting 53, 58
—, intestinal blood flow 35
fatty acid synthesis in ruminants 58
— — synthetase 54, 55, 78, 79ff
— — —, catalytic efficiency 82, 83, 84
— — — in hepatoma 64
— — —, inhibition 83
— — —, quantity 81, 82
— — —, subunits 83
— acids, long chain, enzymatic synthesis 54f, 55
— acyl-CoA derivatives 71
feeding 60, 62, 63, 65, 84
fluid transfer 24
fluorescence histochemical studies 11
— — technique 4, 5, 6, 21
fluorocitrate 69

Gad 111, 124, 127, 129, 130, 135, 137, 138, 139, 140, 141, 143, 149
— inhibitors 101
gall bladder, reflex relaxation 22
ganglion cells, noradrenaline action on 18
ganglionic site of adrenergic inhibitory action 15, 17, 18
gas-liquid chromatography 100
gastric secretion 26
gastrin 26
gastrointestinal see also intestinal
— adrenergic innervation 4
— movement, inhibition of 2f
GABA antagonists 112
—, conformation of 110
— efflux 124
—, histochemical methods 111
— in hippocampus 130
— levels 106, 110, 123, 124, 128, 131, 134, 137, 138, 139, 140, 142, 143, 146, 148, 152, 161
— metabolism 111
—, neuropharmacology of 110—113
— and PAD 160
—, postsynaptic actions 112, 125, 129, 131, 132, 133, 136, 137, 139, 141, 143, 144, 145, 146, 156, 157, 162
— release 102, 112, 137, 142
— shunt 111
— synthesis 101

GABA-T 111, 124, 135, 137, 139, 149
— and tetanus toxin 161
GABA transaminase 111
— as transmitter 98, 101, 126, 128, 132, 133, 135, 136, 137, 138, 139, 142, 144, 145, 147, 159
— transport 111
— uptake 101, 105, 124, 131, 139, 153, 162
— — in cerebellum 129
gelsemine 109, 158
glial cells 118
globus pallidus, GABA in 110
gluconeogenesis in ruminants 58
glucose 6-phosphate dehydrogenase 81
glutamate, enzymes involved 150
— gradient in dorsal root 150
— in hippocampus 130
—, storage of 120
L-glutamate antagonists 119, 122
— diethylester 116, 119, 135, 136, 155
— levels 106, 120, 121, 127, 134, 145, 148, 150, 161, 163
— metabolism 117
— neurochemistry 117
—, postsynaptic action 118, 122, 127, 132, 133, 135, 137, 139, 140, 144, 145, 153
— release 102, 118, 121, 137
— transport 118
— uptake 101, 105, 120, 150, 154
glutamic acid decarboxylase see GAD
L-glutamine 117
— synthetase 118
D-glycerate 107
glycine 126
— antagonists 108, 158
— in cerebellum 129
— decarboxylase 151
— efflux 152
—, end-product inhibition 107
— hyperpolarization, reversal of 156
— levels 106, 107, 122, 123, 128, 142, 148, 151, 161
— metabolism 107, 108, 151
—, neurochemistry of 107—110
—, postsynaptic actions 108, 125, 129, 132, 136, 137, 139, 141, 143, 144, 145, 146, 155, 156—159
— release 108
— as transmitter 126, 129, 138, 145, 156, 157
— transport 108
— uptake 101, 105, 122, 131, 139, 142, 151, 152
glycinergic interneurones 158
glycolaldehyd 108
glycolysis 58
granule cell inhibiton 133
granular vesicles 12, 13
guanethidine 15, 16, 37

haloperidol 109, 112
Henle plexus 11
hepatoma 64
hexamethonium 14, 29
hexose monophosphate shunt dehydrogenase 79, 80, 81, 84
high threshold intestinal reflex 30, 32, 37
hippocampus and amino acids 130
histamine 26
D-homocysteate 105, 118
DL-homocysteate 119, 122, 133, 135, 139, 144, 153
hybrid ping-pong mechanism 68
hydrophobic bonds 69
1-hydroxy-3-amino-pyrrolidone-2 119, 122, 155
hydroxycitrate 57
5-hydroxydopamine 12
6-hydroxydopamine 9, 12
hydroxylamine 111, 141
5-hydroxytryptamine 136, 144
hyoscine 16
hyperpolarization and noradrenaline 19
hypogastric nerves 22, 23
hypoglossal motoneurones 143, 144
hypolipidemic agents 78
hypotaurine 115, 120
hypotension, systemic 34, 35
hypothalamic defense area 32, 34
hypothalamus and amino acids 137

ileo-coecal sphincter 28
ileo-colic sphincter 21, 22, 35
ileus, adynamic 36
—, paralytic 36
imidazole 19, 20
imidazole-4-acetic acid 115, 126, 145
imipramine 109, 112
immunochemical titration 61, 62
inferior colliculus neurones 143
— mesenteric ganglion 3, 4
inhibitory neurones, nonadrenergic 20, 21
— synapses, location of 103
internodal strands 6
intestinal absorption and adrenaline 27
— acidity, regulation 38
— blood flow 24f, 29, 35
— — —, redistribution of 24, 25
— — supply 38
— — vessels 13, 23, 35, 38
— motility and blood flow 27
— tone, inhibition of 28
intestine, relaxation of 2
—, segmental adrenergic control 6
intestino-intestinal reflex, high threshold 17
intramural inhibitory nerves 14
ion pumps 103

isocitrate 68
— dehydrogenase 79
isoprenaline 23, 26
—, action in taenia 20
— and K^+ uptake 18

juglone 118

K^+ see potassium
α-ketoglutarat 56
kynurenate 65

laparatomy 28, 38
laudanosine 109, 158
levorphan 109
linoleic acid 60
lipogenesis, homoeostatic function 53
—, rate of 54
lipogenic enzymes, coordinate response 84
longitudinal muscle 9, 16
LOT inhibition 133
LSD 116, 119, 144

malate permease 57
malic enzyme 79, 80, 81
malonyl CoA 54, 65, 66, 67, 68, 75, 82
masseter motoneurones 143
medial longitudinal fascicle 143
Meissner plexus 11
membrane, conductance and amino acids 144, 145
— — and GABA 112, 129, 141, 159, 163
— — and glutamate 154
— — and glycine 109, 141, 156
— resistance and noradrenaline 19
— stabilization 20
mesenteric ganglia, inferior 22, 28
— ganglion, superior 3, 4, 22
— inhibitory nerves 17
— nerves, afferents 31, 32
— —, denervation of 4
— —, stimulation of 2, 3, 14, 16, 17
— vessels 23
L-methionine-DL-sulphoximine 119, 122, 136, 154, 155
2-methoxyaporphine 119, 122, 136, 155
N-methyl-D-aspartate 105, 117, 118, 135, 154
DL-α-methyl glutamic acid 119, 135
microelectrophoresis 103
mitral cells and GABA 132, 133
morphine 109, 158
motility, intestinal 27, 28
mucosal blood flow 24, 26
— breakdown 35
— glands, adrenergic innervation 11
mucus secretion 26
muscarinic agents 16
muscularis externa, adrenergic innervation 9
— mucosae, adrenergic innervation 11
myenteric plexus 6, 16, 17
— —, adrenergic cells in 5, 7
— —, — innervation 9
myogenic tone and adrenaline 15, 17

NADPH, cytoplasmic 79–81
neurones, GABA content of 124, 143
nicotine 3, 5, 22, 30, 31
noradrenaline 133
—, action on acetylcholine release 14, 18
—, — on excitatory neurones 14, 17, 18
—, autoradiographic localization 7
— diffusion 16, 17
—, discharge in enteric neurones 18
—, effect on sphincters 22, 23
— levels in gut 2, 3, 4
—, site of action 13f
— and K^+ uptake 18
— overflow 16
—, receptor sites for action 18
—, release of upon nerve stimulation 2,3
nucleus ambiguus 143
— supraopticus 137

obesity 59, 61, 62
oculomotoneurones 142
oesophageal sphincter 21, 22, 34
olfactory bulb, amino acids 132–134
— —, mitral cell inhibition 133
opaque vesicles 12
ouabain 114
oxaloacetate 56, 80
— transcarboxylase reaction 68

pacemaker potential generation 19
PAD 153, 160
pain fibres and intestinal motility 34, 38
pallido-nigral path 139
palmityl-CoA 58, 71, 72, 73, 75, 76, 78
palmityl carnitine 71
pancreatic juice, alkaline 38
parasympathetic neurones 153, 159
paravascular adrenergic nerves 6
paraventricular nucleus 137
penicillin 126, 158, 160
pentobarbitone, LOT inhibition 133
pentolinium 16
pentose phosphate cyclus 79
pericellular endings, adrenergic 6
peristalsis in ileus 36, 37
peritonitis 36, 37
phenazine methosulfate 81
phentolamine 16, 19, 37
phosphodiesterase activity 19
phospholipids 72
phosphorylase a 19, 20
physalaemin 153

picrotoxin 104, 105, 112, 113, 126, 132, 133, 136, 137, 139, 141, 142, 143, 144, 145, 146, 147, 155, 158, 162
picrotoxinin 113, 129, 160
plasma, amino acids 106
postganglionic adrenergic neurones 3
— neurones, noradrenaline action upon 14, 18
postsynaptic inhibiton 155, 157, 160
potassium concentration and PAD 160
—, equilibrium potential in taenia 19
— permeability and adrenergic receptors 19
— — in virgin 20
pregnancy and K^+ and Cl^- conductance 20
pressor reflex 33
pressure, intraarteriolar 25
presynaptic inhibition 103, 146, 155, 159, 160
— — of cholinergic neurones in gut 18
primary afferent depolarization see PAD
— — terminals 103
— — —, excitability of 153
— — —, — and GABA 160
proline 116
propanolol 16
protein, hepatic, half-life time 64
protoveratrines 112
PT cells and GABA 125
p-type vesicles 12
Purkinje cell inhibition 141, 142
— cells and GABA 128, 140, 141
— — glutamate 127
puromycin 19, 58, 59, 64, 181, 182
pyloric sphincter 28
pyridoxal phosphate 111, 112
— 5'-phosphate 117
pyruvate carboxylase reaction 68

raphe nuclei 144
α-receptor blocker 23
α-receptors, adrenaline action 19
— and cyclic AMP 20
— in gut muscle 18
— — vessels 25
— and K^+ permeability 19
—, myometrial 20
— in sphincters 22, 23
β-receptor blocker 23
— blockade 37
β-receptors and adenyl cyclase 20
—, adrenaline action 19
— in gut muscle 18
— — vessels 25
— and K^+ permeability 19
— in sphincters 22, 23
rectum, adrenergic innervation 11
recurrent inhibition 157
red nucleus and amino acids 140
reflex, depressor 29, 33
— activity on intestinal motility 29
— —, vasomotor 29
—, high threshold intestino-intestinal inhibitory 30, 32, 33, 37
—, intestino-intestinal inhibitory 30, 31, 34
—, low threshold intestio-intestinal inhibitory 32, 33, 37, 38
—, peristaltic 17
—, spinal intestino-intestinal inhibitory 32, 33
reflexes, descending inhibitory 21, 30
—, vagal 30
remote inhibition 103, 155
Renshaw cells 153, 156, 158, 159
reserpine 3
resistance to flow 25
resonance spectroscopy 107, 110
retina, amino acidas in 131, 132
—, structure 131
reticular activating system and amino acid release 121, 124
reticulo-hypothalamic-hypophyseal paths 137
reticulospinal neurones 144
rhinencephalon, amino acids in 130, 131
rigidity 147
rotenone 81
ruminants and fatty acid synthesis 57, 58

secretin 26
seizures 130
— and glutamate 117
semicarbazide 146, 160
serine 107, 108, 109, 114, 115, 153
— hydroxymethyltransferase 108, 151
— levels 107, 148
shikimin 113
shock, irreversible 35
small intestine, adrenergic innervation 11
— —, vesicles in 13
sodium ions and amino acid transport 109, 118, 121
— — and PAD 160
somatic afferents and intestinal motility 34
sphincter contraction, reflectory 34, 35, 38
— Oddi 21, 22
—, tonic constriction 27, 28
sphincters 38
— of gut 21f, 34, 35
— in ileus 37
spinal anesthesia 36
— cord, amino acids in 147ff
— — cultures 153
— interneurones and aspartate 153
— — and GABA 159, 161
— — and glycine 151, 156f
— reflexes and ileus 36, 37
splanchnic nerve, rhythmic firing 29
— — stimulation 2, 16, 22, 26
— vessels 23

— —, tone in 27, 28, 29
spreading depression 132
— — and amino acids 121
stellate cells, cortical 126
starvation 60, 63, 65, 77, 82, 84
stimulation, supraspinal 32
stomach, adrenergic innervation 11
—, vesicles in 13
strio-nigral path 138
strychnine 104, 105, 109, 114, 125, 126, 129, 131, 132, 133, 136, 137, 139, 141, 142, 143, 144, 145, 146, 155, 157, 158, 159, 162,
submucosa 11
submucosus plexus, adrenergic cells 5, 7
substantia nigra and amino acids 137f
— —, GABA in 110
sucrose gap, double 19
sugars, phosphorylated 82
superior vestibular nucleus, inhibition in 143
sympathectomy 28
sympathetic inhibitory paths 21
— nerves, stimulation of 22
— neurones 153
sympathomimetic amines 2
synapses, adrenergic with enteric neurones 7
—, location of 103
synaptic excitation, ionic mechanism 154
synaptosomal preparations 101, 102
synaptosomes 101, 120
synthetase, fatty acids 54

taenia coli, catecholamine inhibition 19
taurine 109, 111, 120, 126, 129, 153, 157
— actions 136, 145
— conformation 113
— efflux 124
— levels 106, 113, 122, 123, 128, 134
— metabolism 114
— release 102
— synthesis 101
— transport 114
— uptake 105, 131
tetanus toxin 109, 112, 126, 129, 159, 161
tetrodotoxin 14, 118, 119, 154, 160
thalamus, amino acids in 134ff
thebaine 109, 158
theophylline 19, 20
thiosemicarbazide 111, 154
thyroid hormones 59, 81, 84
thyrotoxic state 60
transmitter see also under amino acids
— inactivation 102, 104, 105
—, postsynaptic action 102
—, reversal potential of effect 103, 109
—, storage 100, 101
— —, inhibitors of 101
—, synaptic release 102
— synthesis 100, 101
— —, inhibitors of 101
—, uptake antagonists 102
tricarballylate 57
trigeminal motoneurones 143
trochlear motoneurones 142, 143
d-tubocurarine 113
tutin 113

ulceration 32, 35, 38
urecholine 26

vagal inhibitory paths 21
vagotomy and enterogastric reflex 31
varicosities 6
vasoconstriction, intestinal 35
vasomotor nerves, resting discharge 29
vasopressin 26
vesicles, granular 12, 13
vestibular nuclei and amino acids 140—142
villi, adrenergic innervation 11

xanthurenate 65

Reviews of Physiology, Biochemistry and Experimental Pharmacology

Vol. 64: Neurophysiology and Neurochemistry of sleep and wakefulness
63 figures
VIII, 342 pages. 1972
Cloth DM 98,–
ISBN 3-540-05462-6

Contents: G. Moruzzi: The Sleep-Waking Cycle. – M. Jouvet: The Role of Monoamines and Acetylcholine-Containing Neurons in the Regulation of the Sleep-Waking Cycle.

Vol. 65: 45 figures
1 portrait
VIII, 191 pages. 1972
Cloth DM 68,–
ISBN 3-540-05814-1

Contents: P. Kruhøffer, Chr. Crone: Einar Lundsgaard, 1899-1968. – J.-S. Pitton: Mechanisms of Bacterial Resistance to Antibiotics. – J. B. Stanbury: Some Recent Developments in the Physiology of the Thyroid Gland. – M. A. Bouman, J. J. Koenderink: Psychophysical Basis of Coincidence Mechanisms in the Human Visual System.

Vol. 66: 46 figures
1 portrait
VIII, 296 pages. 1972
Cloth DM 98,–
ISBN 3-540-05882-6

Contents:
H.-J. Schümann: Peter Holtz, 1902-1970. – H. van den Bosch, L. M. G. van Golde, L. L. M. van Deenen: Dynamics of Phosphoglycerides. – N. Emmelin, U. Trendelenburg: Degeneration Activity after Parasympathetic or Sympathetic Denervation. – P. N. Patil, J. B. LaPidus: Stereoisomerism of Adrenergic Drugs.

Vol. 67: 43 figures
IV, 226 pages. 1972
Cloth DM 86,–
ISBN 3-540-05959-8

Contents: K. Koizumi, C. M. Brooks: The Integration of Autonomic Reactions: A Discussion of Autonomic Reflexes, their Control and their Association with Somatic Reactions. – W. Elger: Physiology and Pharmacology of Female Reproduction under the Aspect of Fertility Control. – W. H. Daughaday, L. S. Jacobs: Human Prolactin.

Vol. 68: 12 figures
IV, 128 pages. 1973
Cloth DM 68,–
ISBN 3-540-06238-6

Contents: P. A. Owren, H. Stormorken: The Mechanism of Blood Coagulation. – W. W. Fleming, J. J. McPhillips, D. P. Westfall: Postjunctional Supersensitivity and Subsensitivity of Excitable Tissues to Drugs. – R. S. Turner, M. M. Burger: The Cell Surface in Cell Interactions.

Prices are subject to change without notice

Springer-Verlag
Berlin
Heidelberg
New York
München Johannesburg London New Delhi Paris Rio de Janeiro Sydney Tokyo Wien

Ergebnisse der Physiologie

Biologischen Chemie und experimentellen Pharmakologie

Reviews of Physiology

Biochemistry and Experimental Pharmacology

Reprint from Volume 69

J. B. Furness and M. Costa

The Adrenergic Innervation of the Gastrointestinal Tract

Nicht im Handel

Springer-Verlag Berlin Heidelberg GmbH 1974

Inhalt/Contents

The Adrenergic Innervation of the Gastrointestinal Tract. By J. B. FURNESS and M. COSTA, Victoria/Australia. With 5 Figures

Regulation of Fatty-Acid Synthesis in Higher Animals. By S. NUMA, Kyoto/Japan. With 6 Figures

Amino Acid Transmitters in the Mammalian Central Nervous System. By D. R. CURTIS and G. A. R. JOHNSTON, Canberra City/Australia. With 4 Figures

Author Index

Subject Index

Ergebnisse der Physiologie

Biologischen Chemie und experimentellen Pharmakologie

Reviews of Physiology

Biochemistry and Experimental Pharmacology

Reprint from Volume 69

Shosaku Numa

Regulation of Fatty-Acid Synthesis in Higher Animals

Nicht im Handel

Springer-Verlag Berlin Heidelberg GmbH 1974

Inhalt/Contents

The Adrenergic Innervation of the Gastrointestinal Tract. By J.B. FURNESS and M. COSTA, Victoria/Australia. With 5 Figures

Regulation of Fatty-Acid Synthesis in Higher Animals. By S. NUMA, Kyoto/Japan. With 6 Figures

Amino Acid Transmitters in the Mammalian Central Nervous System. By D. R. CURTIS and G. A. R. JOHNSTON, Canberra City/Australia. With 4 Figures

Author Index

Subject Index

Ergebnisse der Physiologie

Biologischen Chemie und experimentellen Pharmakologie

Reviews of Physiology

Biochemistry and Experimental Pharmacology

Reprint from Volume 69

David R. Curtis and Graham A. R. Johnston

Amino Acid Transmitters in the Mammalian Central Nervous System

Nicht im Handel

Springer-Verlag Berlin Heidelberg GmbH 1974

Inhalt/Contents

The Adrenergic Innervation of the Gastrointestinal Tract. By J. B. FURNESS and M. COSTA, Victoria/Australia. With 5 Figures

Regulation of Fatty-Acid Synthesis in Higher Animals. By S. NUMA, Kyoto/Japan. With 6 Figures

Amino Acid Transmitters in the Mammalian Central Nervous System. By D. R. CURTIS and G. A. R. JOHNSTON, Canberra City/Australia. With 4 Figures

Author Index

Subject Index